T0091897

METHODS IN MOLECULAR BIOLOGY

Series Editor
John M. Walker
School of Life and Medical Sciences
University of Hertfordshire
Hatfield, Hertfordshire AL10 9AB, UK

For further volumes:
http://www.springer.com/series/7651

Phytoplasmas

Methods and Protocols

Edited by

Rita Musetti and Laura Pagliari

Department of Agricultural, Food, Environmental and Animal Sciences, University of Udine, Udine, Italy

Humana Press

Editors
Rita Musetti
Department of Agricultural, Food,
Environmental and Animal Sciences
University of Udine
Udine, Italy

Laura Pagliari
Department of Agricultural, Food,
Environmental and Animal Sciences
University of Udine
Udine, Italy

ISSN 1064-3745   ISSN 1940-6029   (electronic)
Methods in Molecular Biology
ISBN 978-1-4939-8836-5   ISBN 978-1-4939-8837-2   (eBook)
https://doi.org/10.1007/978-1-4939-8837-2

Library of Congress Control Number: 2018957657

© Springer Science+Business Media, LLC, part of Springer Nature 2019
This work is subject to copyright. All rights are reserved by the Publisher, whether the whole or part of the material is concerned, specifically the rights of translation, reprinting, reuse of illustrations, recitation, broadcasting, reproduction on microfilms or in any other physical way, and transmission or information storage and retrieval, electronic adaptation, computer software, or by similar or dissimilar methodology now known or hereafter developed.
The use of general descriptive names, registered names, trademarks, service marks, etc. in this publication does not imply, even in the absence of a specific statement, that such names are exempt from the relevant protective laws and regulations and therefore free for general use.
The publisher, the authors, and the editors are safe to assume that the advice and information in this book are believed to be true and accurate at the date of publication. Neither the publisher nor the authors or the editors give a warranty, express or implied, with respect to the material contained herein or for any errors or omissions that may have been made. The publisher remains neutral with regard to jurisdictional claims in published maps and institutional affiliations.

This Humana Press imprint is published by the registered company Springer Science+Business Media, LLC, part of Springer Nature.
The registered company address is: 233 Spring Street, New York, NY 10013, U.S.A.

Preface

Phytoplasmas are phytopathogenic agents associated with thousands of diseases affecting wild and cultivated plants all over the world. The disease severity and the absence of effective curative strategies result in important economic damages. Palm lethal yellows, Australian grapevine yellows, alfalfa virescence and witches' broom, and grapevine yellows (FD and BN) are some examples of economically important phytoplasma diseases.

The aim of this book is to present a set of modern protocols forming a solid background for those who want to start or improve research program on phytoplasmas.

The first part of the book collects nursery techniques for maintaining phytoplasma collections and producing infected plants. Moreover, some guidelines about the main macroscopic symptoms associated with phytoplasma diseases help the reader in the recognition of infected plants, both on field and in greenhouse. Various methods for the detection and analysis are then described, covering both traditional and innovative protocols, set up for border inspection or to detect different pathogens associated with phytoplasmas in mixed infection, for in-field or remote-sensing analyses.

Because of the lack of several metabolic essential functions, phytoplasmas are intrinsically uncultivable and, thus, it is particularly challenging the distinction between pathogen and plant system. For this reason, we present a new pipeline to produce phytoplasma genome draft (*see* Chap. 16) and protocols for the analysis of phytoplasma gene expression (*see* Chap. 18). Then, some methods for the investigation of the phloem tissue, the privileged host site for phytoplasmas, are defined. Electron and fluorescent microscopy analyses follow, presenting also in vivo observation protocols.

For a wider approach in the study of plant–pathogen interactions, this new edition presents also some chapters about the characterization of the molecular targets of phytoplasma effector, the analysis of volatile organic compound (VOC) pattern, and phytohormone production by phytoplasma-infected plants. The description of producing transgenic lines both in plant and bacteria concludes this part.

Udine, Italy

Rita Musetti
Laura Pagliari

Contents

Contributors

SIMONA ABBÀ • *Institute for Sustainable Plant Protection, CNR, Torino, Italy*

YUSUF ABOU-JAWDAH • *Department of Agriculture, Faculty of Agricultural and Food Sciences, American University of Beirut, Beirut, Lebanon*

PETER ABRAHAMIAN • *Gulf Coast and Research Education Center, Wimauma, FL, USA*

VICKEN AKNADIBOSSIAN • *Department of Agriculture, Faculty of Agricultural and Food Sciences, American University of Beirut, Beirut, Lebanon*

MARILIA ALMEIDA-TRAPP • *Department of Bioorganic Chemistry, Max Planck Institute for Chemical Ecology, Jena, Germany*

H. AL-SADDIK • *Agroecology, Agrosup Dijon, INRA, Univ. Bourgogne Franche-Comté, Dijon, France*

SANJA BARIC • *Faculty of Science and Technology, Free University of Bozen-Bolzano, Bozen-Bolzano, BZ, Italy*

ASSUNTA BERTACCINI • *Phytobacteriology Laboratory, DISTAL, Alma Mater Studiorum, University of Bologna, Bologna, Italy*

D. BOSCO • *Department of Agriculture, Forestry and Food Sciences, University of Torino, Grugliasco, Italy*

KRISTI D. BOTTNER-PARKER • *Molecular Plant Pathology Laboratory, USDA, ARS, Beltsville, MD, USA*

SARA BUOSO • *Department of Agricultural, Food, Environmental and Animal Sciences, University of Udine, Udine, Italy*

STEFANIE VERA BUXA • *Centre for BioSystems, Land Use and Nutrition, Institute of Phytopathology, Justus Liebig University Giessen, Giessen, Germany*

J. CHUCHE • *IFV, Pôle Nouvelle Aquitaine, Blanquefort, France; UMT Seven "Santé des écosystèmes viticoles économes en intrants", Villenave d'Ornon, France*

F. COINTAULT • *Agroecology, Agrosup Dijon, INRA, Univ. Bourgogne Franche-Comté, Dijon, France*

NICOLETTA CONTALDO • *Phytobacteriology Laboratory, DISTAL, Alma Mater Studiorum, University of Bologna, Bologna, Italy*

TANJA DREO • *National Institute of Biology, Ljubljana, Slovenia*

PAOLO ERMACORA • *Department of Agricultural, Food, Environmental and Animal Sciences, University of Udine, Udine, Italy*

NICOLA FIORE • *Department of Plant Health, Faculty of Agricultural Sciences, University of Chile, Santiago, Chile*

GIUSEPPE FIRRAO • *Department of Agricultural, Food, Environmental and Animal Sciences (DI4A), University of Udine, Udine, Italy*

ALEXANDRA C. U. FURCH • *Department of Plant Physiology, Faculty of Biological Science, Matthias-Schleiden-Institute for Genetics, Bioinformatics and Molecular Botany, Friedrich-Schiller-University Jena, Jena, Germany*

LUCIANA GALETTO • *Istituto per la Protezione Sostenibile delle Piante, CNR, Torino, Italy*

JANNICKE GALLINGER • *Laboratory of Applied Chemical Ecology, Institute for Plant Protection in Fruit Crops and Viticulture, Federal Research Centre for Cultivated Plants, Julius Kühn-Institut, Dossenheim, Germany; Plant Chemical Ecology, Technical University of Darmstadt, Darmstadt, Germany*

JÜRGEN GROSS • *Laboratory of Applied Chemical Ecology, Institute for Plant Protection in Fruit Crops and Viticulture, Federal Research Centre for Cultivated Plants, Julius Kühn-Institut, Dossenheim, Germany; Plant Chemical Ecology, Technical University of Darmstadt, Darmstadt, Germany*

BETTINA HAUSE • *Department of Cell and Metabolic Biology, Leibniz Institute of Plant Biochemistry, Halle, Germany*

JENNIFER HODGETTS • *Fera, The National Agri-Food Innovation Campus, York, UK*

KATRIN JANIK • *Functional Genomics, Laimburg Research Centre, Auer/Ora, BZ, Italy*

MAAN JAWHARI • *Department of Agriculture, Faculty of Agricultural and Food Sciences, American University of Beirut, Beirut, Lebanon*

SHIGEYUKI KAKIZAWA • *Bioproduction Research Institute, National Institute of Advanced Industrial Science and Technology (AIST), Tsukuba, Ibaraki, Japan*

TORSTEN KNAUER • *Max-Planck Institute for Chemical Ecology, Jena, Germany*

KERSTIN KRÜGER • *Department of Zoology and Entomology, University of Pretoria, Pretoria, South Africa*

A. LAYBROS • *Agroecology, Agrosup Dijon, INRA, Univ. Bourgogne Franche-Comté, Dijon, France*

ING-MING LEE • *Molecular Plant Pathology Laboratory, USDA, ARS, Beltsville, MD, USA*

ALBERTO LOSCHI • *Department of Agricultural, Food, Environmental and Animal Sciences, University of Udine, Udine, Italy*

IOANNA MALANDRAKI • *Laboratory of Virology, Department of Phytopathology, Benaki Phytopathological Institute, Athens, Greece*

CARMINE MARCONE • *Department of Pharmacy, University of Salerno, Fisciano, SA, Italy*

MARTA MARTINI • *Department of Agricultural, Food, Environmental and Animal Sciences (DI4A), University of Udine, Udine, Italy*

CRISTINA MARZACHÌ • *Istituto per la Protezione Sostenibile delle Piante, CNR, Torino, Italy*

NATAŠA MEHLE • *National Institute of Biology, Ljubljana, Slovenia*

STEFANO MINGUZZI • *Department of Agricultural and Food Sciences (DISTAL), University of Bologna, Bologna, Italy*

AXEL MITHÖFER • *Department of Bioorganic Chemistry, Max Planck Institute for Chemical Ecology, Jena, Germany*

CECILIA MITTELBERGER • *Functional Genomics, Laimburg Research Centre, Auer/Ora, BZ, Italy*

R. MUSETTI • *Department of Agricultural, Food, Environmental and Animal Sciences, University of Udine, Udine, Italy*

RUGGERO OSLER • *Department of Agricultural, Food, Environmental and Animal Sciences, University of Udine, Udine, Italy*

DAVIDE PACIFICO • *Institute of Biosciences and Bioresources, CNR, Palermo, Italy*

L. PAGLIARI • *Department of Agricultural, Food, Environmental and Animal Sciences, University of Udine, Udine, Italy*

SABRINA PALMANO • *Institute for Sustainable Plant Protection, CNR, Torino, Italy*

SAMANTA PALTRINIERI • *Phytobacteriology Laboratory, DISTAL, Alma Mater Studiorum, University of Bologna, Bologna, Italy*

CESARE POLANO • *Department of Agricultural, Food, Environmental and Animal Sciences (DI4A), University of Udine, Udine, Italy*

MAHNAZ RASHIDI • *Citrus Experimental Station, University of Florida, Lake Alfred, FL, USA*

CLAUDIO RATTI • *Department of Agricultural and Food Sciences (DISTAL), University of Bologna, Bologna, Italy*

MARGIT RID • *Laboratory of Applied Chemical Ecology, Institute for Plant Protection in Fruit Crops and Viticulture, Federal Research Centre for Cultivated Plants, Julius Kühn-Institut, Dossenheim, Germany; Institute of Evolutionary Ecology and Conservation Genomics, University of Ulm, Ulm, Germany*

SIMONETTA SANTI • *Department of Agricultural, Food, Environmental and Animal Sciences, University of Udine, Udine, Italy*

J. C. SIMON • *Agroecology, Agrosup Dijon, INRA, Univ. Bourgogne Franche-Comté, Dijon, France*

HAGEN STELLMACH • *Department of Cell and Metabolic Biology, Leibniz Institute of Plant Biochemistry, Halle, Germany*

PATIL TAWIDIAN • *Department of Agriculture, Faculty of Agricultural and Food Sciences, American University of Beirut, Beirut, Lebanon*

D. THIÉRY • *INRA, UMR 1065 Save "Santé et Agroécologie du VignoblE", Villenave d'Ornon, France; UMT Seven "Santé des écosystèmes viticoles économes en intrants", Villenave d'Ornon, France*

MASSIMO TURINA • *Institute for Sustainable Plant Protection, National Research Council of Italy, IPSP-CNR, Turin, Italy*

MARTA VALLINO • *Istituto per la Protezione Sostenibile delle Piante, CNR, Torino, Italy*

AART J. E. VAN BEL • *Centre for BioSystems, Land Use and Nutrition, Institute of Phytopathology, Justus-Liebig University, Giessen, Germany*

CHRISTINA VARVERI • *Department of Phytopathology, Laboratory of Virology, Benaki Phytopathological Institute, Athens, Greece*

NIKON VASSILAKOS • *Department of Phytopathology, Laboratory of Virology, Benaki Phytopathological Institute, Athens, Greece*

MATTHIAS R. ZIMMERMANN • *Department of Plant Physiology, Faculty of Biological Science, Matthias-Schleiden-Institute for Genetics, Bioinformatics and Molecular Botany, Friedrich-Schiller-University Jena, Jena, Germany*

Chapter 1

Phytoplasmas: An Introduction

L. Pagliari and R. Musetti

Abstract

Phytoplasmas are among the most recently discovered plant pathogens. They are wall-less prokaryotes restricted to phloem tissue, associated with diseases affecting several hundred plant species. The impact of phytoplasma diseases on agriculture is impressive and, at the present day, no effective curative strategy has been developed. The availability of rapid and sensitive techniques for phytoplasma detection as well as the possibility to study their relationship with the host plants is a prerequisite for the management of phytoplasma-associated diseases.

Key words Phytoplasmas, Phloem, Disease, Detection, Defense mechanisms

Phytoplasmas are prokaryotic plant pathogens belonging to the class *Mollicutes* (order *Acholeplasmatales*, family Acholeplasmataceae), a group of wall-less microorganisms phylogenetically related to low G+C Gram-positive bacteria [1]. Phytoplasmas were discovered in 1967 [2] and were named mycoplasma-like organisms (MLOs), due to their morphological and ultrastructural similarity to mycoplasmas, already known as aetiologic agents in animal and human diseases. Following the application of molecular technologies MLOs were designed as a coherent, genus-level taxon, named "*Candidatus* Phytoplasma" [3]. In this new clade, groups and subgroups have been defined and many of them are now considered species [4]. The most comprehensive and widely accepted phytoplasma classification system relies on restriction fragment length polymorphism (RFLP) analysis of polymerase chain reaction (PCR)-amplified 16S rDNA [5–8] (*see* Chapters 7 and 8).

Phytoplasmas are similar to bacterial bodies of small dimensions, varying from 200 nm to 800 nm in diameter, delimited by a plasma membrane, but devoid of the cell wall [9]. The absence of a rigid cell wall allows them to be highly pleomorphic and to change shape adapting to the environment. This feature is probably associated with the fact that, as the other *Mollicutes*, phytoplasmas are obligate parasites. The durable adaptation to life as obligated

Rita Musetti and Laura Pagliari (eds.), *Phytoplasmas: Methods and Protocols*, Methods in Molecular Biology, vol. 1875, https://doi.org/10.1007/978-1-4939-8837-2_1, © Springer Science+Business Media, LLC, part of Springer Nature 2019

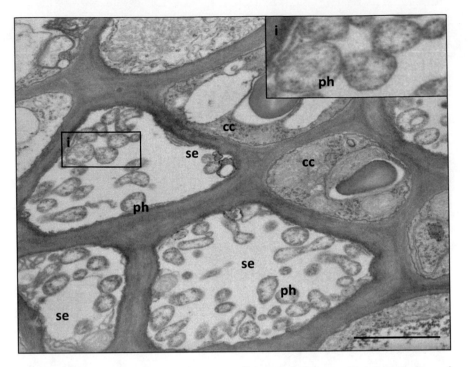

Fig. 1 TEM micrograph showing Chrysanthemum yellows-phytoplasmas in infected sieve elements of *Arabidopsis thaliana*. Phytoplasmas show a typical pleomorphic shape, delimited by an electron-dense membrane. Bar corresponds to 1 μm. *cc* companion cell, *ph* phytoplasma, *se* sieve element

parasite is also demonstrated by the strong host-specific, the tissue-specific correlation, the extremely difficulty to cultivate them in vitro [10] and the lack of several pathways for the synthesis of compounds considered to be necessary for the cell metabolism [11] (Fig. 1).

Phytoplasmas live inside the cells of plants and insect vectors, and, with a unique life cycle, they replicate intracellularly in both [12]. Through insect nutrition activity on infected plants, phytoplasmas enter the vector. Phytoplasmas are transmitted by phloem-feeding insect species within the Order *Hemiptera*, such as *Cicadellidea* (leafhoppers), *Fulgoridea* (cicada), and *Psyllidae* (psyllids) [13]. The insects must feed for an extended period of time (called acquisition access period) to acquire a sufficient titer of phytoplasmas to establish infection. During a latent period in the vector, phytoplasmas pass from the alimentary canal through the midgut into the hemolymph, they invade salivary gland cells, multiply and are incorporated into saliva. Then they are transmitted to a new host plant by injection into phloem tissue during insect feeding, during the so-called inoculation access period [12, 14–16]. Inside the sieve elements the phytoplasmas move systemically through the plant. Phytoplasma spread into the plant cannot be explained solely by assimilate flow [17–19]. On the other hand, considering the fact that phytoplasmas have no gene coding for cytoskeleton elements or flagella, their active movement seems unlikely [15]. Phytoplasma life-cycle can be replicated in controlled conditions in the laboratories, with the use of infective insects collected in field (*see* Chapter 4) or with vector-rearing and infection (*see* Chapter 3). Alternatively, it is possible to transmit phytoplasma by vegetative propagation (grafting and cuttings), as proposed in Chapter 2.

Phytoplasmas are reported to be associated with plant diseases in several hundreds of plant species, including many important vegetable and fruit crops, ornamental and timber plants, causing an impressive impact on agriculture [20, 21]. Although not all infections are necessarily deleterious, the great majority of phytoplasma diseases causes stunting of overall plant growth, general decline, loss of productivity, and, in some cases, plant death [22] (for an in-depth symptom description *see* Chapter 5). The detection of these micro-organisms is a prerequisite for the management of phytoplasma-associated diseases and, for this reason, the development of sensitive detection (*see* Chapters 6, 7, and 8) and quantification (*see* Chapters 9 and 10) protocols has been a continuous effort in the last decades. Over recent years, there has been a drive toward simpler and quicker detection methods that can be performed for border inspection or multiple pathogens, in field (*see* Chapters 11, 12, 13, 14, and 15) or can even imply remote-sensing technique (*see* Chapter 17).

Symptoms in infected plants suggest a profound disturbance of the hormonal balance and interference with the phloem mass flow [9, 23], nevertheless phytoplasma research is actually facing the absence of a clear comprehension of phytoplasma and infected-plant physiology. Considering that phytoplasmas are strictly associated with the host tissue, it is particularly challenging the study of the basic details of phytoplasma biology and their pathogenic behavior. In this new edition, we present a new pipeline to produce phytoplasma genome draft (*see* Chapter 16) and protocols for the analysis of phytoplasma gene expression (*see* Chapter 18).

Pathogen presence and activity, as well as its recognition by the host plant, drive to many biochemical changes indicating the activation of plant defence response. Phytoplasma infection induces Ca^{2+} influx into the sieve elements, leading to sieve-tube blockage [24, 25]. In fact, transmission electron microscopy (TEM) images revealed sieve-element filament formation and agglutination and callose deposition at sieve plate level [24, 26]. In addition, Ca^{2+} signals are decoded and relayed by signaling molecules, generating various intracellular cascades leading to changes in metabolism and gene expression [27–29]. Phytoplasma infection can lead to the involvement of other important signal and defence molecules such as hydrogen peroxide (H_2O_2) [30, 31], and different phytohormones [32–34]. Different authors reported also a downregulation of photosynthetic proteins in phytoplasma-infected plants [35–37], accompanied by a reduction of total chlorophyll content [38–41]. The limited expression of the photosynthetic proteins induces alteration in the photosynthetic whole chain (mainly affecting photosystem II activity), compromising the whole photosynthetic process. This inhibition seems to have an impact on the carbohydrate metabolism, particularly on the accumulation of soluble carbohydrates and starch, as observed in source leaves of plants infected by phytoplasmas [35, 40–43]. It has been hypothesized that photosynthesis impairment and the following decrease in synthesis of ribulose-1, 5-biphosphate carboxylase (RuBPC), the major soluble protein of the leaf [45], could be related to the reduction in total soluble proteins observed in many phytoplasma-infected plants [38, 39, 46]. Despite the decrease in the total protein content, following pathogen infection, a lot of proteins are produced by the host plant. Most of them are related to the plant-defense mechanisms and belong to the pathogenesis-related protein (PR-protein) families [24, 40, 46–48]. PR proteins accumulate locally in the infected leaves and are also induced systemically, dealing to the development of systemic acquired resistance (SAR) [50].

Most ultrastructural and biochemical changes in infected plants mainly involve phloem tissue, deeply described in Chapter 19. For this reason, in comparison with the last version of this volume, we added a new part on site-specific analyses (*see* Chapters 20, 21, and 22). Furthermore, a protocol for studying the interactions between the phytoplasma immunodominant membrane proteins (IMP) and vector proteins is here presented (*see* Chapter 23). Phytoplasmas interact with insect cytoskeleton, imposing actin reorganization [51, 52]. For a wider approach in the study of plant-pathogen interactions, this new edition presents also chapters about the characterization of the molecular targets of phytoplasma effector (*see* Chapter 24) and the analysis of volatile organic compound (VOC) patterns emission (*see* Chapter 25) and phytohormone production (*see* Chapter 26) by phytoplasma-infected plants.

References

1. Weisburg WG, Tully JG, Rose DL et al (1989) A phylogenetic analysis of the mycoplasmas: basis for their classification. J Bacteriol 171 (12):6455–6467

2. Doi Y, Teranaka M, Yora K et al (1967) Mycoplasma- or PLT group-like microorganisms found in the phloem elements of plants infected with mulberry dwarf, potato witches' broom, aster yellows or paulownia witches' broom (in Japanese with English summary). Ann Phytopath Soc Japan 33:259–266

3. IRPCM P, Spiroplasma, WTPTG (2004) Candidatus Phytoplasma', a taxon for the wall-less, nonhelical prokaryotes that colonize plant phloem and insects. Int J Syst Evol Microbiol 54(Pt 4):1243

4. Bertaccini A, Duduk B (2009) Phytoplasma and phytoplasma diseases: a review of recent research. Phytopathol Mediterr 48 (3):355–378

5. Lee IM, Gundersen-Rindal DE, Davis RE et al (2004) 'Candidatus Phytoplasma asteris', a novel phytoplasma taxon associated with aster yellows and related diseases. Int J Syst Evol Microbiol 54(4):1037–1048

6. Lee M, Martini M, Marcone C et al (2004) Classification of phytoplasma strains in the elm yellows group (16SrV) and proposal of

'Candidatus Phytoplasma ulmi' for the phytoplasma associated with elm yellows. Int J Syst Evol Microbiol 54(2):337–347

7. Marcone C, Lee IM, Davis RE et al (2000) Classification of aster yellowsgroup phytoplasmas based on combined analyses of rRNA and tuf gene sequences. Int J Syst Evol Microbiol 50(5):1703–1713

8. Zhao Y, Davis RE (2016) Criteria for phytoplasma 16Sr group/subgroup delineation and the need of a platform for proper registration of new groups and subgroups. Int J Syst Evol Microbiol 66(5):2121–2123

9. Lee IM, Davis RE, Gundersen-Rindal DE (2000) Phytoplasma: Phytopathogenic Mollicutes 1. Annu Rev Microbiol 54(1):221–255

10. Contaldo N, Bertaccini A, Paltrinieri S et al (2012) Axenic culture of plant pathogenic phytoplasmas. Phytopathol Mediterr 51 (3):607–617

11. Marcone C, Neimark H, Ragozzino A et al (1999) Chromosome sizes of phytoplasmas composing major phylogenetic groups and subgroups. Phytopathology 89(9):805–810

12. Hogenhout SA, Loria R (2008) Virulence mechanisms of gram-positive plant pathogenic bacteria. Curr Opin Plant Biol 11(4):449–456

13. Weintraub PG, Beanland L (2006) Insect vectors of phytoplasmas. Annu Rev Entomol 51:91–111

14. Bosco D, Galetto L, Leoncini P et al (2007) Interrelationships between "Candidatus Phytoplasma asteris" and its leafhopper vectors (Homoptera: Cicadellidae). J Econ Entomol 100(5):1504–1511

15. Christensen NM, Axelsen KB, Nicolaisen M et al (2005) Phytoplasmas and their interactions with hosts. Trends Plant Sci 10 (11):526–535

16. Oshima K, Ishii Y, Kakizawa S et al (2011) Dramatic transcriptional changes in an intracellular parasite enable host switching between plant and insect. PLoS One 6(8):e23242

17. Schaper U, Seemüller E (1984) Recolonization of the stem of apple proliferation and pear decline-diseased trees by the causal organisms in spring. Z Pflanzenkrankh Pflanzenschutz 91:608–613

18. Marcone C, Weintraub PG, Jones P (2009) Movement of Phytoplasmas and the development of disease in the plant. In: Genomes, plant hosts and vectors, vol 114

19. Pagliari L, Buoso S, Santi S et al (2017) Filamentous sieve element proteins are able to limit phloem mass flow, but not phytoplasma spread. J Exp Bot 68(13):3673–3688

20. Bertaccini A, Duduk B, Paltrinieri S et al (2014) Phytoplasmas and phytoplasma diseases: a severe threat to agriculture. Am J Plant Sci 5:763–1788

21. Valiunas V, Wang HZ, Li L et al (2015) A comparison of two cellular delivery mechanisms for small interfering RNA. Physiol Rep 3 (2):e12286

22. Seemüller E, Garnier M, Schneider B (2002) Mycoplasmas of plants and insects. In: Molecular biology and pathogenicity of mycoplasmas. Springer, New York, pp 91–115

23. Osler R, Carraro L, Loi N et al (1996). Le più importanti malattie da fitoplasmi nel Friuli-Venezia Giulia: atlante. Edito da Ente regionale per la promozione e lo sviluppo dell'agricoltura del Friuli-Venezia Giulia

24. Musetti R, Buxa SV, De Marco F et al (2013) Phytoplasma-triggered Ca^{2+} influx is involved in sieve-tube blockage. MPMI 26(4):379–386

25. Musetti R, Favali MA (2003) Calcium localization and X-ray microanalysis in Catharanthus roseus L. infected with phytoplasmas. Micron 34:387–393

26. Lherminier J, Benhamou N, Larrue J et al (2003) Cytological characterization of elicitin-induced protection in tobacco plants infected by Phytophthora parasitica or phytoplasma. Phytopathology 93(10):1308–1319

27. Kudla J, Batistič O, Hashimoto K (2010) Calcium signals: the lead currency of plant information processing. Plant Cell 22(3):541–563

28. McAinsh MR, Pittman JK (2009) Shaping the calcium signature. New Phytol 181 (2):275–294

29. van Bel AJ, Furch AC, Will T et al (2014) Spread the news: systemic dissemination and local impact of Ca^{2+} signals along the phloem pathway. J Exp Bot 65:1761–1787

30. Musetti R, Sanità di Toppi L, Martini M et al (2005) Hydrogen peroxide localization and antioxidant status in the recovery of apricot plants from European stone fruit yellows. Eur J Plant Pathol 112(1):53–61

31. Sánchez-Rojo S, López-Delgado HA, Mora-Herrera ME et al (2011) Salicylic acid protects potato plants-from phytoplasma-associated stress and improves tuber photosynthate assimilation. Am J Pot Res 88(2):175–183

32. Minato N, Himeno M, Hoshi A et al (2014) The phytoplasmal virulence factor TENGU causes plant sterility by downregulating of the jasmonic acid and auxin pathways. Sci Rep 4:1399

33. Punelli F, Al Hassan M, Fileccia V et al (2016) A microarray analysis highlights the role of tetrapyrrole pathways in grapevine responses to "stolbur" phytoplasma, phloem virus infections and recovered status. Physiol Mol Plant Path 93:129–137

34. Zimmermann MR, Schneider B, Mithöfer A et al (2015) Implications of Candidatus Phytoplasma Mali infection on phloem function of apple trees. Endocytobiosis Cell Res 26:67–75

35. Ji X, Gai Y, Zheng C et al (2009) Comparative proteomic analysis provides new insights into mulberry dwarf responses in mulberry (*Morus alba* L.). Proteomics 9(23):5328–5339

36. Hren M, Nikolić P, Rotter A et al (2009) 'Bois noir'phytoplasma induces significant reprogramming of the leaf transcriptome in the field grown grapevine. BMC Genomics 10 (1):1

37. Taheri F, Nematzadeh G, Zamharir MG et al (2011) Proteomic analysis of the Mexican lime tree response to "Candidatus Phytoplasma aurantifolia" infection. Mol BioSystems 7 (11):3028–3035

38. Bertamini M, Grando MS, Muthuchelian K et al (2002a) Effect of phytoplasmal infection on photosystem II efficiency and thylakoid membrane protein changes in field grown apple (Malus pumila) leaves. Physiol Mol Plant Path 61(6):349–356

39. Bertamini M, Nedunchezhian N, Tomasi F et al (2002b) Phytoplasma [Stolbur-subgroup bois noir-BN] infection inhibits photosynthetic pigments, ribulose-1, 5-bisphosphate carboxylase and photosynthetic activities in field grown grapevine (*Vitis vinifera* L. cv. Chardonnay) leaves. Physiol Mol Plant Path 61 (6):357–366

40. Junqueira A, Bedendo I, Pascholati S (2004) Biochemical changes in corn plants infected by the maize bushy stunt phytoplasma. Physiol Mol Plant Path 65(4):181–185

41. Lepka P, Stitt M, Moll E et al (1999) Effect of phytoplasmal infection on concentration and translocation of carbohydrates and amino acids in periwinkle and tobacco. Physiol Mol Plant Path 55(1):59–68

41. Zafari S, Niknam V, Musetti R et al (2012) Effect of phytoplasma infection on metabolite content and antioxidant enzyme activity in lime (*Citrus aurantifolia*). Acta Physiol Plant 34 (2):561–568

42. Maust BE, Espadas F, Talavera C et al (2003) Changes in carbohydrate metabolism in coconut palms infected with the lethal yellowing phytoplasma. Phytopathology 93(8):976–981

43. Pagliari L, Martini M, Loschi A et al (2016) Looking inside phytoplasma-infected sieve elements: a combined microscopy approach using Arabidopsis thaliana as a model plant. Micron 89:87–97

44. Bertamini M, Grando MS, Nedunchezhian N (2003) Effects of phytoplasma infection on pigments, chlorophyll-protein complex and photosynthetic activities in field grown apple leaves. Biol Plant 47(2):237–242

45. Favali MA, Sanità di Toppi L, Vestena C et al (2001) Phytoplasmas associated with tomato stolbur disease. Acta Hortic 551:93–99

46. Margaria P, Palmano S (2011) Response of the Vitis vinifera L. cv. 'Nebbiolo' proteome to Flavescence dorée phytoplasma infection. Proteomics 11(2):212–224

47. Santi S, Grisan S, Pierasco A et al (2013) Laser microdissection of grapevine leaf phloem infected by stolbur reveals site-specific gene responses associated to sucrose transport and metabolism. Plant Cell Environ 36 (2):343–355

48. Zhong BX, Shen YW (2004) Accumulation of pathogenesis-related Type-5 like proteins in Phytoplasma infected garland chrysanthemum Chrysanthemum coronarium. Acta Biochim Biophys Sin 36(11):773–779

49. van Loon LC, van Strien EA (1999) The families of pathogenesis-related proteins, their activities, and comparative analysis of PR-1 type proteins. Physiol Mol Plant Path 55 (2):85–97

50. Boonrod K, Munteanu B, Jarausch B et al (2012) An immunodominant membrane protein (imp) of 'Candidatus Phytoplasma mali' binds to plant actin. Mol Plant-Microbe Interact 25(7):889–895

51. Galetto L, Bosco D, Balestrini R et al (2011) The major antigenic membrane protein of "Candidatus Phytoplasma asteris" selectively interacts with ATP synthase and actin of leafhopper vectors. PLoS One 6(7):e22571

Part I

Preparing Plant Material

Chapter 2

Micro-Tom Tomato Grafting for Stolbur-Phytoplasma Transmission: Different Grafting Techniques

Sara Buoso and Alberto Loschi

Abstract

Tomato plant, being a model system in scientific research, is widely used to study plant-phytoplasma interaction. Grafting is the faster and most effective method to obtain infected plants. This chapter describes the greenhouse culture of tomato, cv. Micro-Tom, and different herbaceous grafting techniques for efficient stolbur-phytoplasma transmission.

Key words Grafting, Greenhouse maintenance, Micro-Tom, Phytoplasma, Stolbur, Tomato

1 Introduction

Tomato plant (*Solanum lycopersicum* L.), besides being an economically important crop, is a model system in different scientific research [1, 2]. In fact, tomato has many interesting features that other model plants, such as *Arabidopsis*, do not have: fleshy fruit, a sympodial shoot, and compound leaves. The sequencing of tomato genome in 2012 [3] has generated useful biological information and enhanced the use as model plant, especially in relation to the studies about plant-pathogen interactions. Tomato, in fact, is naturally affected by a diversity of diseases, associated with different pathogens. Moreover, resistance to virus and other microorganisms has been largely investigated [4].

Among the different tomato cultivars, Micro-Tom [5] is particularly indicated for scientific investigation due to its small size, high-density culture, and rapid growth [6, 7]. Large collections of Micro-Tom mutants, produced by gamma-ray irradiation and ethylmethanesulfonate (EMS), are available from the National BioResource Project (NBRP) Tomato in Japan via the "TOMA-TOMA" database [8]. Moreover, Shikata and Ezura [6] have developed an efficient *Agrobacterium*-mediated transformation protocol for Micro-Tom. Successful application of Crispr/Cas9 system in

Rita Musetti and Laura Pagliari (eds.), *Phytoplasmas: Methods and Protocols*, Methods in Molecular Biology, vol. 1875, https://doi.org/10.1007/978-1-4939-8837-2_2, © Springer Science+Business Media, LLC, part of Springer Nature 2019

Micro-Tom for genome editing has been reported in few articles [9–11]. Using this technology, an efficient and site-directed mutagenesis has been achieved to investigate plant functional genomics and crop improvement, without the laborious and time-consuming screening process characterized by traditional mutagenesis methods.

Candidatus Phytoplasma solani ('*Ca*. P. solani', group 16SrXII-A) is an A2 quarantine pathogen in Europe (EPPO, European and Mediterranean Plant Protection Organization) (*see* **Note 1**) and is naturally hosted by a wide range of crops including *Solanaceae* [12] and grapevine, inducing a disease known as stolbur. Therefore, tomato has been used in the study on '*Ca*. P. solani'--plant interaction [13–17], much more than other test plants, such as *Catharanthus roseus* (L.) G. Don, *Vicia faba* (L.), and *Arabidopsis thaliana* [18, 19].

'*Ca*. P. solani' is transmitted by vegetative propagation (grafting and cuttings) and by several insect vector species. In experimental conditions, grafting is the most rapid and effective method. Thus, in this chapter we describe different herbaceous grafting techniques for an efficient stolbur-phytoplasma transmission in tomato cv. Micro-Tom.

All the grafting methods illustrated below can be performed also for the maintenance of phytoplasma in *C. roseus*, the test plant generally used for this purpose. Moreover, some of the described methods can be used in heterologous grafting for the transmission of the phytoplasma from different plant sources to *C. roseus* and tomato plants, as described by Aryan et al. [17].

2 Materials

Considering that '*Ca*. P. solani' is listed as quarantine pest for Europe (*see* **Note 1**), infected plants should be maintained in insect-proof rooms, in confined greenhouse and every experiment should be carried out under safety conditions and according to the local current phytosanitary rules. To reduce the chance of insect-vector casual introduction, few precautions can be adopted such as the use of (1) white net (fine mesh) to protect the entrance of the chamber, (2) chromotropic traps to monitor insect presence, and (3) periodic insecticide treatments.

2.1 Plant Culture

1. Potting substrate mix: peat and perlite (10–15%) (*see* **Note 2**) eventually added with compost.
2. Fertilizers: slow release fertilizers with macro and microelements.
3. Plastic plateaux for seedling.
4. Plastic pots (squared 7 cm × 7 cm, 7 cm high, or bigger).

5. White insect-proof net.

6. Artificial lighting system (lamp *metal-halide* or *light-emitting diodes*).

7. Plastic film.

8. Bamboo or plastic plant stakes (ca. 30 cm high) (*see* **Note 3**).

9. *Solanum lycopersicum* cv. Micro-Tom seeds (*see* **Note 4**).

10. Sodium hypochlorite solution 1–1.5% (v/v).

11. Broad spectrum fungicide and insecticide (*see* **Note 5**).

2.2 Grafting

1. Healthy tomato cv. Micro-Tom plants.

2. Stolbur-infected tomato.

3. Transparent plastic bags (approximately 20 cm × 15 cm or at least the double height of the scion).

4. Plant ties.

5. Razor blades, scalpels, and cutters (*see* **Note 6**).

6. 90–100% Ethanol.

7. Parafilm.

8. Grafting clip (plastic or silicon).

9. Labels.

10. Pencil.

11. Hole punch.

12. Tubes.

3 Methods

3.1 Growth of Healthy Plants from Seeds

The greenhouse conditions are the same for both healthy and phytoplasma-infected plants. The plants are grown under a long-day photoperiod, with 14–16 h light. During the day, if natural irradiance decreases below 4000 lx, supplementary lighting must be provided (preferably with automatic activation). The daytime temperature should be between 21 and 27 °C, with a night minimum temperature 17 °C. Heating and cooling systems should be automatically activated if the temperature in the greenhouse decreases or increases below the settings. Presence of a data logger for temperature monitoring is recommended.

1. Sterilize seeds soaking in sodium hypochlorite 1–1.5% solution for 3–5 min, then rinse with distilled water.

2. Fill the plastic plateaux with the mix substrate (*see* **Note 7**) and pour out it until saturation with water.

3. Sow seeds with a minimum distance of 1 cm (*see* **Note 8**) from each other, cover with half cm of substrate, and pour out gently (*see* **Note 9**).

4. Cover the plateaux with the plastic film until the plant emerging to maintain high humidity (*see* **Note 10**). Then begin to pierce the film gradually and after 4–5 days cover it off.

5. After ca. 15–20 days from sowing, the plants are ready for transplanting. Prepare some plastic pots (recycled pots should be sterilized with sodium hypochlorite) filled with substrate and transplant the plants individually. After a month, transplant them another time in bigger sterilized pots (*see* **Note 11**).

6. Check daily the plants condition, water gently, and fertilize every 10–15 days switching from N and P rich fertilizers to K and Ca. Pay attention also to the eventual appearance of phytosanitary problems, such as mites, insects, powdery mildew, and *Botrytis*.

3.2 Grafting for Phytoplasma Transmission

About 2 months from the sowing, plants are ready to be grafted for phytoplasma transmission (*see* **Note 12**). The choice of the herbaceous grafting type depends on the kind of experiment to be performed (*see* **Note 13**) and on the available material (healthy and infected plant). Healthy plants, used as rootstock, must have a good vegetative development, lack of visible diseases (*see* **Note 14**) and be cultivated in a controlled area, avoiding insect vector presence. Infected plants, from which scion is taken, must show all the typical disease symptoms (Fig. 1), but not be too old (woody tissues are not suitable for grafting).

It is important to stress the fact that not only phytoplasmas, but also other endophytic microorganisms, may move from the infected scion to the healthy part. Moreover, in grafted plants,

Fig. 1 Phenotypes of healthy and fully symptomatic plants at 90 days after sowing (**a**) normal growth in healthy tomato; (**b**) stolbur-infected tomato with leaf yellowing, flower abnormalities, stunting and reduced leaf area

phytohormone-, long-distance protein-, and small RNA-movement may result altered. For this reason, healthy plant grafted with healthy scion should be used as control in the experiments.

3.2.1 Side Graft

This kind of grafting guarantees the best success both for scion survival and phytoplasma transmission (roughly 95%). Moreover, the grafted plant shows harmonic growth and clear symptom development. The infected scion and the stem of the healthy rootstock need to have approximately the same diameter, to obtain a tight anatomic connection between the tissues.

1. Every cut must be made with razor blades, scalpels, and cutters, sterilized with ethanol. The cutting must be precise, linear, and clear, without remaining lacerated tissues.

2. Cut vertically a portion of stem, with a variable length (from 0.5 cm to 2 cm), in the middle of healthy rootstock (Fig. 2a, b). To better stabilize the scion, it is possible to prepare a little pocket at the end of the cut (Fig. 2c).

Fig. 2 Side graft stages; (**a–c**) vertical cutting of stem in the healthy rootstock; (**d**) oblique cutting in infected scion; (**e**) completed graft; (**f**) successful rootstock-scion connection 1 month after grafting

3. For the scion, choose a symptomatic shoot from the phytoplasma-infected tomato plant, approximately 7–10 cm long (*see* **Note 15**).

4. In the final part of the scion, make an oblique cut of almost the same size of the cut made on the rootstock plant (Fig. 2d).

5. Insert the scion into the cut of the rootstock plant.

6. Wrap firmly the two parts with parafilm (or grafting clips) to fix the graft (Fig. 2e).

7. Treat with fungicide to avoid the development of *Botrytis*. Place a transparent plastic bag (*see* **Note 16**) over the scions or around the plant (sustained by the stakes) to maintain high humidity.

8. Label all the plants with the phytoplasma name and the grafting date.

9. Keep plants protected from direct light for at least a week; a panel (e.g., Styrofoam™) placed 20–30 cm over the grafted plants could protect them from direct sunlight.

10. Check daily the grafting status to prevent the development of fungal pathogens and, when necessary, open the bag, treat the plants with fungicide, and then close immediately.

11. After 15 days open the bag gradually, for instance cutting the edge corners. Leave the bag open on the plant for other 3 days to permit the scion acclimatization to the environmental conditions.

12. 4–5 weeks after grafting, the symptoms of phytoplasma infections will appear.

3.2.2 Apical Wedge Graft   This grafting technique is very simple to execute and guarantees roughly the complete success of phytoplasma transmission. On the other hand, the harmonic development of the plant will be impaired, so it is preferentially recommended for phytoplasma-maintenance purposes.

Compared to the side graft described here above, the apical wedge graft changes only in the preparation of the rootstock plant:

1. Cut off the top of the main stem of the healthy plant, then make a vertical cut in the middle of the stem (Fig. 3a, b).

2. In the final part of the scion, make an oblique cut on both the sides, to obtain a wedge of almost the same size of the cut made on the rootstock plant (Fig. 3c, d). Insert the scion and wrap firmly to the receiving stem with parafilm (Fig. 3e, f) or grafting clip.

3. Treat with fungicide to avoid the development of *Botrytis* and cover with a transparent plastic bag as described above (Fig. 3g).

Fig. 3 Apical wedge graft stages; (**a**) removal of the apical stem in healthy rootstock; (**b**) vertical cut of the stem for scion insertion; (**c** and **d**) oblique cutting on both the sides in the infected scion; (**e**) insertion of the scion into the rootstock; (**f**) fixed graft by parafilm; (**g**) graft covering with a transparent plastic bag

4. Label all the grafted plants with the phytoplasma name and the grafting date.

5. After 4–5 weeks, the symptoms of phytoplasma infections will appear.

3.2.3 Leaf Grafting

This type of grafting requires precision and care during the execution, because it is necessary that both scion and rootstock midribs fit perfectly together. The success of this kind of grafting is very low but is useful when a poor amount of infected material is available, or when it is recommended to reduce the damage produced by the impact of the previous described grafting techniques.

1. Cut a disc from the midrib section of the infected leaf with a hole punch. The leaf must be well developed but not too old (*see* **Note 17**).

2. Cut the healthy leaf with the hole punch and discard the leaf disk (Fig. 4a).

3. Take the infected disc and put it on the hole of the healthy leaf (Fig. 4b). The midribs should be aligned as best as possible,

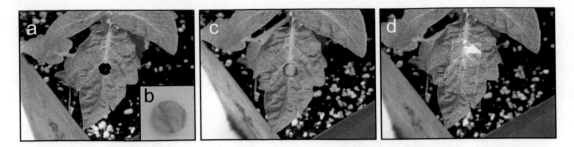

Fig. 4 Leaf grafting stages; (**a**) cutting of healthy leaf; (**b**) infected disc; (**c**) insertion of infected disk in the healthy leaves; (**d**) sealing with tape

and the disc should carefully match the hole. Put a piece of adhesive tape at the bottom of the receiving leaf to hold the scion leaf disc in place while necessary adjustments are made.

4. Take another piece of tape of similar length and place it above the leaf, then press it firmly (Fig. 4c).

5. Treat with fungicide to avoid the development of *Botrytis* or other fungal disease and place a transparent plastic bag over the leaf. Fix it around the plant or the stick.

6. Label plant with phytoplasma name and date of grafting.

7. After 1–2 weeks will be possible to determine the disc scion survival, as dead leaf will turn into browny colour. Symptoms occur 4–5 weeks after grafting.

3.2.4 Approach Grafting

This type of grafting is characterized by the use of a scion that remains attached to its own root system at the time of grafting. Approach grafting should be used in heterologous grafting for transmission of the stolbur phytoplasma from different plant sources to tomato plants. Unlike all other methods, the scion is less prone to become water stressed, resulting in a high probability of success. Alternatively, the scion could be cut off from its own root and put in a tube with water (Fig. 5c).

1. Cut vertically a portion of stem with the same length in healthy and infected plants (Fig. 5a).

2. Tying the two stems together at cut site with parafilm (Fig. 5b).

3. After 4–5 weeks, the symptoms of phytoplasma infections will appear.

4   Notes

1. For more information refer to EPPO website (https://www.eppo.int/QUARANTINE/quarantine.html), which every year provides updated lists of quarantine pests within the European and Mediterranean region.

Fig. 5 Approach grafting stages; (**a**) vertical cutting of stem in the healthy and infected plant; (**b**) complete graft union; (**c**) approach grafting with scion cut and insertion in a tube for hydration

2. A good substrate for tomato growth must ensure a good soil aeration and structural stability (low slumping effect). The ideal pH should be between 5.7 and 6.5. Commercial horticulture mixes for professional use can guarantee high quality substrates.

3. The plastic support stakes are to be preferred to woody ones, to avoid mold development. Nevertheless, woody stakes can be treated with an appropriate fungicide.

4. Micro-Tom wild-type seeds can be purchased in the "TOMA-TOMA" database (http://tomatoma.nbrp.jp/), where a rich collection of mutant lines is also provided. Seeds can be produced and collected by fully developed plants. When plants reach the anthesis phase (fully open flower), shake the flower individually to replace the natural self-pollination by insect or wind. At red ripe stage, harvest the fruits and collect seeds in a tissue net (ca. 0.5 mm mesh), close it with tie and submerge it in a water bath to remove locular tissue. Then proceed with a 5–10 min wash in hypochlorite solution (1–1.5%) and rinse in water to eliminate the remaining locular tissue. Dry the seeds overnight on a clean net tissue. Transfer the dried seeds to a paper bag and store in dry and cool conditions.

5. If you use a new active ingredient, check the potential harmful effect on few plants before spraying it on the test plants and use pesticides according to label information.

6. The blades must be very sharp to obtain plain cut surfaces and to minimize the tissues damages. Before every use, sterilize the blades with ethanol.

7. For sowing use a fine substrate which provides optimum condition for seed germination and root growth.

8. The distance between the seeds must be suitable for the next transplanting operation and to facilitate the right development of the roots and the plant in general.

9. If the watering is too violent, it could disturb the regular seed germination.

10. The maintenance of humidity is essential to guarantee seed germination. A too high level of humidity can lead to the development of root pathogens (*Phythium* spp. and *Phythophthora* spp.). Therefore, treatments with fungicide are recommended to avoid the development of root rots.

11. Tomato plants can be also grown in hydroponic condition [20]. Hydroponic system may be helpful to study the interactions among phytoplasmas and specific nutrients or to survey plant response at root level.

12. When request by the experimental conditions, it is possible to use a younger healthy plant as rootstock. Younger plants ensure quicker symptom development and clearer symptoms. On the other hand, the younger is the plant, the more difficult will be the grafting operation, because of the tight diameter of the stem.

13. An experiment planning is mandatory for research success. Some experimentation requires the use of healthy plants deriving from seeds produced by a single plant.

14. The sanitary status of the plants must be checked by symptom appearance and molecular detection analyses, also to exclude mixed infections with viruses or other phytoplasmas (*Cfr* Chapter 5).

15. The stem scion of infected plant must match in diameter with the stem of rootstock plant.

16. The plastic bag is used to maintain the scion hydrated to ensure the grafting success. The plastic bag must be at least twice bigger than the scion. It is possible to cover the whole plant, even if it is not recommended for a proper plant development.

17. Leaves of healthy plants must be well developed and not too young; the use of young leaves makes difficult to cut and manipulate the discs. For a successful match of tissues, the punched discs of the scion need to be slightly bigger than the rootstock plant hole diameter.

References

1. Kimura S, Sinha N (2008) Tomato (Solanum lycopersicum): a model fruit-bearing crop. CSH Protoc 2008(11):pdb-emo105

2. Zorzoli R, Pratta GR, Rodríguez GR, Picardi LA (2007) Advances in biotechnology: tomato as a plant model system. Funct Plant Sci Biotechnol 1(1):146–159

3. The Tomato Genome Consortium (2012) The tomato genome sequence provides insights into fleshy fruit evolution. Nature 485:635–641

4. Arie T, Takahashi H, Kodama M, Teraoka T (2007) Tomato as a model plant for plant-pathogen interactions. Plant Biotechnol 24(1):135–147

5. Scott JW, Harbaugh BK (1989) Micro-tom. A miniature dwarf tomato, vol S-370. FL Agric Exp Sta Circ, Florida, pp 1–6

6. Shikata M, Ezura H (2016) Micro-tom tomato as an alternative plant model system: mutant collection and efficient transformation. In: Botella JR, Botella MA (eds) Plant signal transduction. Methods in molecular biology, vol 1363. Humana Press, New York

7. Meissner R, Jacobson Y, Melamed S, Levyatuv S, Shalev G, Ashri A, Elkind Y, Levy A (1997) A new model system for tomato genetics. Plant J 12:1465–1472

8. Saito T, Ariizumi T, Okabe Y, Asamizu E, Hiwasa-Tanase K, Fukuda N, Mizoguchi T, Yamazaki Y, Aoki K, Ezura H (2011) TOMA-TOMA: a novel tomato mutant database distributing micro-tom mutant collections. Plant Cell Physiol 52:283–296

9. Ueta R, Abe C, Watanabe T et al (2017) Rapid breeding of parthenocarpic tomato plants using CRISPR/Cas9. Sci Rep 7:507

10. Pan C, Ye L, Qin L et al (2016) CRISPR/Cas9-mediated efficient and heritable targeted mutagenesis in tomato plants in the first and later generations. Sci Rep 6:24765

11. Brooks C, Nekrasov V, Lippman ZB, Van Eck J (2014) Efficient gene editing in tomato in the first generation using the clustered regularly interspaced short palindromic repeats/CRISPR-associated9 system. Plant Physiol 166(3):1292–1297

12. Quaglino F, Zhao Y, Casati P, Bulgari D, Bianco PA, Wei W, Davis RE (2013) "Candidatus Phytoplasma solani", a novel taxon associated with stolbur and bois noir related diseases of plants. Int J Syst Evol Microbiol 63:2879–2894

13. Pracros P, Renaudin J, Eveillard S, Mouras A, Hernould M (2006) Tomato flower abnormalities induced by stolbur phytoplasma infection are associated with changes of expression of floral development genes. Mol Plant-Microbe Interact 19(1):62–68

14. Pracros P, Hernould M, Teyssier E, Eveillard S, Renaudin J (2007) Stolbur phytoplasma-infected tomato showed alteration of SlDEF methylation status and deregulation of methyl-transferase genes expression. B Insectol 60(2):221–222

15. Machenaud J, Henri R, Dieuaide-Noubhani M, Pracros P, Renaudin J, Eveillard S (2007) Gene expression and enzymatic activity of invertases and sucrose synthase in Spiroplasma citri or stolbur phytoplasma infected plants. B Insectol 60(2):219–220

16. Buxa SV, Degola F, Polizzotto R et al (2015) Phytoplasma infection in tomato is associated with re-organization of plasma membrane, ER stacks and actin filaments in sieve elements. Front Plant Sci 6:650

17. Aryan A, Musetti R, Riedle-Bauer M, Brader G (2016) Phytoplasma transmission by heterologous grafting influences viability of the scion and results in early symptom development in periwinkle rootstock. J Phytopathol 164(9):631–640

18. Choi YH, Tapias EC, Kim HK et al (2004) Metabolic discrimination of Catharanthus roseus leaves infected by phytoplasma using 1H-NMR spectroscopy and multivariate data analysis. Plant Physiol 135(4):2398–2410

19. Riedle-Bauer M, Sára A, Regner F (2008) Transmission of a stolbur phytoplasma by the Agalliinae leafhopper Anaceratagallia ribauti (Hemiptera, Auchenorrhyncha, Cicadellidae). J Phytopathol 156(11–12):687–690

20. Motohashi R, Enoki H, Fukazawa C, Kiriiwa Y (2015) Hydroponic culture of 'micro-tom' tomato. Bio-Protoc 5(19):e1613

Chapter 3

Phytoplasma Transmission: Insect Rearing and Infection Protocols

L. Pagliari, J. Chuche, D. Bosco, and D. Thiéry

Abstract

Phytoplasmas are obligate pathogens and thus they can be studied only in association with their plants or insect hosts. In this chapter, we present protocols for rearing some phytoplasma insect vectors, to obtain infected insects and plants under controlled environmental conditions. We focus on *Euscelidius variegatus* and *Macrosteles quadripunctulatus* that can infect *Arabidopsis thaliana*, and *Hyalesthes obsoletus* and *Scaphoideus titanus*, that can infect grapevine.

Key words *Euscelidius variegatus*, *Macrosteles quadripunctulatus*, *Hyalesthes obsoletus*, *Scaphoideus titanus*, Infection protocol

1 Introduction

Since their discovery, the study of phytoplasmas has been hampered by the impossibility to culture them in vitro, although Contaldo et al. [1] reported some positive results, so far not confirmed by other laboratories. In fact, they lack several pathways for the synthesis of compounds considered to be necessary for the cell metabolism [2], exhibiting a strong host-specific correlation. This made necessary to study phytoplasmas associated with their hosts, plants, or insect vectors. Pathosystems developed under controlled environmental conditions allow us to focus on host-pathogen interaction, excluding unpredictable factors related to field condition. In this chapter we present protocols for rearing some phytoplasma vectors, to obtain infected insects and plants and, ultimately, to maintain phytoplasma strains by insect transmission. We focus on rearing protocols of two insects, *Euscelidius variegatus* and *Macrosteles quadripunctulatus*, that can infect *Arabidopsis thaliana* [3–5], that is a model plant of growing interest also for the studies of phytoplasma-plant interactions. Moreover, we present two important case studies for European viticulture, *Scaphoideus titanus* vector of flavescence dorée (FD) phtoplasma [6], and *Hyalesthes*

Rita Musetti and Laura Pagliari (eds.), *Phytoplasmas: Methods and Protocols*, Methods in Molecular Biology, vol. 1875,
https://doi.org/10.1007/978-1-4939-8837-2_3, © Springer Science+Business Media, LLC, part of Springer Nature 2019

obsoletus vector of Bois noir [7]. We provide a data sheet for each vector species with details on rearing and acquisition and transmission techniques.

2 Materials

2.1 *Insect Rearing and Infection*

1. Growth chambers. All phytoplasma-infected insects and plants should be kept in controlled growth rooms. Main requirements include sealed access, timer-controlled fluorescent lighting, temperature, and humidity control.

2. Glasshouse for growing plants for maintenance of insect colonies. In some countries, contained facilities for quarantine phytoplasma-infected plants (for *S. titanus* insects infected by FD phytoplasma) are required.

3. Large plexiglass and net cages for insect colony development (in case of *E. variegatus* and *M. quadripunctulatus*, and *H. obsoletus*) or egg hatchings (for *S. titanus*) (Fig. 1). To avoid fungi infection on herbaceous plants, cages should be properly cleaned with sanitizing products and should have large windows protected by insect-proof net.

4. Host plants (*see* Subheadings 3.1.3 and 3.2.3 for further details), proper pots, sterilized with sodium hypochlorite, and proper soil for each kind of plant.

5. Aspirator for carefully handling the insects. A simple mouth aspirator can be built with a 50 ml centrifuge tube, two pieces of 50 cm of transparent plastic tube (with a diameter of 5–7 mm) and a piece of gauze, to be used as a filter. Cut the tube lid, to create a hole where fixing one piece of the plastic transparent tube: it will be used for insect aspiration. At the same extremity of the tube, stick a small piece of gauze (which will avoid the accidental insect swallowing by the operator). Cut the bottom of the tube to create a hole where fixing the other piece of the plastic transparent tube: it will be inserted in the cage to capture the insects and to collect them inside the tube (*see* **Note 1**).

6. Insecticides and fungicide chemicals.

2.2 *Host Plant Infection*

1. For most of the host plants here described, a plexiglass cage can be used.

2. For Arabidopsis infection, plants should be exposed to infective insects (*see* **Note 2**). For this purpose, gauze cage or plexiglass tubes can be used.

Fig. 1 Example of a plastic cage for insect colony development (in case of *E. variegatus* and *M. quadripunctulatus*) and egg hatchings (for *S. titanus* and *H. obsoletus*)

2.3 Vector Infection Via Abdominal Microinjection

1. Autoclaved ceramic pestle and mortars.

2. Injection buffer. Prepare 30 μl of buffer per insect as follows: 300 mM glycine, 30 mM $MgCl_2$, pH 8.0 [8].

3. 0.45 μm sterile filters, glass capillaries, needle-puller device, or a Cell Tram Oil microinjector (Eppendorf).

4. CO_2 flush.

2.4 Artificial Feeding

1. 1.5 ml microcentrifuge tubes, cotton wool, and Parafilm.

2. Feeding solution: 5% sucrose in TE [8, 9], or 5% sucrose, 10 mM Tris/Cl, 1 mM EDTA, pH 8.0 [10, 11].

3 Methods

3.1 Euscelidius variegatus and Macrosteles quadripunctulatus

3.1.1 Insect Description

E. variegatus is commonly found in weeds, lawns, pastures in Europe, Asia, northern Africa, and the Western United States [12, 13]. It has a light brown color with numerous fuscous markings on body (Fig. 2a). Nymphs can be recognized by the absence of wing and marked transverse stripes (Fig. 2b, c). Adults are characterized by a medium size, ranging from 3.90–4.50 mm (male) to 4.10–5.50 mm (female). Nymph development requires five instar stages, each of them lasting approximately 1 week [14]. *E. variegatus* is a known vector of phytoplasmas of clover phyllody [15], aster yellows [16], X-disease [17], Chrysanthemum yellows [18], and FD [19], although it cannot acquire this phytoplasma from infected grapevines [8, 20].

M. quadripunctulatus is widely distributed from western Europe to central Russia, south to Cyprus, Iraq, and Kashmir [21]. It inhabits dry climatic regions, preferring scant, disperse vegetation and dry, well-drained soils. Among the plant species, it prefers *Medicago sativa* (L.), *Trifolium repens* (L.), *Agropyron repens* (L.) Beauv., *Poa pratensis* (L.), and *Digitaria sanguinalis* (L.) [22]. It has a greenish yellow color, with two pairs of black spots at the vertex and short black longitudinal bands between eyes and ocelli and ocelli reddish [21]. Length (including tergmen) varies from 2.9–3.3 mm (male) to 3.2–3.7 mm (female) [23] (Fig. 2d–f). Its transmission capacity has been demonstrated for aster yellows (AY) phytoplasma in lettuce and carrot [24], Kok-saghyz yellows [25], and chrysanthemum yellows (CY) phytoplasma [18, 26, 27].

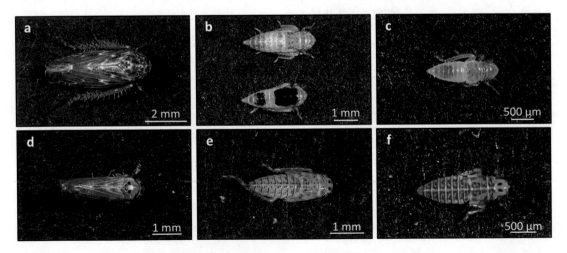

Fig. 2 *E. variegatus* development stages: adult (**a**), late instar nymphs (**b**), early instar nymphs (**c**). *M. quadripunctulatus* development stages: adult (**d**), late instar nymphs (**e**), early instar nymphs (**f**). Late instar nymphs (**b, e**) are transferred to CY-infected daisy plants

3.1.2 Rearing	Both leafhoppers can be reared in plexiglass cages on oat (*Avena sativa* L.) (*see* **Note 3**) at 20/22 °C and long-day conditions (16 h light/8 h dark period). Since they are multivoltine insects and thus they breed continuously, an age-structured rearing is advised. An oviposition chamber where adults lay eggs on host plants for a short period (from few days to 1 week) is recommended. After the oviposition, host plants are moved to new cages where the eggs give rise to the nymphs. Considering that, a complete egg hatching needs from 3 to 4 weeks [14], after a month oat plants should be replaced with fresh ones, to allow the nymphs to feed on good quality plants.

3.1.3 Host Plant Cultivation

Chrysanthemum carinatum (Schousboe) (daisy) plants are grown from seed in greenhouse at 20/22 °C, under long-day conditions (16 h light/8 h dark period), at $100\ \mu Em^{-2}\ s^{-1}$ light intensity (with Plant Growth fluorescent lamps) and 50–70% humidity. After about 10 days from seedling emergence, plants are transplanted into 8 cm pots. Roughly 40 days after germination, plants at the 6–8 leaf stage are exposed to infective leafhoppers.

A. thaliana plants are grown at 20/22 °C, under short-day conditions (9 h light/15 h dark period) (*see* **Note 4**), at $120–150\ \mu Em^{-2}\ s^{-1}$ light intensity (with Plant Growth fluorescent lamps) and 50–70% humidity. Seeds are hydrated on wet blotting paper at room temperature for roughly 3 h. Soaked seeds are posed in soil and vermiculite mixture and pots are maintained at 4 °C, for the so-called stratification period, to improve germination rate and synchrony. After 3 days, pots are placed in the growth room in high-humidity conditions, which can be easily reached covering the pots with a close-fitting clear plastic dome. After 5 days, the dome is slightly displaced to reduce relative humidity gradually. After a few days of acclimation, the dome can be removed entirely. 20-day-old plants are than transplanted in single 8 cm pots (*see* **Note 5**).

3.1.4 Transmission Protocol

As explained above, both leafhoppers can acquire and transmit different phytoplasmas. Here, we present the protocol for the transmission of CY phytoplasma that can be easily transmitted to *A. thaliana*.

Late instar nymphs (Fig. 2b, e) (*see* **Note 6**) are taken from healthy colonies grown on oats and transferred to CY-infected daisy plants (*see* **Note 7**) for a 7-day acquisition-feeding period. Twenty (for *M. quadripunctulatus*) or thirty days (for *E. variegatus*) after nymph transfer, 45-day-old *A. thaliana* plants [corresponding to the 3.50 growth stage [28]] are individually exposed to 3 infective insects. Control plants are exposed to healthy insects. At the end of the 7-day inoculation feeding period, insects are manually removed and/or plants are treated with insecticide (*see* **Note 8**). Infected and control plants are maintained in two separated insect-proof

Fig. 3 Healthy (**a**) and infected (**b**) daisies. Infected daisies are characterized by leaf yellowing, reduced growth, and a clear curvature of the apical part. Healthy (**c**) and infected (**d**) Arabidospis. Infected Arabidopsis are yellowing and stunted, with shorter leaves, with a thick main vein and a smaller petiolar area

plexiglass cagesuntil symptom development [roughly 20 days after inoculation, corresponding to the 3.90 growth stage [28]].

Purcell et al. [12] noticed rod-shaped bacteria in the hemolymph of *E. variegatus*, designated as BEV. Even if the presence of endosymbionts and parasitic bacteria in insects may influence transmission ability [29], it was demonstrated that in *E. variegatus* BEV do not interfere with CY transmission [30].

3.1.5 Symptom Development

Within 20 days from inoculation, common phytoplasma disease symptoms appear on plants exposed to infective insects. Considering that CY is transmitted by single *M. quadripunctulatus* and *E. variegatus* with high efficiency (on daisy plants with transmission rates of 100% and 82%, respectively), almost all the plants exposed to infective insect should develop symptoms [31].

In contrast to healthy plants (Fig. 3a), infected daisies are characterized by leaf yellowing and reduced growth (Fig. 3b). The apical part of the plant shows clear curvature (of roughly 30°) in comparison to the plant growth vertical axis, with shorten internodes and short and thick leaves.

In Arabidospis, yellowing and general stunting are accompanied by a decrease of ~40% in growth [5]. In contrast to healthy plants (Fig. 2c), leaves having emerged after phytoplasma inoculation were shorter, with a thick main vein and a smaller petiolar area (Fig. 2d).

3.1.6 Vector Infection Via Abdominal Microinjection

Microinjection is a useful tool for studying transmission mechanisms by vectors as it allows us to overcome the barrier of salivary glands, reduce latency period and is highly efficient [30]. Moreover, with this technique it is possible to test for transmission insects with different feeding habits or insects that could not feed properly on the source plants used. This technique, first set by Black [32] to deliver phytoplasma extracts into leafhoppers, was used by [8] to assess vector specificity of FD phytoplasma in several species of Cicadellidae [10] and to assess the role of midgut barrier in phytoplasma colonization of the insect [30]. The same technique is applied to leafhopper vectors of spiroplasmas [33].

Phytoplasma suspension for microinjection is prepared by crushing infected *E. variegatus* in ice-cold filter sterilized injection buffer. The extract is centrifugated (10 min, 800 g, 4 °C) and the supernatant filtered through 0.45 μm sterile filters. Newly emerged healthy *E. variegatus* adults are anesthetized by CO_2 flushing for few seconds and, immediately, about 0.2 μl of solution is injected between two abdominal segments under a stereomicroscope. The phytoplasma suspension must be maintained on ice and used within 4 h [10, 30]. This technique requires some skills from the operator to avoid high mortality rates.

3.1.7 Artificial Feeding: A Test for the Transmission Ability of Candidate Insect Vectors

The transmission ability of candidate insect vectors is normally tested by PCR only after the inoculation of the target plant, because of the destructive feature of this technique. A rapid and nondestructive method is the use of artificial feeding assays, that is successfully applied with several Hemiptera species and was developed also for leafhoppers and planthoppers [8, 9, 11, 21, 34]. Moreover, this technique can be adopted when the host plant species is unknown or test plants are poor hosts for the potential vector.

1.5–2 ml microcentrifuge tubes can be used as insect chambers. Caps (*see* **Note 9**) are filled with 200 μl of 5% sucrose in TE [8, 9] and sealed with Parafilm. The bottom ends of tubes are cut, to introduce an insect. Finally, the cut-ends are sealed with cotton wool and each tube is kept at 23 to 25 °C for 48 to 72 h in a vertical position with the cap facing a light source to attract the insects to the feeding medium. At the end of the inoculation period (*see* **Note 10**), the artificial feeding buffer is gently aspirated and processed for phytoplasma detection (*see* **Note 11**).

3.2 Scaphoideus titanus and Hyalesthes obsoletus

3.2.1 Insect Description

Scaphoideus titanus (Hemiptera: Cicadellidae) (Fig. 4) is the main vector of FD phytoplasma and widespread in most European vineyards and was reported in most American states and Canadian provinces [6]. Nymph color varies according to the nymphal stage. At hatching they are almost translucent, go through a milky white, then become ivory white at the end of the second instar. In the third instar, the nymph become an ivory yellow more and more accentuated as they age. Finally, the fourth and fifth instars are characterized by the appearance of light to dark brown irregular spots and the appearance of wing and elytral drafts [35]. Nymphs have two black spots arranged symmetrically in the dorso-lateral position at the posterior end of the abdomen [35]. Adult females are larger (5.5–5.8 mm) than males (4.8–5 mm) and there are three brown transverse bands at the vertex level for females, compared to only one for males [35]. Nymph development requires five instar

Fig. 4 Nymphs (**a**) and adult (**b**) of *Scaphoideus titanus*

Fig. 5 Nymphs (**a**) and adult (**b**) *Hyalesthes obsoletus* on *Urtica dioica* in a rearing cage

stages, each of them lasting approximately 10 days [36]. This leaf-hopper is mostly recorded on *Vitis vinifera* in Europe, while in North America, *V. labrusca* and *V. riparia* are reported as the preferred host plant [37, 38]. Five nymphal instars lead to adults in about 50 days in laboratory conditions [36]. Nymphal instars can be distinguished using the key of Della Giustina et al. [39]. Longevity of males can reach 55 days while females can live up to 80 days, but infected individuals have a reduced lifespan [40]. *S. titanus* is the main vector of phytoplasmas responsible of FD [35]. However, *S. titanus* can also transmit phytoplasma from the 16SrI-C [41, 42] and 16SrI-B groups [43] and it was introduced in Europe probably as a consequence of the massive importation of American rootstocks after the phylloxera crisis [6].

H. obsoletus (Hemiptera: Cixiidae) (Fig. 5) is a Paleartic species and the main natural vector of "*Candidatus* Phytoplasma solani," (16SrXII-A genetic group) which is associated with diseases of several crops such as grapevine, lavender, maize, and potato. Nymphs are white to brown, with red eyes, and the abdomen is bearing a lot of wax secreted by wax plates. one to third larval instars do not have compound eyes [44]. Adults are dark with red eyes two and females have wax glands and are often bearing wax on the end of their abdomen. Length is about 4 mm with females that are bigger than males and host plant having an effect on the adult size. This mesophilic species is monovoltin in Europe but bivoltin in Israel [45]. Females lay their eggs in the soil near the basis of different host plants species. After hatching, the five nymphal instars live underground and feed on the roots. Acquisition of "*Ca.* P. solani" can be achieved by nymphs feeding on infected root phloem, while transmission to cultivated and wild plants is done by flying adults during summer. Because most of the crops affected by "*Ca.* P. solani" are dead-end hosts phytoplasma acquisition occurs on wild plants. This insect is found on a great diversity of plants, mainly wild. About 19 species belonging to 10 different plant families are known to harbor both nymphs and adults but adults can be observed on more species [46]. Multiple hosts enhance the opportunity for genetic local adaptations and led for *H. obsoletus* to the existence of host races which specialize on different host plants [47]. As a consequence, a *H. obsoletus* from a host plant cannot be reared on another host plant.

3.2.2 Rearing

S. titanus is a univoltine species that is difficult to rear in captivity from egg to egg. It is possible to make all the life cycle in controlled conditions, but it is time consuming and no one has succeeded yet in obtaining more than one generation a year. Nymphs can be obtained from eggs laid in the field by collecting two-year-old (or older) grapevine woody canes during winter [48]. Woody canes should be collected in vineyards with high populations of the leafhopper (yellow sticky traps can estimate the population level during summer). After collection, the woody canes can be checked to see if they are bearing eggs, and then are kept in a cold room at 5 ± 1 °C and 85–90% humidity until egg incubation. Egg hatchings are obtained by placing woody canes inside plastic hatching cages ($50 \times 38 \times 36$ cm) in a climatic chamber under a 16 h light/ 8 h dark period, at 23 ± 1 °C, and 65–70% humidity. To avoid eggs desiccation a 1 cm layer of vermiculite is placed below the eggs and is kept humid. To harvest neonate nymphs, food is provided by placing grapevine leaves maintained in a glass tube with water, ca. 20 days after the canes with eggs are removed from the cold room. Leaves must be replaced when they began to wither. Nymphs can be reared individually or in group on grapevine cutting [49]. Nymphs and adults can also be kept alive by placing them in a small container in which a grapevine leaf disk was laid over a 1 cm layer of technical agar solution [0.8% (wt/vol)] at the bottom and is replaced twice a week [50]. Alternatively, woody canes can be placed in larger cages together with small potted grapevine and broad bean plants: hatched nymphs can survive very well under these conditions and can be collected for experiments when needed. In order to extend the period of egg hatchings and perform transmission experiments over the year, grapevine wood with eggs can be stored in cold chamber at 4–6 °C for some months.

H. obsoletus can be reared on different species, e.g., *Lavandula angustifolia* (lavender), *Urtica dioica* (nettle), and *Salvia sclarea* (clary sage) [44, 51, 52]. Because it is quite impossible to collect eggs in the soil, rearing should start with nymphs from uprooted plants or adults collected in the field. Females lay their eggs at the basis of seedlings or just under the soil surface. It is impossible to know the sanitary status of captured nymphs and adults. Thus, after egg laying and before first hatchings, eggs should be gently transferred close to the roots of a new plant obtained from seed to initiate a phytoplasma-free rearing. This step can be avoided to obtain diseased plants further used for phytoplasma acquisition.

Hereafter we describe the rearing on nettle. Plants are kept at 25 ± 1 °C under a 16 h light/8 h dark period and 80% relative humidity. The soil used in the pots should be loose to allow nymphs movements and access to roots, for example a mixture of 50% peat and 50% small gravel, ca. 0.5 mm diameter [44]. Pots are watered every 2 days by pouring water into the saucers in which each pot was standing. Care is taken to prevent overwatering to avoid egg and nymphal drowning. After about 1 month, nymphs hatched and developed above ground at the base of the plant. Adult emergence starts about 2 months later, and they are gradually transferred to a new cage to initiate the next generation. Only a small proportion of eggs give rise to adults due to a high nymphal mortality rate.

3.2.3 Host Plant Cultivation

Vitis vinifera (grapevine) can be obtained from cuttings or in vitro plantlets. Cuttings are grown in a potting compost mix (Substrate 5; Klasmann-Deilmann, Geeste, Germany) at 22 ± 2 °C, under long-day conditions (16 h light/8 h dark period) and irrigated twice a week. Plantlets are planted into 12 cm pots filled with same compost mix as cuttings [53]. Plantlets are grown at 25 ± 2 °C, and 50–80% humidity, under illumination of 50 $\mu Em^{-2} s^{-1}$ for a 16 h light/8 h dark period (Osram, Lumilux), and under 400 W high pressure sodium lamps for a 14 h light/10 h dark period, in vitro and in the greenhouse, respectively. Since no *Vitis* plant is known to be immune to FD phytoplasma (FDP), in theory all *Vitis* species/cv can be used in transmission experiments. However, since infected rootstocks are generally non-symptomatic and cultivated varieties differ a lot in FDP susceptibility [53, 54], the use of a very susceptible variety is advised to obtain a better estimation of the vector infectivity (*see* **Note 12**).

Vicia faba (broadbean) plants are grown from seed under the same conditions than grapevine cuttings.

L. angustifolia, *S. sclarea*, and *U. dioica* plants are grown from seeds in pots (12 cm diameter) filled with a peat-based standard substrate containing a slow-release fertilizer in a heated greenhouse at 25 ± 1 °C under a 16 h light:8 h dark photoperiod and 80% relative humidity.

3.2.4 Transmission Protocol	Because FD is listed in Quarantine Pests for Europe (*see* **Note 13**), all experiments using FDP, grapevine, and *S. titanus* should be done in confined greenhouse and under safety rules. FDP can be maintained by continuous broadbean to broadbean transmissions, using *E. variegatus* as an alternative vector [20]. When needed, *S. titanus* can be fed on FDP-infected broad beans, which ensure a higher acquisition rate compared to infected grapevines [48]. Third to fifth instar nymphs are caged on FD—broadbeans to allow FDP acquisition. After 1 week of acquisition access period, insects are transferred onto grapevine cuttings for a 3–4 weeks latency period. Then, a variable number of adults are caged on a grapevine plant and removed after 1 week. Plants are treated with insecticide (*see* **Note 8**) and maintained in confined greenhouse at 25 °C constant, which is the optimum temperature for the multiplication of FDP [55]. The transmission rate of FDP is higher in males than in females [56], probably due to the different feeding behavior observed between both sexes [57]. Test plants for FDP transmission assays can be cuttings (grafted or not), which generally require 1 year to be scored for infection, or herbaceous ex vitro plantlets, that are very susceptible to phytoplasma infections and develop disease symptoms and can be scored by infection by PCR in some weeks [53].

Due to the existence of host plant races, phytoplasma acquisition by *H. obsoletus* should not be done on any plant species. One possible way is to rear two distinct strains, a healthy and an infected one. The main difficulty is avoiding cross-contamination, by keeping the rearings in separate chambers/greenhouses but this system can produce infected vectors anytime. It is also possible to transfer third and fourth instar nymphs from a healthy plant to an infected one and collect the adult that would be infective 4–5 weeks later. In laboratory/greenhouse conditions, nymphs can develop on the aerial parts of the plants.

3.2.5 Symptom Development	At 5 and 10 weeks post-transmission, the FD symptoms and the number of contaminated grapes can easily be scored, when using ex vitro plantlets [53]. Since evaluation by symptoms is not reliable enough on grapevine plantlets, PCR assays of leave and/or root tissues for FDP presence are needed. Test cuttings (grafted or not) generally requires 1 year to be scored for infection. The first visible

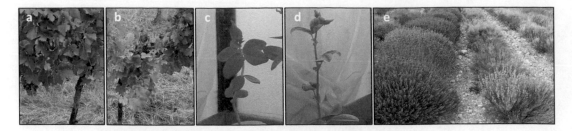

Fig. 6 Healthy (**a**) and Flavescence doree symptomatic (**b**) *Vitis vinifera* cv. Cabernet Sauvignon in a vineyard. Healthy (**c**) and Flavescence doree symptomatic (**d**) *Vicia faba*. Healthy (left) and lavender decline symptomatic (right) *Lavandula angustifolia* in the field (**e**)

symptoms on grapevine cuttings are on leaves that roll downward and become yellowish for white cultivars or reddish for red ones. The new shoots then become weeping shaped due to a lack of lignification and some black punctuations can be observed on the petioles. Hydric stress increases the symptoms expression [58] (Fig. 6a, b).

Symptomatic broadbeans are stunted and had small leaves with upward-curled edges. Flowers are not affected (Fig. 6c, d).

Symptoms of lavender decline, the diseased caused by "*Ca*. P. solani*," are yellowing and either standing up or rolling down of the leaves, and reduction and abortion of inflorescences [59]. Infected clary sages show typical symptoms of stolbur, such as stunted and very small leaves [52] (Fig. 6e).

4   Notes

1. When sucking insects with a mouth aspirator, the operator should shake the plants to push the insects out of the plants and collect only those ones that are along the cage walls: insects set on plants could have their stylet inserted in the leaf tissue, and thus being damaged by aspiration. Moreover, leaves can be contaminated by thrips that can be transferred from one cage to another together with vectors.

2. Efficient transmission can be obtained with just one insect, nevertheless, when one wishes to maximize transmission, the use of batches of infective insects per plant is advisable.

3. *E. variegatus* and *M. quadripunctulatus* can be reared also on barley (*Hordeum vulgare*), wheat (*Triticum* spp.), and perennial ryegrass (*Lolium perennae*) ([12], **Bosco, personal observation**). *M. quadripunctulatus* can be reared also on barley (*Hordeum vulgare*) and perennial ryegrass (*Lolium perennae*).

4. Short-day conditions enhance vegetative growth, necessary for the following steps of the infection protocol.

5. Extensive information about *A. thaliana* cultivation can be found in "101 ways to try to grow Arabidopsis" in Horticulture Department of Purdue University website (https://ag. purdue.edu/hla/Hort/Greenhouse/Pages/101-Ways-to-Grow-Arabidopsis.aspx), which describes various aspects of Arabidopsis growing.

6. While in *M. quadripunctulatus* the acquisition efficiency of CYP is not influenced by nymph life stage, in *E. variegatus* late instar nymphs are more efficient in acquiring CY phytoplasma when compared to younger stages [18].

7. *Catharantus roseus* (L.) can be used as an alternative CY-source plant, even if both leafhoppers show better acquisition efficiency on daisy and suffer high mortality on this plant species [18].

8. Different insecticides can be used, but in case the inoculated plants have to be further used as inoculum source for insects, non-residual insecticides should be used.

9. White microcentrifuge tubes are indicated to guarantee maximum transparency, nevertheless, a plastic yellow transparent film can be placed on the top of the Eppendorf tubes to attract insect to the feeding medium.

10. Despite that insects are handled with the maximum care, some of them do not survive to this treatment. Because of the sucrose-TE diet, survival of *E. variegatus* (in the fourth and fifth weeks after the beginning of acquisition) is lower than 50% [9]. Dead insects can be collected and PCR-tested for phytoplasma presence.

11. Briefly, phytoplasma cells are pelleted by centrifugation at $12,000 \times g$ for 15 min and diluted in 10 µl of 0.5 M NaOH and 20 µl of 1 M Tris–HCl (pH 8.0), containing 1% sodium dodecyl sulfate and 20 mM EDTA. After a 15 min incubation at 65 °C, genomic DNA is precipitated with two volumes of ethanol and dissolved in 30 µl of TE. For other detail, please refer to [11].

12. Highly susceptible cultivars, as Baco 22A, Chardonnay, Barbera, Cabernet Sauvignon, are suggested.

13. For any information about quarantine pests in Europe please refer to EPPO website (https://www.eppo.int/QUARAN TINE/quarantine.html). EPPO is an intergovernmental organization responsible for cooperation and harmonization in plant protection within the European and Mediterranean region. EPPO lists are reviewed every year by the Working Party on Phytosanitary Regulations and approved by Council.

Acknowledgment

The authors would like to thank Alberto Loschi (University of Udine, Italy), Filippo Bujan (University of Udine, Italy), and Dr. Stefano Demichelis (University of Torino, Italy) for their help in taking pictures.

References

1. Contaldo N, Bertaccini A, Paltrinieri S et al (2012) Axenic culture of plant pathogenic phytoplasmas. Phytopathol Mediterr 51:607–617

2. Bai X, Zhang J, Ewing A et al (2006) Living with genome instability: the adaptation of phytoplasmas to diverse environments of their insect and plant hosts. J Bacteriol 18:3682–3696

3. Cettul E, Firrao G (2011) Development of phytoplasma-induced flower symptoms in *Arabidopsis thaliana*. Physiol Mol Plant Pathol 76:204–211

4. Pacifico D, Galetto L, Rashidi M et al (2015) Decreasing global transcript levels over time suggest that phytoplasma cells enter stationary phase during plant and insect colonization. Appl Environ Microbiol 81(7):2591–2602

5. Pagliari L, Buoso S, Santi S et al (2017) Filamentous sieve element proteins are able to limit phloem mass flow, but not phytoplasma spread. J Exp Bot 68(13):3673–3688

6. Chuche J, Thiéry D (2014) Biology and ecology of the Flavescence dorée vector *Scaphoideus titanus*: a review. Agron Sust Devel 34:381–403

7. Maixner M (1994) *Hyalesthes obsoletus* (Auchenorrhyncha: Cixiidae). Vitis 33:103–104

8. Bressan A, Clair D, Semetey O et al (2006) Insect injection and artificial feeding bioassays to test the vector specificity of flavescence Doree phytoplasma. Phytopathology 96(7):790–796

9. Tanne E, Boudon-Padieu E, Clair D et al (2001) Detection of phytoplasma by polymerase chain reaction of insect feeding medium and its use in determining vectoring ability. Phytopathology 91:741–746

10. Rashidi M, Galetto L, Bosco D et al (2015) Role of the major antigenic membrane protein in phytoplasma transmission by two insect vector species. BMC Microbiol 15(1):193

11. Bosco D, Tedeschi R (2013) Insect vector transmission assays. In: Dickinson M, Hodgetts J (eds) Phytoplasma: methods in molecular biology (methods and protocols), vol 938. Humana Press, Totowa, NJ

12. Purcell AH, Steiner T, Mégraud F et al (1986) In vitro isolation of a transovarially transmitted bacterium from the leafhopper *Euscelidius variegatus* (Hemiptera: Cicadellidae). J Invertebr Pathol 48(1):66–73

13. Reis F, Aguin-Pombo D (2003) *Euscelidius variegatus* (Kirschbaum, 1858), a new leafhopper record to Madeira archipelago (Hemiptera, Cicadellidae). In: Vieraea, vol 31, pp 27–31

14. Caudwell A, Larrue J (1977) La production de cicadelles saines et infectieuses pour les épreuves d'infectivité chez les jaunisses à Mollicutes des végétaux. L'élevage de *Euscelidius variegatus* KBM et la ponte sur mousse de polyuréthane. Ann Zool Ecol Anim 9:443–456

15. Giannotti J (1969) Transmission of clover phyllody by a new leafhopper vector, *Euscelidius variegatus*. Plant Dis Rep 53:173

16. Severin HHP (1947) Newly discovered leafhopper vectors of California Aster-yellows virus. Phytopathology 37(5):364

17. Jensen DD (1969) Comparative transmission of western X-disease virus by Colladonus montanus, C. Geminatus, and a new leafhopper vector, *Euscelidius variegatus*. J Econ Entomol 62(5):1147–1150

18. Palermo S, Arzone A, Bosco D (2001) Vector-pathogen-host plant relationships of chrysanthemum yellows (CY) phytoplasma and the vector leafhoppers *Macrosteles quadripunctulatus* and *Euscelidius variegatus*. Entomol Exp Appl 99(3):347–354

19. Lefol C Lherminier J, Boudon-Padieu E et al (1994) Propagation of Flavescence dorée MLO (mycoplasma-like organism) in the leafhopper vector *Euscelidius variegatus* Kbm. J Invertebr Pathol 63(3):285–293

20. Caudwell A, Kuszala C, Larrue J et al (1972) Transmission de la Flavescence dorée de la Fève à la Fève par des cicadelles des genres Euscelis et Euscelidius. Intervention possible de ces insectes dans l'épidémiologie du Bois noir en Bourgogne. Ann Phytopathol 1:181–189

21. Zhang J, Miller S, Hoy C et al (1998) A rapid method for detection and differentiation of aster-yellows phytoplasma-infected and

inoculative leafhoppers. Phytopathology 88 (Suppl):S84

22. Kirby P (2000) Some records of *Macrosteles quadripunctulatus* (Kirschbaum) (Hemiptera: Cicadellidae). Br J Entomol Nat History 13 (1):67–68

23. Kwon YJ (1988) Taxonomic revision of the leafhopper genus' Macrosteles' fieber of the world (Homoptera: Cicadellidae) Doctoral dissertation, University of Wales, College of Cardiff

24. Orenstein S, Franck A, Kuznetzova L et al (1999) Association of phytoplasmas with a yellows disease of carrot in Israel. J Plant Pathol 81:193–199

25. Brcak J (1979) Leafhopper and planthopper vectors of plant disease agents in central and southern Europe. In: Maramorosch K, Harris KF (eds) Leafhopper vectors and plant disease agents. Academic Press, London, pp 97–146

26. Minucci C, Boccardo G (1997) Genetic diversity in the stolbur phytoplasma group. Phytopathol Mediterr 36(1):45–49

27. Alma A, Conti M, Boccardo G (2000) Leafhopper transmission of a phytoplasma of the 16Sr-IB group [Chrysanthemum yellows (CY)] to grapevine [Vitis vinifera L.]. Petria (Italy)

28. Boyes DC, Zayed AM, Ascenzi R et al (2001) Growth stage–based phenotypic analysis of Arabidopsis a model for high throughput functional genomics in plants. Plant Cell 13 (7):1499–1510

29. Beard CB, Durvasula RV, Richards FF (1998) Bacterial symbiosis in arthropods and the control of disease transmission. Emerg Infect Dis 4 (4):581

30. Galetto L, Nardi M, Saracco P et al (2009) Variation in vector competency depends on chrysanthemum yellows phytoplasma distribution within *Euscelidius variegatus*. Entomol Exp Appl 131(2):200–207

31. Bosco D, Galetto L, Leoncini P et al (2007) Interrelationships between 'Candidatus Phytoplasma asteris' and its leafhopper vectors (Homoptera: Cicadellidae). J Econ Entomol 100:1504–1511

32. Black LM (1940) Mechanical transmission of aster yellows virus to leafhoppers. Phytopathology 30:2–3

33. Foissac X, Danet JL, Saillard C et al (1997) Mutagenesis by insertion of Tn4001 into the genome of Spiroplasma citri: characterization of mutants affected in plant pathogenicity and transmission to the plant by the leafhopper vector Circulifer haematoceps. Mol Plant Microbe In 10(4):454–461

34. Ge Q, Maixner M (2003) Comparative experimental transmission of grapevine yellows phytoplasmas to plants and artificial feeding medium. Pages 109–110. 14th Meeting of the International Council for the Study of Virus and Virus-Like Diseases of the Grapevine (ICVG), Locorotondo, Italy, 12–17 Sept 2003

35. Schvester D, Moutous G, Carle P (1962) Scaphoideus littoralis Ball. (Homopt. Jassidae) cicadelle vectrice de la Flavescence dorée de la vigne. Rev Zool Agr Appl 12:118–131

36. Boudon-Padieu E (2000) Cicadelle vectrice de la flavescence dorée, *Scaphoideus titanus* Ball, 1932. In: Stockel J (ed) Ravageurs de la vigne. Féret, Bordeaux, pp 110–120

37. Vidano C (1964) Scoperta in Italia dello Scaphoideus littoralis Ball cicalina americana collegata alla «Flavescence dorée» della Vite. L'Italia agricola 101:1031–1049

38. Maixner M, Pearson RC, Boudon-Padieu E et al (1993) *Scaphoideus titanus*, a possible vector of grapevine yellows in New York. Plant Dis 77:408–413

39. Della Giustina W, Hogrel R, Della Giustina M (1992) Description des différents stades larvaires de *Scaphoideus titanus* Ball (Homoptera, Cicadellidae). Bull Soc Entomol Fr 97:269–276

40. Bressan A, Girolami V, Boudon-Padieu E (2005) Reduced fitness of the leafhopper vector *Scaphoideus titanus* exposed to Flavescence dorée phytoplasma. Entomol Exp Appl 115:283–290

41. Caudwell A, Larrue J, Kuszala C et al (1971) Pluralité des jaunisses de la vigne. Ann Phytopathol 3:95–105

42. Boudon-Padieu E, Larrue J, Caudwell A (1990) Serological detection and characterization of grapevine Flavescence dorée MLO and other plant MLOs. IOM Letters 1:217–218

43. Alma A, Palermo S, Boccardo G et al (2001) Transmission of chrysanthemum yellows, a subgroup 16SrI-B phytoplasma, to grapevine by four leafhopper species. J Plant Pathol 83:181–187

44. Sforza R, Bourgoin T, Wilson ST et al (1999) Field observations, laboratory rearing and descriptions of immatures of the planthopper *Hyalesthes obsoletus* (Hemiptera: Cixiidae). Eur J Entomol 96:409–418

45. Sharon R, Soroker V, Wesley SD et al (2005) Vitex agnus-castus is a preferred host plant for *Hyalesthes obsoletus*. J Chem Ecol 31:1051–1063

46. Riolo P, Minuz R, Anfora G et al (2012) Perception of host plant volatiles in *Hyalesthes*

obsoletus: behavior, morphology, and electrophysiology. J Chem Ecol 38:1017–1030

47. Johannesen J, Lux B, Michel K et al (2008) Invasion biology and host specificity of the grapevine yellows disease vector *Hyalesthes obsoletus* in Europe. Entomol Exp Appl 126:217–227

48. Caudwell A, Kuszala C, Bachelier JC et al (1970) Transmission de la Flavescence dorée de la vigne aux plantes herbacées par l'allongement du temps d'utilisation de la cicadelle Scaphoideus littoralis BALL et l'étude de sa survie sur un grand nombre d'espèces végétales. Ann Phytopathol 2:415–428

49. Chuche J, Thiéry D (2009) Cold winter temperatures condition the egg-hatching dynamics of a grape disease vector. Naturwissenschaften 96(7):827–834

50. Mazzoni V, Lucchi A, Cokl A et al (2009) Disruption of the reproductive behaviour of *Scaphoideus titanus* by playback of vibrational signals. Entomol Exp Appl 133:174–185

51. Kessler S, Schaerer S, Delabays N et al (2011) Host plant preferences of *Hyalesthes obsoletus*, the vector of the grapevine yellows disease 'bois noir', in Switzerland. Entomol Exp Appl 139:60–67

52. Chuche J, Danet JL, Rivoal JB et al (2018) Minor cultures as hosts for vectors of extensive crop diseases: does *Salvia sclarea* act as a pathogen and vector reservoir for lavender decline? J Pestic Sci 91(1):145–155

53. Eveillard S, Jollard C, Labroussaa F et al (2016) Contrasting susceptibilities to Flavescence dorée in *Vitis vinifera*, rootstocks and wild Vitis species. Front Plant Sci 7:12

54. Roggia C, Caciagli P, Galetto L et al (2014) Flavescence dorée phytoplasma titre in field-infected Barbera and Nebbiolo grapevines. Plant Pathol 63(1):31–41

55. Salar P, Charenton C, Foissac X et al (2013) Multiplication kinetics of Flavescence dorée phytoplasma in broad bean. Effect of phytoplasma strain and temperature. Eur J Plant Pathol 135:371–381

56. Schvester D, Carle A, Moutous G (1969) Nouvelles données sur la transmission de la Flavescence dorée de la vigne par Scaphoideus littoralis Ball. Ann Zool Ecol Anim 1:445–465

57. Chuche J, Sauvion N, Thiéry D (2017b) Mixed xylem and phloem sap ingestion in sheath-feeders as normal dietary behavior: evidence from the leafhopper *Scaphoideus titanus*. J Insect Physiol 102:62–72

58. Caudwell A (1964) Identification d'une nouvelle maladie à virus de la vigne, la «Flavescence dorée». Etude des phénomènes de localisation des symptômes et de rétablissement. Ann Epiphyties 15(1):193

59. Boudon-Padieu E, Cousin MT (1999) Yellow decline of Lavandula hybrida rev and L. vera DC. Int J Trop Plant Dis 17:1–34

Chapter 4

Sampling Methods for Leafhopper, Planthopper, and Psyllid Vectors

Kerstin Krüger and Nicola Fiore

Abstract

To reduce the spread of phytoplasmas in a crop or in a certain geographic area, epidemiological studies are of crucial importance in determining which insect species transmit these pathogens. In this chapter, we describe methods of capturing the insect vectors of phytoplasmas and the criteria for choosing the method(s) according to the objective to be achieved.

Key words Leafhoppers, Planthoppers, Psyllids, Survey, Traps, Vacuum sampling, Sampling strategies, Potential vectors of phytoplasmas, Transmission trials

1 Introduction

Phytoplasmas are spread via vegetative propagation of infected plant material and in nature through insect vectors in the families Cicadellidae (leafhoppers), Psyllidae (psyllids), and the superfamily Fulgoroidea (planthoppers), which feed on the phloem sap of infected plants [1–3]. In addition to increasing the metabolic activity of infected plants, phytoplasmas reduce or enhance the fitness of insect vectors [4, 5]. In a few insect species/phytoplasma combinations, transovarial transmission was also demonstrated [6–9].

Vectors of phytoplasmas are most likely to be found on leaves, flowers, and fruit of host plants during the growing season, but they may also occur beneath the bark of woody plants where some of them overwinter. Nymphs of Cixiidae (Fulgoroidea) that vector phytoplasmas, e.g., *Haplaxius crudus* van Duzee (previously *Myndus crudus*) transmitting lethal yellowing disease in palm trees [10] and *Hyalesthes obsoletus* Signoret, a vector of stolbur phytoplasma in grapevine [11], are an exception. Adult females lay eggs in the soil and nymphs complete their development subterraneously, feeding on the roots of their host plants [10, 12].

Rita Musetti and Laura Pagliari (eds.), *Phytoplasmas: Methods and Protocols*, Methods in Molecular Biology, vol. 1875, https://doi.org/10.1007/978-1-4939-8837-2_4, © Springer Science+Business Media, LLC, part of Springer Nature 2019

The control of phytoplasma diseases is mainly based on preventing the infection. Thus, once the presence of a disease has been detected in a specific crop and area, it is essential to know which insect vectors are responsible for the pathogen spread. A variety of methods are available to sample insect vectors of phytoplasmas. Commonly used methods to collect insects include sweep netting, beat sampling, vacuum sampling, yellow sticky traps, light traps, and malaise traps [13, 14]. The choice of method is determined by the insect taxon, by live stage of concern, and by the purpose of the study. In fact, insect can be sampled for establishing cultures or for surveys such as taxonomic or molecular studies, the determination of the species occurring in an area or the identification of potential vectors of phytoplasmas. Surveys can be either qualitative or quantitative, depending on the objective. In contrast, insect monitoring is used to obtain information on population size, density, and composition, to detect variations in insect abundance and to provide decision support regarding management of insect pests.

It is unlikely that a single sampling method is sufficient. Often two or three methods have to be used together if there is little prior knowledge concerning the insect vector(s) (e.g., [15–17]). For example, identification of potential vectors of phytoplasmas usually entails a combination of visual inspection of plants, sweep netting (e.g., leafhoppers) or beat sampling (e.g., psyllids), and a trapping method. Sampling should not be restricted to the crop in question but should also extend to weeds and other plants because these could serve as alternative hosts for the pathogen and the insect vector(s) and influence the presence and abundance of vectors [13]. Sampling of insects from trees and shrubs may require different methods than collection of insects from herbaceous plants. In addition, methods can be sex biased [18]. All the methods for collecting insects can be divided into active (e.g., visual inspection, sweep netting, vacuum collection) and passive (trapping) collection.

2   Materials

Most collecting materials (Figs. 1–3), such as sweep nets or beating sheets, are available from specialized entomological suppliers or can be self-assembled.

1. Field recording sheets (Fig. 1a).

2. Hand-held Global Positioning System (GPS) device.

3. Camera.

4. Glass vials or plastic jars, microtubes, "zip-lock" plastic bags for insect or plant specimens, tissue paper (Fig. 1b, c).

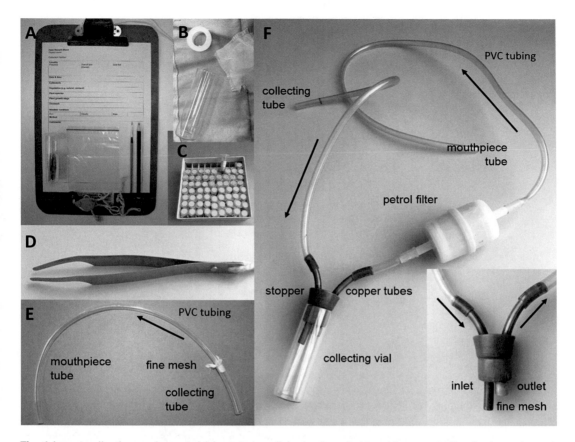

Fig. 1 Insect collecting equipment. (**a**) Insect recording sheet, pocket hand lens, and other basic equipment. (**b**) Collecting vial. (**c**) Microtubes for preserving insects in ethanol. (**d**) Soft forceps. (**e**) Simple aspirator (pooter) consisting of a mouthpiece tube and a slightly wider tube that serves as a collecting container and fits over the narrower mouthpiece tube. To prevent inhalation of insects, gauze is fitted over the smaller tube before inserting it into the larger tube. (**f**) Aspirator with collecting vial (pooter). A petrol filter in the middle and gauze at the inner end of the mouthpiece tube provide filters to prevent inhalation of insects and small particles. Insects are collected in the vial. The arrows indicate the direction of airflow

5. Pocket hand lens (×10 magnification or higher), soft forceps (Fig. 1d), scissors, pruning shears, pocket knife, pencil, ethanol-resistant permanent markers, sticky labels, paper clips, fine paint brush, vials with precut labels, vial with rubber bands, lighter, string, cotton wool, clear tape.

6. 75% or absolute ethanol for insect preservation.

7. Aspirator (pooter).

    (a) Simple aspirator (Fig. 1e): PVC tubing (long tube ca. 6 mm in diameter and ca. 40 cm long; short tube ca. 7 cm long, short PVC tube to fit over longer tube), gauze (2 × 2 cm). Connect the longer piece of PVC tubing with the shorter piece tightly to fit over the longer tube. To prevent inhaling of insects the small piece of

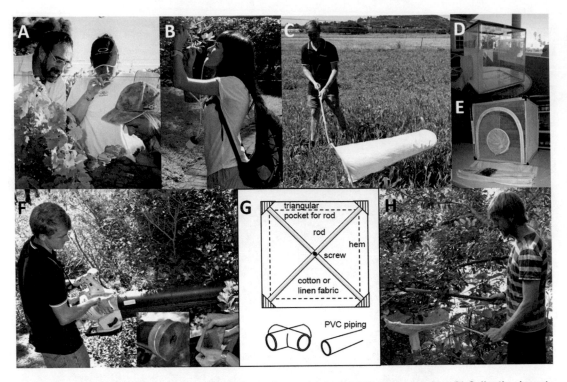

Fig. 2 Common methods used for sampling of phytoplasma vectors. (**a**) Visual inspection. (**b**) Collecting insects with an aspirator (pooter). (**c**) Sweep netting (Photo: D. J. van Wyk). (**d**) Glass insect cage for selectively collecting insects of interest. (**e**) Collapsible insect cage made of gauze and fabric and clear plastic. (**f**) Insect collection with a leaf blower (Photo: D.J. van Wyk). (**g**) Example of a self-made beating sheet. (**h**) Knockdown using beat sampling. A jar is attached to the beating sheet to collect dislodged insects

gauze is fitted over the end of the shorter tube that is pushed over one end of the longer tube. The other end of the longer PVC tube is used as the mouthpiece.

(b) Aspirator with collecting vial (Fig. 1f): PVC tubing (5–8 mm diameter, at least 80 cm long, depending on insects to be collected and preference of collector), metal tubing (two pieces ca. 8 cm long, slightly bent at one end), rubber or cork stopper to fit collecting vial, gauze (2 × 2 cm), glass or plastic vial. Make two holes in the stopper to fit the two metal tubes through the stopper with their straight section. Fit gauze with a small piece of PVC tubing to hold it in place over the outlet of the metal tube (the gauze is used as a filter and the end of the mouth piece connection pointing into the collecting jar to prevent inhalation of trapped insects). Attach the PVC tubing to the outer ends of the metal tubes; cut mouth piece tube and insert petrol filter to collect small particles and prevent their inhalation. Add vial to stopper to collect trapped insects. If necessary, a small piece of tissue paper

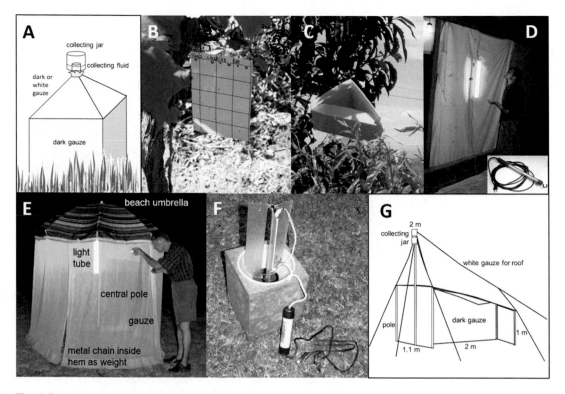

Fig. 3 Common traps used for sampling of phytoplasma vectors. (**a**) Emergence trap (cage). (**b**) Yellow sticky trap. **C**. *Delta* trap. (**d**) Light trap (white sheet). (**e**) Light trap (beach umbrella) (**f**) Portable Heath light trap. (**g**) Malaise trap

can be added to the vial to absorb moisture. Do not add ethanol or any other chemical to the vial because the fumes will be inhaled.

8. Backpack or any other suitable bag; portable tool box for small equipment.

9. Sweep net (Fig. 2c). Fit the net over a round metal rim with a diameter of ca. 30 to 40 cm. The bag of the net should consist of fine netting (gauze) reinforced with sturdier material (e.g., cotton, linen) where it folds over the rim to reduce wear. The netting should be fine enough for allowing air to flow through to reduce resistance during sweeping. The length of the bag should be more than double the diameter to be able to be flipped over the frame after a sweep to trap insects in the tip of the bag. The sweep net has to be sturdy enough to avoid damage when dragged through vegetation, hitting branches, or stones. Attach a wooden or light aluminum handle, ca. 1 m long, to the rim.

10. Beating sheet (Fig. 2g). Connect two rods with a screw in the middle or use plastic piping and a cross connector to form an X; the length of the rods/plastic pipes is determined by the size of

the beating sheet. Add a washer when connecting the rods with a screw so that the rods can be collapsed for transport. The X structure provides support for a sheet made of white (black) sturdy cotton or linen fabric (ca. 1 m square, including hem) that is fitted over the frame with small pockets consisting of triangles sown onto the edges of the cloth. A useful size for the beating sheet is a 70 cm square or smaller, depending on the vegetation to be sampled. A ca. 1 m long wooden stick is used for beating.

11. Cordless hand-held leaf blower that can blow and vacuum for vacuum (suction) sampling; nylon knee high stockings and strong rubber bands that fit over the tube (Fig. 2f). Leaf blowers are available from garden stores. Commercial portable vacuum samplers (e.g., D-Vac, Vortis, backpack aspirators) are available from specialized suppliers. Several designs are in use (*see* **Note 1**).

12. Insect cage (Fig. 2d, e). Collapsible or any other type of insect cage.

13. Trapping, e.g., emergence traps, yellow sticky traps, bait traps, light traps, malaise traps.

    (a) Emergence traps (Fig. 3a). Fine dark (black, dark green) mesh gauze for a square or rectangular cage with a pyramidal structure on top; rods (wood, plastic, or light metal for the frame); collecting jar to trap insects. The gauze is draped over the cage and the collecting jar fitted on top.

    (b) Sticky traps (Fig. 3b). Double-sided plastic sheets (usually bright yellow; *see* **Note 2**); the size may vary, e.g., 20 cm × 10 cm, but should be standardized within a study; non-drying insect glue (e.g., polyisobutylene); large paper clips. Clear plastic bags (cut open at two sides) or clear plastic to cover sticky traps for viewing of specimens. The traps are available commercially.

    (c) Bait traps (Fig. 3c). Chemical lures or attractants (pheromones, plant volatiles (volatile organic compounds (VOCs) for bait traps are commercially available or can be synthesized [14]. The chemicals are released via a dispenser, sol-gel formulations, etc. Several trap designs are available and some have been designed for specific taxa [18].

    (d) Light traps.
    - White fabric sheet (Fig. 3d).
    - "Beach umbrella" light trap (Fig. 3e). Beach umbrella, fine netting (e.g., mosquito net). Attach the net to the rim of the umbrella. Use linen or tough cotton fabric for the hem. Add a metal chain to the inside of the hem as a weight to stabilize the net in windy conditions.

For both traps use a fluorescent tube fitting with two black (near-ultraviolet, UV-A) or any other white light; black lights should not be looked for extended periods of time [19]; use a string to attach the light source to a preferred fixture in front of the white sheet or in the center of the inside of the beach umbrella trap.

- Portable light trap (Fig. 3f). A bucket or other container; a funnel large enough to fit upside-down into the container with an opening that allows insect to pass through but small enough to prevent insects from escaping; a light bulb (e.g., mercury-vapor bulb (MV-bulb)) suspended above the bucket with the funnel.

For all traps: Extension cords; portable generator or battery and converter if no electricity is available; a headlamp. Several light trap models are available from commercial suppliers.

(e) Malaise trap (Fig. 3g). Netting (gauze) for the sides and the net spanned vertically between two poles of different heights (usually dark green or black), four equal-sized poles (two each to support netting on each side); one large center pole; gauze for the roof (usually white), string; tent pegs; collecting jar. The size of malaise traps is variable. A simple design for a bidirectional trap is shown in Fig. 3g. For different designs *see* van Achterberg [20].

3   Methods

A large variety of sampling methods are available (*see* **Note 2**). The methods mentioned below are commonly used for sampling insect vectors of phytoplasmas.

3.1   Basic Indications for Insect Collection

1. Fill field recording sheets for noting the unique collection number, details of a sampling locality (town, farm (owner(s)' name), grid reference), date and time, name(s) of collectors, vegetation (e.g., natural, orchard), plant species, plant growth stage, plant disease, weather conditions (e.g., sun, cloudy, rain), and other observations made (Fig. 1a). It is useful to take a photo of the recording sheet so that those subsequently taken to record observations can be linked to a specific locality or sampling event when visiting more than one locality during a field trip or the same locality over different days.

2. Record the position of the sampling locality using a hand-held Global Positioning System (GPS) device.

3. Use collecting containers, e.g., microtubes and vials, for keeping insects alive or preserving them in, e.g., ethanol (Fig. 1b, c). Dry dead insects can be transported in zip lock plastic bags; for freshly killed insects add tissue paper to absorb moisture. Plastic bags with tissue paper can also be used for plant samples to confirm host plants.

4. Label, using labels and pencil, in and on containers with the basic collecting information, should be written in pencil because water (rain) and ethanol can render labels written with a ball point pen or ink unreadable when coming into contact with it (*see* **Note 3**).

5. Collect small and fragile trapped insects using an aspirator (Fig. 1e, f).

   How to use an aspirator: hold mouthpiece and approach insect(s), hold collecting tube over the insect(s) and inhale.

   (a) Simple aspirator (Fig. 1e): inhale through the longer PVC tube to trap insects in the shorter piece; once an insect has been trapped place a finger over the opening of the shorter tube to prevent the insect from escaping; gently exhale and blow insect into collecting container.

   Advantage: easy to build.
   Disadvantage: the number of insects that can be collected simultaneously is limited.

   (b) Aspirator with collecting vial (Fig. 1f): mark mouthpiece or insect collecting piece to avoid confusion; after collection of insects remove vial and close with lid (ventilated if insects are to be kept alive). To prevent insects from escaping when removing the vial, it can be gently shaken so that the insects accumulate at the bottom. The vial is then quickly detached and closed with a lid; add new vial as stopper (*see* **Note 4**).

   Advantage: several insects can be collected at once.
   Disadvantage: more complex to build than a simple aspirator.

6. During the sampling, it is advisable to use a backpack or other suitable bag to carry the equipment in the field. Very useful are the toolboxes (e.g. pencils, permanent marker pent, labels, plastic bags, spare collecting jars, paper clips, pruning shears, etc.)

3.2 Active Sampling

Active sampling methods can be used for all live stages of insect vectors of phytoplasmas (Table 1).

3.2.1 Visual Inspection

Visual inspection is suitable for both observing and collecting insects and it is frequently used as a supporting method (Fig. 2a). Observing insects provides information on their biology, e.g., behavioral observations or identifying host plant species, which

Table 1

Comparison of common field collecting methods for insect vectors of phytoplasmas

Collecting method	Advantage	Disadvantage
Active sampling		
Visual inspection	• Provides information on host plant species • Species can be collected selectively • Insects are collected alive • Useful for surveys	• Labor intensive • Inconspicuous insects may be overlooked • Efficiency varies with time of the day/insect activity • Efficiency depends on observer • Not suitable for estimating absolute abundance
Sweep netting	• Easy to use • Relatively large numbers of insects can be captured • Both nymphs (leafhoppers and planthoppers) and adults are collected • Insects can be extracted selectively from samples • Insects are collected alive • A relatively large area can be covered • Can provide information on host plant if sweeping a single plant species • Useful for surveys	• Information on host plant is unreliable if sweeping is done in vegetation with different plant species (insects may be trapped incidentally, for example in orchards with undercover growth or mixed vegetation) • Insects may drop from plants or fly away after the first sweep • Depends on vertical distribution of insects • Cannot be used when the vegetation is wet, very dense or thorny • Depends on the time of day and weather conditions (e.g., sunny vs cloudy, wind) • Difficult to standardize sampling for estimating absolute abundance
Beat sampling	• Same as for sweep netting	• Difficult to use when vegetation is wet • Depends on the time of day • Not suitable for dense vegetation or narrow rows of herbaceous crops (alternatives: Sweep netting, vacuum sampling) • Difficult to standardize sampling for estimating absolute abundance
Vacuum sampling	• Relatively large numbers of insects can be captured • Both nymphs (leafhoppers and planthoppers) and adults are collected • Insects are collected alive • Can be used in dense vegetation and vegetation with thorns • More accurate than sweep netting for standardized sampling (equipment has to be calibrated for standardized sampling to obtain estimates of absolute abundance)	• Delicate insects may be damaged • Depends on the time of the day • Not suitable for wet vegetation

(continued)

Table 1
(continued)

Collecting method	Advantage	Disadvantage
Passive sampling (collection of actively moving insects)		
Emergence trap	• Soil dwelling insects can be collected, i.e. nymphs of Cixiidae • Insects can be trapped dead or alive	• Only a small area is covered, posing a problem for patchily distributed insects such as nymphs of cixiids in the soil [12] • May provide information on host plant if placed over a single plant species • Difficult to standardize sampling for estimating abundance
Sticky trap	• Many species are collected • Can be used for surveys, monitoring flight activity and determining relative abundance	• No information on host plant(s) • Insects are killed and may be damaged • Only winged insects are recorded • Samples difficult to process, especially if specimens have to be removed from traps
Bait trap	• Insects can be trapped dead or alive • Can be used for mass-trapping • Can be used for monitoring flight activity and abundance	• Does not provide information on host plant(s) • Only species attracted to the bait, sometimes enhanced by color, are recorded
Light trap	• Collects nocturnal insects • Different species can be collected • Species can be collected selectively • Insects can be collected dead or alive • Useful for surveys	• Does not provide information on host plant(s) • Catches can be very variable depending on, e.g., type of trap, habitat and weather (e.g., rain, wind, ambient temperature) • Sampling insects directly from sheets/nets is labor intensive • Trap catches very low during periods of full moon (the light of the moon competes with the light of the trap)
Malaise trap	• A large number of insects and species can be collected • Useful for surveys	• No information on host plant(s) • Does not provide estimates for absolute abundance or species composition

can be useful when establishing a laboratory culture of an insect vector. However, an insect species may be observed feeding on a specific plant species but may not necessarily use it for breeding. This method can be used for all insect vector taxa of phytoplasmas, especially eggs of insect vectors and psyllid nymphs. Insects detected on plants can often simply be collected with an aspirator (Fig. 2b), trapped in a collecting jar or vial or, in case of eggs and psyllid nymphs, collected with the plant material. Adult phytoplasma vectors (leaf- and planthoppers, psyllids) and nymphs of leaf- and planthoppers are highly mobile and quick action is required when collecting insects with small equipment. Taking photographs of insects and plants may be useful as supporting documentation for insect and host plant records, sampling area, vegetation, etc. Plant samples with eggs or nymphs attached, or plant samples collected for host plant or phytoplasma identification can be transferred to plastic bags with tissue paper added to absorb moisture.

3.2.2 Sweep Netting

Sweep nets constitute one of the most widely used methods to collect insects from vegetation (Fig. 2c). The method is particularly useful for capturing mobile insects such as leaf- and planthoppers and adult psyllids. Insects are captured by moving the net back and forth in front of the collector. Technique is important: sweeps should be carried out relatively forcefully and a fair amount of speed to avoid highly mobile insects from escaping. Insects may occur on different parts of plants, e.g., on outside leaves or hiding inside. Many insects may be missed when collecting from the outside or top of plants only. Care should therefore be taken that sweeps cover also the inside or base of plants. At the end of a sweep the net is quickly moved back and forth to accumulate the insects at the base of the net, and the bottom section of the net is then folded over the rim or closed with a hand to prevent insects from escaping. Insects of interest can be collected selectively with an aspirator and the remainder released again, or the entire sample can be transferred to a collecting jar. Lighter sweep nets can also be used for collecting airborne insects, although this is usually not a useful technique for vectors of phytoplasmas.

3.2.3 Vacuum Sampling

Vacuum sampling is useful for leafhoppers, planthoppers, and psyllids. The method is recommended for insects that are difficult to catch with methods such as sweep netting or beat sampling. For vacuum sampling a hand-held leaf blower is reversed so that it "vacuums" insects. A nylon knee-high stocking is placed over the opening of the inlet tube to trap insects. The stocking can be held in place with strong rubber bands (Fig. 2f). After the collection of insects, the stocking is removed to transfer the sample to collecting jars with a preservative or to freeze insects for later sorting, or insects can be kept alive and released into ventilated containers or cages.

3.2.4 Beat Sampling

A beating sheet is useful for sampling slow-moving insects that are easily dislodged from trees or shrubs, e.g., adult psyllids (Fig. 2g). The beating sheet is held or placed beneath branches of trees or shrubs to be sampled. A stick is used to shake the plant vigorously from above to dislodge insects which will fall onto the sheet. For collecting insects from herbaceous plants or crops, the sheet can be placed on the ground beneath the plant or between two rows and branches shaken or hit downward by hand. A sweep net (a collecting jar can be attached at the bottom; Fig. 2h) or a white umbrella can be used as an alternative to a beating sheet. Insects that have fallen onto the sheet can then be collected with an aspirator.

3.3 Passive Sampling

The two basic approaches to passive sampling with traps are based on attraction and interception. Attraction traps use color, light, or odor (pheromones, allelochemicals) to attract and trap insects. Interception traps obstruct the path of insects that are then trapped. The two methods are often combined. Passive sampling is used for adult leafhoppers, planthoppers and psyllids, as well as nymphs of leafhoppers and planthoppers (Table 1). Insects are frequently trapped on a sticky surface (*see* **Note 5**) or a colleting jar filled with a preservative (*see* **Note 6**).

3.3.1 Emergence Cage Traps

Emergence traps (Fig. 3a) can be used for collecting soil-dwelling insects, such as nymphs of Cixiidae (Fulgomorpha). The trap uses light to attract insects (positive phototactic response). An open-bottomed cage is placed over vegetation on the ground; tall vegetation can be cut prior to placing the trap. Soil-dwelling insects move upwards toward the light and fall into a collecting jar at the top. Insects can be collected dead in a liquid preservative added to the collecting jar, or alive. Various compositions of preservatives can be used (*see* **Note 6**). The liquid is usually replaced weekly.

3.3.2 Sticky Traps

Many leafhoppers, planthoppers, and psyllids are attracted to yellow. Depending on the insect(s) to be trapped, sticky traps can be placed in the plant canopy, the canopy-weed interface, or close to the weed layer beneath the canopy (Fig. 3b). They can be fastened to wire frameworks with paper clips and can be enclosed with wire mesh to prevent leaves from getting stuck to the traps [21]. Traps are usually replaced weekly or fortnightly. Traps for insects required for DNA extraction can be exposed for a week [22, 23] but should, preferably, be removed sooner although they can be stored for longer at room temperature [24]. After removal, labeled traps can be transferred to clear plastic bags (cut open at two sides for ease of handling) or plastic sheets to view insects (*see* **Note 7**).

3.3.3 Bait Traps

The use of bait traps for sampling insect vectors of phytoplasmas has been described in detail in Weintraub and Gross [14]. Bait traps make use of intraspecific pheromones (sex pheromones, aggregation pheromones) involved in insect communication or interspecific allelochemicals such as kairomones, which are chemical cues used for orientation and host finding [14]. The chemicals are frequently available commercially. Delta traps in combination with a lure are commonly used in orchards for monitoring specific insects (Fig. 3c). The trap is folded in a tent-like fashion. The lure is suspended from the top inside the trap. Insects are trapped on a sticky surface at the bottom.

3.3.4 Light Traps

Light trapping is useful for collecting many leaf- and planthopper species. Several trap designs are available and are described in Southwood and Henderson [18]. A simple way of using light to attract insects is to use a white sheet draped over a clothes line or attached to another suitable fixture. The light source is suspended in front of the cloth (Fig. 3d). The direction of the placement of the sheet is of importance. Another simple nondirectional light trap is a "beach umbrella" with fine netting (e.g., mosquito netting material) attached to the rim of the umbrella and a light source placed in the center (Fig. 3e). A headlight is recommended for collecting and processing insects in the dark. Both methods require regular active sampling of attracted insects during the trapping period. The beach umbrella light trap has the advantage that insects not collected during the night and trapped insight the tent can be collected the next day. For both types of traps insects can be collected selectively with an aspirator or collecting jar. Portable light traps, e.g., the Heath light trap, can be left in the field (Fig. 3f). Insects are trapped and fall into a collecting container. This type of trap can be emptied daily or weekly, depending on the purpose of the study and the size of the catch. The positioning of light traps is crucial for efficient insect collection.

3.3.5 Malaise Traps

This type of trap consists of a tent-like interception structure usually placed on the ground across flight paths for collecting flying insects, e.g., leafhoppers and planthoppers (Fig. 3g). Several models are in use [20]. Flying insects are intercepted by a dark (black or dark green) fabric barrier spanned vertically between two poles of different heights in the middle of an open-sided tent. Like the emergence cage, the bidirectional trap makes use of the phototactic response of insects which move upward to the corner of the longer pole of the sloping roof (often white) where they accumulate and pass into a collecting jar containing a preservative (*see* **Note 6**). Accurate positioning and orientation of the trap is essential for efficient insect collection, e.g., correct positioning in the flight path of insects [18]. Depending on catch size, jars may have to be emptied daily or after several days.

3.4 Processing and Preservation of Samples

If live insects are required, collecting vials with live insects can be transported directly to the laboratory, in a cooler box if temperatures are high, or trapped insects can be released into cages at the field site. Releasing insects into a cage is useful to prevent overheating and death of specimens in vials in hot weather conditions or to avoid overcrowding. Plant material can be added to cages to offer hiding places and allow insects to feed during transport.

Collapsible cages made of gauze and soft material are useful for transport and storage (Fig. 2d). Cages made of glass or clear non-static plastic on the other hand allow for better viewing and are useful for extracting specific insects from the samples (Fig. 2e). Insects can be collected selectively from the cages with an aspirator.

Samples used for morphological identification can be stored in 75% ethanol, or absolute ethanol if specimens are also required for molecular work, e.g., species identification or determining presence of phytoplasmas.

Apart from storing samples in ethanol, dead insects can be directly transferred to plastic bags together with tissue paper to absorb excess moisture and kept cool or frozen for later sorting in the laboratory. However, insects can become squashed during transport and sorting insects may be time consuming if debris has been collected together with the samples. Insects can be sorted using soft forceps (Fig. 1d) or a fine paint brush.

4   Notes

1. Comparisons of different hand-held vacuum samplers were made by Thomas [25], Zou et al. [26], and Cherrill et al. [27].

2. A number of publications deal with various aspects of sampling insects for phytoplasma studies. Weintraub and gross [14] provide an overview of sampling insect vectors of phytoplasmas with an emphasis on using novel techniques for trapping insect vectors with infochemical lures. For a full description of insect collecting methods, *see* Southwood and Henderson [18], Schauff [28], and Grootaert et al. [29]. For collecting insects from palm trees, *see* Howard et al. [30].

3. Sticky labels can be used for labeling plastic bags, jars, and microtubes for easy sorting and processing. Labels on microtubes should be secured with clear tape. Writing directly on microtubes with an ethanol-resistant permanent marker may become unreadable. In addition, paper labels with the collecting information must be added to plastic bags, jars, and tubes in case the outside labels become dislodged or unreadable, as well as for later processing of samples.

4. Insects collected can be more easily observed through glass vials. However, glass is prone to breakage and clear plastic vials provide an alternative, but these may cause a problem for smaller insects which may get stuck to the tube due to static electricity.

5. Many species that transmit phytoplasmas are attracted to yellow. There are exceptions, however. For example, it is recommended that the planthopper *H. crudus*, a vector of lethal yellow disease, should be monitored with white or blue sticky

traps [31]. Another example is the planthopper vector of flavescence dorée, *Scaphoideus titanus* Ball, where more individuals were trapped on red than on yellow, blue, or white sticky traps [32].

6. The type of collecting fluid used is important because insects need to be preserved for morphological or molecular identification, or for later storage in collections. The choice depends on the frequency of emptying the collecting containers and the purpose, e.g., collecting insects for molecular studies. Common preservatives or insect collecting fluids are 20% ethylene glycol, 70% propylene glycol, or 70% ethanol [33].

7. Insects can be removed from sticky traps or surfaces and cleaned with organic solvents, e.g., ethyl acetate, hexanes, benzene, and xylene [14, 34]. Specimens have to be dried well before DNA extraction to remove the solvent.

References

1. Bertaccini A, Davis RE, Lee I-M (1992) *In vitro* micropropagation for maintenance of mycoplasmalike organisms in infected plant tissues. Hort Science 27:1041–1043

2. Jarausch W, Lansac M, Dosba F (1996) Long-term maintenance of nonculturable apple-proliferation phytoplasmas in their micropropagated natural host plant. Plant Pathol 45:778–786

3. Bertaccini A (2007) Phytoplasmas: diversity, taxonomy, and epidemiology. Front Biosci 12:673–689

4. Sugio A, Kingdom HN, MacLean AM et al (2011) Phytoplasma protein effector SAP11 enhances insect vector reproduction by manipulating plant development and defense hormone biosynthesis. Proc Natl Acad Sci U S A 108:E1254–E1263

5. Bertaccini A, Duduk B, Paltrinieri S et al (2014) Phytoplasmas and phytoplasma diseases: a severe threat to agriculture. AJPS 5:1763–1788

6. Alma A, Bosco D, Danielli A et al (1997) Identification of phytoplasmas in eggs, nymphs and adults of *Scaphoideus titanus* ball reared on healthy plants. Insect Mol Biol 6:115–121

7. Kawakita H, Saiki T, Wei W et al (2000) Identification of mulberry dwarf phytoplasmas in the genital organs and eggs of leafhopper *Hishimonoides sellatiformis*. Phytopathology 90:909–914

8. Hanboonsong Y, Choosai C, Panyim S et al (2002) Transovarial transmission of sugarcane white leaf phytoplasma in the insect vector *Matsumuratettix hiroglyphicus* (Matsumura). Insect Mol Biol 11:97–103

9. Tedeschi R, Ferrato V, Rossi J et al (2006) Possible phytoplasma transovarial transmission in the psyllids *Cacopsylla melanoneura* and *Cacopsylla pruni*. Plant Pathol 55:18–24

10. Gurr GM, Johnson AC, Ash GJ et al (2016) Coconut lethal yellowing diseases: a phytoplasma threat to palms of global economic and social significance. Front Plant Sci 7:1521

11. Maixner M (2010) Phytoplasma epidemiological systems with multiple plant hosts. In: Weintraub PJ, Jones P (eds) Phytoplasmas: genomes, plant hosts and vectors. CABI International, Wallingford

12. Kaul C, Seitz A, Maixner M et al (2009) Infection of bois-noir tuf-type-I stolbur phytoplasma in *Hyalesthes obsoletus* (Hemiptera: Cixiidae) larvae and influence on larval size. J Appl Entomol 133:596–601

13. Weintraub PG, Beanland L (2006) Insect vectors of phytoplasmas. Annu Rev Entomol 51:91–111

14. Weintraub P, Gross J (2013) Capturing insect vectors of phytoplasmas. In: Dickinson M, Hodgetts J (eds) Phytoplasma. Methods in molecular biology (methods and protocols). Springer Science, New York

15. Tedeschi R, Bosco D, Alma A (2002) Population dynamics of *Cacopsylla melanoneura* (Homoptera: Psyllidae), a vector of apple proliferation phytoplasma in northwestern Italy. J Econ Entomol 95:544–551

16. Koji S, Fujinuma S, Midega CO et al (2012) Seasonal abundance of *Maiestas banda* (Hemiptera: Cicadellidae), a vector of phytoplasma, and other leafhoppers and planthoppers (Hemiptera: Delphacidae) associated with

Napier grass (*Pennisetum purpureum*) in Kenya. J Pestic Sci 85:37–46

17. Trivellone V, Pollini Paltrinieri L, Jermini M et al (2012) Management pressure drives leafhopper communities in vineyards in southern Switzerland. Insect Conserv Divers 5:75–85

18. Southwood TRE, Henderson PA (2000) Ecological methods. Blackwell Science, Oxford

19. Sliney DH, Gilbert DW, Lyon T (2016) Ultraviolet safety assessments of insect light traps. J Occup Environ Hyg 13:413–424

20. van Achterberg K (2009) Can Townes type malaise traps be improved? Some recent developments. Entomologische Berichten 69:129

21. Mpunami A, Tymon A, Jones P et al (2000) Identification of potential vectors of the coconut lethal disease phytoplasma. Plant Pathol 49:355–361

22. Weber A, Maixner M (1998) Survey of populations of the planthopper *Hyalesthes obsoletus* sign. (Auchenorrhyncha, Cixiidae) for infection with the phytoplasma causing grapevine yellows in Germany. J Appl Entomol 122:375–381

23. Orenstein S, Zahavi T, Nestel D et al (2003) Spatial dispersion patterns of potential leafhopper and planthopper (Homoptera) vectors of phytoplasma in wine vineyards. Ann Appl Biol 142:341–348

24. Bertin S, Bosco D (2013) Molecular identification of phytoplasma vector species. In: Dickinson M, Hodgetts J (eds) Phytoplasma. Methods in molecular biology (methods and protocols). Springer Science, New York

25. Thomas DB (2012) Comparison of insect vacuums for sampling Asian citrus psyllid (Homoptera: Psyllidae) on citrus trees. Southwest Entomol 37:55–60

26. Zou Y, van Telgen MD, Chen J et al (2016) Modification and application of a leaf blower-vac for field sampling of arthropods. J Vis Exp 114:54655

27. Cherrill A, Burkmar R, Quenu H et al (2017) Suction samplers for grassland invertebrates: the species diversity and composition of spider and Auchenorrhyncha assemblages collected with Vortis™ and G-vac devices pages. Bull Insectology 70:283–290

28. Schauff ME (2001) Collecting and preserving insects and mites: techniques & tools. Systematic Entomology Laboratory, USDA, National Museum of Natural History, Washington, DC

29. Grootaert P, Pollet M, Dekoninck W et al (2010) Sampling insects: general techniques, strategies and remarks. Manual on field recording techniques and protocols for all taxa biodiversity inventories and monitoring. Abc Taxa, Belgium

30. Howard FW, Moore D, Giblin-Davis RM et al (2001) Insects on palms. CABI Publishing, Wallingford

31. Cherry RH, Howard FW (1984) Sampling for adults of the planthopper *Myndus crudus* a vector of lethal yellowing of palms. Trop Pest Manag 30:22–25

32. Lessio F, Alma A (2004) Dispersal patterns and chromatic response of *Scaphoideus titanus* ball (Homoptera Cicadellidae), vector of the phytoplasma agent of grapevine flavescence dorée. Agric For Entomol 6:121–128

33. Stewart AJA (2002) Techniques for sampling Auchenorrhyncha in grasslands. Denisia 4:491–512

34. Murphy WL (1985) Procedure for the removal of insect specimens from sticky-trap material. Ann Entomol Soc Am 78:881–881

Chapter 5

Symptoms of Phytoplasma Diseases

Paolo Ermacora and Ruggero Osler

Abstract

Phytoplasmas are associated with diseases in several hundreds of cultivated herbaceous and woody plants. Their impact in agriculture and the periodical outbreak of worrying epidemics make very important, besides precise laboratory-based diagnosis, the direct in-field recognition of phytoplasma disease symptoms. Even if some symptoms are typical of this kind of pathogens, in-field diagnosis requires the knowledge of the host plant, strong field experience, and awareness of the symptom variability of the various organs of the plant during different seasons and under various environmental conditions. It is therefore very important to be familiar with factors like environmental conditions, agronomical features, and disease progression that influence symptom expression. Therefore, a satisfactory diagnosis should be based on repeated and complete observations scored over the entire plant and across different times of the year. A more suitable diagnosis is possible if the observer is able to recognize and distinguish the symptoms of other biotic or abiotic diseases. A general rule is to observe three different symptoms, at least, and to seek input from the grower about the initial development, frequency, diffusion, and particular characteristics of the disease.

After a short introduction the following symptoms are presented: the most common and representative symptoms caused by phytoplasmas; the most common symptoms of phytoplasma diseases occurring in particular plant organs, with some references to specific diseases; phytoplasma symptoms on the model plant periwinkle (*Vinca rosea* or *Catharanthus roseus*); the main factors influencing phytoplasma symptoms expression; and several practical procedures that should be followed for suitable diagnosis. A series of original photos have been included to illustrate typical symptoms.

Key words Chlorosis, Decline, Plant diseases, Virescence, Witches' broom

1 Introduction

Phytoplasmas are bacteria-like organisms. Their genome is small (680–1600 kb) [1], when compared with their ancestral walled bacteria, and lacks several metabolic pathways for the synthesis of compounds indispensable for their survival and multiplication. For example, phytoplasmas lack genes for the biosynthesis of amino acids and fatty acids, the tricarboxylic acid (TCA) cycle, and oxidative phosphorylation (production of ATP) [2, 3]. As a consequence, trophic substances must be obtained from host plants [4], making phytoplasmas obligate parasites, strictly dependent

Rita Musetti and Laura Pagliari (eds.), *Phytoplasmas: Methods and Protocols*, Methods in Molecular Biology, vol. 1875,
https://doi.org/10.1007/978-1-4939-8837-2_5, © Springer Science+Business Media, LLC, part of Springer Nature 2019

on the host. In plants, phytoplasmas are localized exclusively in the sieve tubes where they multiply actively and move systemically within the host [5]. Their distribution in sieve tubes can be erratic and their concentration is often low or particularly low, especially in woody hosts [6].

Phytoplasma presence causes in the plant the development of visible symptoms suggesting a profound disturbance status. It is possible to refer the symptoms occurring in the infected plant to the following major inductive causes:

(a) Interference to the hormonal system leading to a series of disturbances in the balance of growth regulators. This is essentially the reason why phytoplasmas are considered typical "auxonic diseases." In fact, phytoplasmas are known to cause dramatic changes in plant development [7], resulting in malformations of the various organs of the plant.

(b) The progressive partial or total blockage of phloem flux. Disturbance of the phloem causes accumulation of organic solutes (e.g, amino acids and sugar) that originate from higher up in the plant but can no longer continue downward. Indeed, it is the extraordinary sugar accumulation in leaves that causes the cascade of symptoms described below [8].

(c) A serious side effect of the block of phloem flux is a marked reduction in essential storage compounds in sink organs, such as roots that can show a distinctive phenotype if compared to roots belonging to uninfected plants [9]. Especially during the last phases of the disease the downstream symptoms seem very similar to those caused by a typical xylematic necrotic disease, such as diffuse progressive withering, serious and broad necrosis, plant decline, and death of the whole plant.

The most common and representative symptoms caused by phytoplasmas:

(a) Leaf yellowing is one of the most common symptoms associated with the presence of phytoplasmas: it is thought to be due to modifications in both carbohydrate synthesis and transportation [3].

(b) Phyllody is another common symptom resulting from phytoplasma infection. In this case, the plant produces leaf-like structures instead of flowers. Generally, the flowers are completely sterile or the seeds do not germinate. Evidence suggests that the phytoplasma effectors interfere with regulation of genes involved in petal formation [10].

(c) Virescence is the development of green flowers due to the loss of pigment in the petal cells [11]. A phytoplasma effector protein (SAP54) has been identified as inducing symptoms of virescence and phyllody when expressed in plants [12].

(d) Witches' broom or proliferation is due to changes in the normal growth patterns of infected plants. These are mainly related to the loss of apical dominance causing the proliferation of axillary shoots. Proliferation can be associated with decreased internode length (dwarfism). A virulence factor (i.e., effector) was identified from a phytoplasma causing yellowing of onions; the active protein was named "tengu-su inducer" (TENGU). TENGU induces characteristic symptoms in infected plants including witches' broom and dwarfism. Notably, ongoing studies are unravelling the molecular mechanisms of phytoplasma symptoms [10].

(e) Heavy leaves with thick laminas, edges rolled up or down, stiff to the touch and brittle. These symptoms are determined by an accumulation of abnormal amounts of carbohydrates in mature leaves [2].

(f) Small malformed crinkled leaves.

(g) Thick bark above the phloem interruption point induced by phytoplasmas (this symptom is possibly confused with different causes such as linear insect bites along the bark, girdling, mechanical bark damages, tight rope around branches, bacterial tumors).

(h) Phloem necrosis; or vein necrosis, since this symptom is mainly macroscopically visible in leaf veins.

(i) Leaf veins that are pale or purple, prominent, and winding.

(j) Leaf petioles that are shorter and thicker than regular leaves.

(k) Small round fruits with long petioles that look like cherries. These fruits are pale or green, with oily skin, poor in sugar and acidity.

(l) Basal suckers, even visible from a distance.

(m) Rosetting occurring in shoot apices. This symptom could be confused with zinc deficiency, mainly in woody plants.

(n) Symptoms that are considered to be characteristic of the general group of phytoplasma diseases, even if nonexclusive, are virescence, phyllody, enlarged malformed stipules, uneven lignification, out of season flowering, and proliferation. Other symptoms are definitely more generic (for groups of pathogens that are not phytoplasmas) such as chlorosis, necrosis, flower abortion, small fruits, stunting, decline, etc.

2 The Most Common Symptoms of Phytoplasma Diseases Occurring in Particular Plant Organs, with Some References to Specific Diseases

2.1 Foliar Symptoms

Early symptomatic leaves on infected plants can be reduced in size with distorted laminas. In some cases, the size of the leaves can be extremely reduced. The initial chlorotic color later turns yellow or

Fig. 1 Foliar symptoms of phytoplasma diseases. (**a**) Grapevine yellows (GY) with typical chlorotic triangle shaped leaves in infected plants (white variety). (**b**) Reddish irregular areas involving veins in cv. Merlot infected with GY. (**c**) Rosettes of yellowing-reddish leaves with enlarged stipules in apple infected with Apple Proliferation. (**d**) Summer leaf rolling of apricot leaves infected with European stone fruit yellows: not to be confused with water stress

reddish and necrotic areas (including the veins) (Fig. 1a–c). Over time this may change the shape of the leaves. For example, in grapevine the leaf edges roll downward. Leaf rolling often results in the leaf having a triangle shape, while in apricot there can be the upward folding of the lamina similar to water stress (Fig. 1d). Sometimes, the symptomatic leaves may undergo to premature senescence and detach; in the case of grapevine and FD, the lamina may detach but the petiole remains attached to the vine branch.

2.2 Symptoms on Flowers

Symptoms in flowers are among the most striking in many phytoplasma diseases. Floral alterations may involve petal color, with various degrees of virescence, or morphological alterations of the

Fig. 2 Symptoms on flowers. (**a**) An inflorescence of Phytolacca Americana infected with PhyV, (16Sr XII-A) on the right with symptoms of virescence and phyllody; inflorescence from a healthy plant on the left. (**b**) Virescence and phyllody on an inflorescence of *Cichorium intybus* infected with Chicory Phyllody (ChiP, 16SrIX-C). (**c**) *Rubus fruticosus* cv. Loch Ness with flowers showing virescence and phyllody symptoms (right). (**d**) *Zinnia elegans* infected with Clp (16SrIII-B) (right) and a healthy inflorescence (left)

petals and reversion into leaves (phyllody) [11] (Fig. 2). In some cases, alterations to flowers are especially serious with severe malformations and frequent sterility. Phytoplasma infection can also misregulate the normal flowering time and induce severe or general necrosis. In particular, the necrosis of inflorescence caused by FD phytoplasma in grapevine could be confused with boron deficiency.

2.3 Symptoms on Fruits

Phytoplasma disease on fruit trees results in quantitative and qualitative losses. In Apple Proliferation (AP)-infected plants, fruit weight is often reduced by 30–60% and organoleptic characteristics are poor [13]. The small fruits are carried on long pedicels and fruit ripening can be delayed (Fig. 3a). In stone fruits the internal pulp shows corked areas (Fig. 3b). Some branches on infected trees may appear normal and produce regular fruit, whereas other branches may show symptoms such as malformed fruits or no production at

Fig. 3 Symptoms on fruits. (**a**) The small apple fruits of cv Florina with elongated petioles are from a tree infected with Apple Proliferation; on the left is an uninfected control. (**b**) Malformed fruits of Japanese plum cv Ozark Premier with internal corky necrotic areas (centre and right); on the left a control from a healthy plant. (**c**) Necrosis on a white variety grape cluster. (**d**) Diffuse shrivelling of berries (cv Chardonnay) in a small cluster compared the normal berries in the bottom cluster during the ripening phase

all. In phytoplasma-infected grapevines, fruit set is reduced and shrivelled grape bunches are quite common. Premature berry drop occurs in some cultivars and it is not unusual for completely necrotic clusters to be present. When infections are established early, complete detachment of entire clusters may occur (Fig. 3c, d). Additionally, the fruit of herbaceous horticultural plants can be affected by phytoplasmas. In infected potatoes, the number of subterranean tubers can be reduced, and the production of aerial tubers often occurs.

2.4 Vegetative Habits   In many cases phytoplasma infections alter the normal growth patterns of plants: as mentioned, dwarfing, bushiness, or witches' broom are common traits that are mainly related to the loss of apical dominance (Fig. 4a, b). Proliferation/witches' broom normally develops on vigorous shoots and the secondary symptoms are: spindly shoots with a reduced-angle of insertion on the principal

Fig. 4 (**a**) Winter witches broom with typical serrated sprouts in an Apple Proliferation-infected tree. (**b**) Bushy and proliferated vegetation from an Elm yellows-infected tree. (**c**) A zigzag cane with irregular reduced internodes in GY-infected grapevine cv Chardonnay. (**d**) Premature early leafing of a Japanese plum infected with European stone fruit yellows (on the right). Normal flowering tree on the left

axis; reduced internode elongation that results in rosettes; and in grapevine the shoots show characteristic zigzag growth and shortened internodes (Fig. 4c). Further, due to incomplete lignification the shoots fail to harden and become flexible and rubbery with a weeping appearance. Early break in dormancy is reported on *Prunus* spp. ESFY infected causing highly susceptibility of affected plants to frost (Fig. 4d).

2.5   Other Symptoms   Phytoplasmas infected trees often show a basal regrowth from the rootstock (Fig. 5a). Phloem necrosis in some *Prunus* species infected by ESFY is a very typical symptom as attested by the name "Plum leptonecrosis" given to the disease in the past

Fig. 5 (a) European stone fruit yellows-infected trees often show basal regrowth from the rootstock: here regrowth of the peach rootstock is shown. **(b)** Phloem necrosis in an apricot tree infected with ESFY. **(c)** Irregular lignification of grapevine canes infected with GY. **(d)** Leafroll is a common symptom of phytoplasma and viruses diseases (e.g. Grapevine leafroll-associated virus). In Grapevine yellows infected leaves **(d)** the veins became chlorotic or reddish and subsequently necrotic in the second case the veins remain green; in **(e)**

(Fig. 5b). In phytoplasma-infected grapevines, due to uneven lignification, the diseased shoots have a weeping appearance (Fig. 5c) and become very susceptible to frost damage during cold winters. The young diseased *V. vinifera* shoots are weak and necrosis of their terminal buds is possible. Bolting (growth of elongated stalks) is another type of symptom that may manifest in cultivated ornamental plants (e.g., *Tagetes sp.*).

3   Phytoplasma Symptoms on Periwinkle (*Vinca rosea* or *Catharanthus roseus*)

Although phytoplasmas are generally linked to specific genera or to a restricted number of plant genera, periwinkle plants are unusual in their ability to be infected by a large number of different phytoplasmas, irrespective of their genetic interrelationships. A notable characteristic of this extraordinary host is its relative capacity to produce different symptoms depending on the different phytoplasma (Fig. 6).

Fig. 6 Differentially symptoms expressed by periwinkle infected with phytoplasmas. (**a**) A bushy *C. roseus* plant infected with Apple Proliferation (AP15, 16SrX-A) showing diffuse yellowing on small leaves. (**b**) Flowers with virescence and mature leaves with distorted laminas and prominent veins. (**c**) Phyllody and rosetting with elongated leaves. (**d**) Small sized flower on periwinkle infected with 16SrX phytoplasma (right) and control flowers (left)

Various phytoplasmas of herbaceous or woody plants were transmitted and maintained to *C. roseus* by using different species of dodder (*Cuscuta* sp.) or insect vectors [14–17]. In this way it was possible to use periwinkle as a model to test plants for phytoplasmas, particularly during the early studies when molecular or serological tools for proper identification and characterisation were not yet available [18, 19]. For example, two distinct groups of related symptoms have been distinguished in our laboratory on *V. rosea* infected by a large number of herbaceous or perennial plant—phytoplasmas (Table 1).

Group 1: flowers with phyllody and virescence present (Fig. 6b, c). Belonging to this group are different 16Sr DNA groups and subgroups (e.g., 16Sr I, Aster yellows group; 16Sr II, Peanut witches' broom; 16Sr III, X-disease group; 16Sr XII, Stolbur group).

Table 1
Differential reactions of *C. roseus* test plants when infected with phytoplasmas belonging to different clusters

Floral symptoms	Isolate (16Sr group-subgroup)	Original host	Geographic origin
Floral reversion (phyllody and virescence)	ACH (16SrI-C)	*Achillea millefolium* L.	Italy (FVG)
	CA (16SrI-C)	*Daucus carota* L.	Italy (FVG)
	LEO (16SrI-C)	*Leontodon hyspidus* L.	Italy (FVG)
	PnWB (16SrII-A)	*Arachis hypogaea* L.	Taiwan[a]
	CP (16SrIII-B)	*Trifolium repens* L.	Italy
	ER (16SrIII-B)	*Erigeron annuus* L.	(Lombardy)
	MA (16SrIII-B)	*Chrysanthemum*	Italy (FVG)
	ChiP (16SrIX-C)	*leucanthemum* L.	Italy (FVG)
	BA (16Sr XII-A)	*Cichorium intybus* L.	Italy (FVG)
	PhyV (16Sr XII-A)	*Catharanthus roseus* G. Don	Italy (FVG)
	SI (16Sr XII-A)	*Phytolacca americana* L.	Italy (FVG)
		Silene alba (Mill.)	Italy (FVG)
Flower size reduction	AP15 (16SrX-A)	*Malus x domestica* Borkh	Italy (FVG)
	LNp (16SrX-B)	cv. Golden d.	Italy (FVG)
	LNS1 (16SrX-B)	*Prunus salicina* Lindl.	Italy (FVG)
		Prunus salicina Lindl.	

[a]Kindly supplied by Dr. I. M. Lee

Group 2: a pronounced reduction in flower size but maintaining the original color (Fig. 6d). Rapid and severe necrosis occurs on leaves that are also small and malformed. Apple proliferation and European Stony Fruit Yellows, which are among the woody plant phytoplasmas, are placed in this group. Recently, Liu and coworkers [20] were able to distinguish five categories of floral malformations in *C. roseus* infected by Peanut Witches broom.

3.1 Symptom Occurrence Over Time

For annual herbaceous plants, the symptoms expression is rarely followed by the development of new asymptomatic vegetation. In contrast, in perennial plants the symptom expression after the first appearance occurs irregularly year by year. Recovery, defined as the spontaneous remission from symptoms in previously symptomatic plants, has been reported for phytoplasma diseases such as AP, European stone fruit yellows (ESFY), Pear decline (PD), and Grapevine yellows (GY) [21–25]. Recovery can be transient or permanent and is influenced by the plant pathosystem in combination with environmental conditions. In case of ESFY and apricot, a high percentage of trees can permanently recover. Similarly, grapes with GY and AP infected apple-trees often start to recover and show just faint foliar symptoms and may eventually become nonsymptomatic for one or more years. In some cases, during the year after first appearance, infected plants undergo progressive decline until death, such as Palm Lethal yellowing [26].

3.2 Factors Influencing Phytoplasma Symptom Expression

Several factors like agronomic and environmental conditions and genetic characteristics of the pathogen and the host can influence symptom expression and severity. In some fruit tree phytoplasma titer and symptom intensity can also be influenced by the rootstock. For example, in apricot cultivars susceptible to ESFY that are grafted on ESFY-tolerant rootstocks like myrobalan, plums show only faint symptoms compared to the same cultivars grafted on ESFY-sensitive rootstocks like peach selections [27]. Agronomic and climatic conditions can modulate symptoms intensity, practices and conditions that enhance the vigor of plants also generally promote symptoms severity. Moreover, symptom expression could be related to differences in phytoplasma strains, for example for "*Candidatus* Phytoplasma *prunorum*," the causal agent of ESFY, several strains have been reported to differ in virulence and in the ability to induce characteristic symptoms [28, 29].

3.3 Symptom Monitoring Systems

For an experienced plant pathologist, it is possible to visually detect a disease generically attributable to phytoplasmas. This is especially the case if the pathologist is also familiar with symptoms caused by other factors such as fungi, viruses and bacteria, or abiotic factors like mineral deficiency, toxic agents, environmental conditions, and erroneous agronomic practices. However, it is not possible to differentiate between strains of a phytoplasma by analyzing the symptoms expressed. In this case, the use of molecular tools is necessary for a precise diagnosis. For example, different diseases caused by different phytoplasma are known to provoke the same symptoms in the common host, such in the case of FD and Grape Bois Noir (BN).

Being aware about the complexity of the problem, we invite to follow the following practical procedures for a reliable diagnosis:

- Base the diagnosis on at least three of the most typical known symptoms of the phytoplasma disease;
- Extend the observations to several plants with similar symptomatology;
- Observe the entire plant from different sides and inside the crown (there exist symptoms that develop better in the shade, or are simply hidden by the new vegetation);
- Analyze all the organs, including the roots if necessary;
- Repeat the diagnosis at least three times per year, also including the dormant season;
- Follow the progress of symptoms during the entire vegetative season;
- Repeat plant observations from top to bottom (include old and new vegetation);

- Observe widespread chlorosis phenomena related to iron or other microelement deficiencies (symptoms caused by phytoplasmas are often associated with nutrient deficiency) [30];

- Discuss with the grower about the initial development, frequency, diffusion, and particular characteristics of the disease.

3.4 Timing for Phytoplasma Symptom Monitoring

As mentioned above, scouting for symptomatic identification of phytoplasma infected plants in the field must be undertaken at the appropriate time of the year and repeated at different times during the year to cover the full range of typical symptoms. For example, to monitor AP symptoms, at least two surveys for every growing season are necessary. A survey in late summer aims to identify general plant decline, foliar alterations like yellowing or reddening, abnormal stipules, small fruit size with long pedicles, and the presence of witches' broom. In addition, during winter it is suggested that a survey be done to monitor the presence of witches' broom (the only AP symptoms visible in the canopy of dormant plants), which is easier without the leaf interference. Because PD usually induces a late general decline in plants and irregular, reddish and premature autumnal leaf fall, the most reliable field inspections are at the end of summer-early autumn. Concerning the pathosystem *Prunus*/European stone fruit yellows (ESFY), the first appropriate monitoring is at the end of winter to identify the presence of premature break of leaf buds and out-of-season blooms. A second visit in summer is required to identify leaf yellowing or reddened leaf rolls. In the case of GY disease, and similarly to Flavescence dorée (FD) and Bois noir (BN), a single survey just before harvest is usually enough to identify the main characteristic symptoms. For herbaceous annual plants the symptoms can be expressed at different stages of growth during the year but also depend on the timing of the infection. In such situations, the flowering period is optimal for diagnosis of phytoplasma symptoms.

Remote sensing based on multispectral analyses (*see* Chapter 17) has been theorized as a tool in plant pathology to acquire information about crop status since the 1970s [31] and is mainly applied to extensive crops [32]. For example, Gurr et al. [26] reported that aerial surveillance by drones would greatly assist in large-scale surveys in Ghana for coconut lethal yellowing disease. Recently Albetis et al. [33] discriminated grapevines symptomatic for Flavescence dorée from the asymptomatic ones using unmanned aerial vehicles and cameras for multispectral imagery and attained the best results with red cultivars.

Some of the symptoms are specific for phytoplasmas, whereas others may be confused with virus-induced symptoms, nutritional disorders, or other causes. In general, the characteristic symptoms of phytoplasmas include yellowing, stunting, proliferation, witches' broom, and floral alterations such as virescence and phyllody

[34, 35]. The severity of the symptoms depends on several factors, one being the grade of virulence of the specific strain of the phytoplasma [36].

Several studies have shown uneven phytoplasma distribution in host plants and seasonal fluctuations of the pathogens in woody hosts.

For some deciduous woody plants it has been suggested that phytoplasmas disappear from the aerial parts of trees during the winter and survive in the root system to re-colonize the stems and branches in spring [13].

Finally, it is worth noting that phytoplasma-diseased plants can recover from symptoms. Recovery can be temporary or permanent. Moreover, recovered plants can acquire a SAR (Systemic Acquired Resistance) [37, 38] that is a type of induced immunity, and very recently this phenomenon was connected with epigenetic processes [39].

References

1. Bertaccini A, Duduk B, Paltrinieri S et al (2014) Phytoplasmas and phytsoplasma diseases: a severe threat to agriculture. Am J Plant Sci 5:1763–1788. https://doi.org/10.4236/ajps.2014.512191

2. Bertamini M, Nedunchezhian N (2001) Effects of phytoplasma [stolbur-subgroup (Bois-noir-BN)] on photosynthetic pigments, saccharides, ribulose 1,5-bisphosphate carboxylase, nitrate and nitrite reductases, and photosynthetic activities in field grown grapevine (Vitis vinifera L. cv. Chardonnay) leaves. Photosynthetica 39:119–122

3. Bertaccini A, Duduk B (2009) Phytoplasma and phytoplasma diseases: a review of recent research. Phytopathol Mediterr 48:355–378

4. Bai X, Zhang J, Ewing E et al (2006) Living with genome instability: the adaptation of phytoplasmas to diverse environments of their insect and plant hosts. J Bacteriol 188:3682–3696. https://doi.org/10.1128/JB.188.10.3682-3696.2006

5. Christensen NM, Nicolaisen M, Hansen M et al (2004) Distribution of phytoplasmas in infected plants as revealed by real-time PCR and bioimaging. Mol Plant-Microbe Interact 17:1175–1184. pmid:15553243

6. Berges R, Rott M, Seemüller E (2000) Range of phytoplasma concentration in various plant hosts as determined by competitive polymerase chain reaction. Phytopathology 90:1145–1152

7. Arashida R, Kakizawa S, Ishii Y et al (2008) Cloning and characterization of the antigenic membrane protein (Amp) gene and in situ detection of Amp from malformed flowers infected with Japanese hydrangea phyllody phytoplasma. Phytopathology 98:769–775

8. Pagliari L, Buoso S, Santi S et al (2017) Filamentous sieve element proteins are able to limit phloem mass flow, but not phytoplasma spread. J Exp Bot 68(13):3673–3688. https://doi.org/10.1093/jxb/erx199

9. Guerriero G, Giorno F, Cicotti AM et al (2012) A gene expression analysis of cell wall biosynthetic genes in Malus × domestica infected by 'Candidatus Phytoplasma mali'. Tree Physiol 32:1365–1377. https://doi.org/10.1093/treephys/tps095

10. Sugio AM, MacLean HN, Kingdom VM et al (2011) Diverse targets of phytoplasma effectors: from plant development to defense against insects. Annu Rev Phytopathol 49:175–195

11. Lee IM, Davis RE, Gundersen-Rindal DE (2000) Phytoplasmas: phytopathogenic mollicutes. Annu Rev Microbiol 56:1593–1597

12. MacLean AM, Sugio A, Makarova OV et al (2011) Phytoplasma effector SAP54 induces indeterminate leaf-like flower development in Arabidopsis plants. Plant Physiol 157 (2):831–841. https://doi.org/10.1104/pp.111.181586

13. Seemüller E, Carraro L, Jarausch W et al (2011) Apple proliferation phytoplasma. In: Hadidi A, Barba M, Candresse T, Jelkmann W (eds) Virus and virus-like diseases of pome and stone fruits. APS Press, St. Paul, MN, pp 67–75

14. Bennett CW (1967) Plant viruses: transmission by dodder. In: Maramorosch K, Koprowski H (eds) Methods in virology, vol 1. Academic, New York, pp 393–401

15. Marwitz R, Petzold H, Ozel M (1974) Untersuchungen zur ubertragbarkeit des moglichen erregers der triebsucht des apfels auf einen krautigen wirt. Phytopathol Z 81:85–91

16. Carraro L, Osler R, Loi N et al (1991) Transmission characteristics of the clover phyllody agent by dodder. J Phytopathol 133:15–22

17. Přibylová J, Špak J (2013) Dodder transmission of phytoplasmas. In: Dickinson M, Hodgetts J (eds) Phytoplasma. Humana Press, pp 41–46

18. Chikowski LN, Sinha RC (1990) Differentiation of MLO diseases by means of symptomatology and vector transmission. In: Stanek G, Cassel G, Tully JG, Whitcomb RF (eds) Recent advances in mycoplasmology. In: Proceedings of the 7th Congress of the International Organization for Mycoplasmology, Vienna, 1988. Gustav Fisher Verlag, Stuttgard, New York, pp 280–287

19. Marwitz R (1990) Diversity of yellows disease agents in plant infections. In: Recent advances in mycoplasmology, Proceedings of the 7th Congress of the International Organization for Mycoplasmology, Baden near Vienna, 1988 pp 431–434

20. Liu LY, Tseng HI, Lin CP et al (2014) High-throughput transcriptome analysis of the leafy flower transition of Catharanthus roseus induced by peanut witches'-broom phytoplasma infection. Plant Cell Physiol 55:942–957

21. Caudwell A (1961) A study of black wood disease of vines: its relationship to flavescence dorée. Annales des Epiphyties 12(3):241–262

22. Schmid G (1965) Five and more years of observations on the proliferation virus of apples in the field. Zastita Biljia 85:285–289

23. Seemüller E, Kunze L, Schaper U (1984) Colonization behaviour of MLO and symptom expression of proliferation-diseased apple trees and decline-diseased pear trees over a period of several years. J Plant Dis Protect 91:525–532

24. Osler R, Loi N, Carraro L, et al (1999) Recovery in plants affected by phytoplasmas. In: Proceedings of the 5th Congress of the European Foundation for Plant Pathology, pp 589–592

25. Carraro L, Ermacora P, Loi N et al (2004) The recovery phenomenon in apple proliferation-infected apple trees. J Plant Pathol 86:141–146

26. Gurr GM, Johnson AC, Ash GJ et al (2016) Coconut lethal yellowing diseases: a phytoplasma threat to palms of global economic and social significance. Front Plant Sci 7:1521. https://doi.org/10.3389/fpls.2016.01521

27. Marcone C, Jarausch B, Jarausch W (2010) 'Candidatus Phytoplasma prunorum', the causal agent of European stone fruit yellows: an overview. J Plant Pathol 92:19–34

28. Dosba F, Lansac M, Mazy K et al (1991) Incidence of different diseases associated with mycoplasmalike organisms in different species of Prunus. Acta Hortic 283:311–320

29. Kison H, Seemüller E (2001) Differences in strain virulence of the European stone fruit yellows phytoplasma and susceptibility of stone fruit trees on various rootstocks to this pathogen. J Phytopathol 149:533–541

30. Sharbatkhari M, Bahar M, Ahoonmanesh A (2008) Detection of the phytoplasmal agent of pear decline in Iran, Isfahan province, using nested-PCR. Int J Plant Prod 2:167–173

31. Bauer ME, Swain PH, Mroczynski RP, et al (1971) Detection of southern corn leaf blight by remote sensing techniques. In: Proceedings of the 7th International Symposium on Remote Sensing of Environment. University of Michigan, Ann Arbor, Michigan

32. Huang W, Luo J, Zhang J, et al (2012) Crop disease and pest monitoring by remote sensing. Remote sensing—applications. In: Escalante B (ed) InTech, doi: https://doi.org/10.5772/35204. https://www.intechopen.com/books/remote-sensing-applications/crop-disease-and-pest-monitoring-by-remote-sensing

33. Albetis J, Duthoit S, Guttler F et al (2017) Detection of Flavescence dorée grapevine disease using unmanned aerial vehicle (UAV) multispectral imagery. Remote Sens 9(4):308

34. Bertaccini A (2007) Phytoplasmas: diversity, taxonomy, and epidemiology. Front Biosci 12:673–689

35. Hogenhout SA, Oshima K, Ammar E et al (2008) Phytoplasmas: bacteria that manipulate

plants and insects. Mol Plant Pathol 9 (4):403–423

36. Seemüller E, Schneider B (2007) Differences in virulence and genomic features of strains of 'Candidatus Phytoplasma mali', the apple proliferation agent. Phytopathology 97:964–970

37. Musetti R, Sanità di Toppi L, Ermacora P et al (2004) Recovery in apple trees infected with the apple proliferation phytoplasma: an ultrastructural and biochemical study. Phytopathology 94:203–208. https://doi.org/10.1094/PHYTO.2004.94.2.203

38. Osler R, Borselli S, Ermacora P et al (2016) Transmissible tolerance to European stone fruit yellows (ESFY) in apricot: cross-protection or a plant mediated process? Phytoparasitica 44:203–211

39. Leljak-Levanic D, Jezic M, Cesar V et al (2010) Biochemical and epigenetic changes in phytoplasma-recovered periwinkle after indole-3-butyric acid treatment. J Appl Microbiol 109:2069–2078

Part II

Molecular Analyses

Chapter 6

Comparison of Different Procedures for DNA Extraction for Routine Diagnosis of Phytoplasmas

Carmine Marcone

Abstract

This chapter presents five different procedures for extracting DNA from phytoplasma-infected plants and insect vectors suitable for PCR assays. One of these procedures enriches phytoplasma DNA through differential centrifugation and is effective in producing highly purified DNA from fresh tissues from a wide variety of herbaceous and woody plants. Although the DNA yield is less than those of other known total DNA extraction procedures, a major advantage of the presented phytoplasma-enriched procedure is that a substantial proportion of the isolated DNA is from phytoplasmas. The other four procedures here presented involve treatments with CTAB-based buffer to lyse cells and purify DNA followed by deproteination and recovery of DNA. These procedures work well for extracting total DNA from fresh, frozen, or lyophilized tissues from a wide variety of plant hosts as well as insect vectors. Because few manipulations are required, the CTAB-based procedures are faster and easier to perform than the phytoplasma-enrichment protocol. In addition, they result in very high yields and provide DNA that is less pure but of suitable quality for the use in standard molecular biological techniques including PCR assays.

Key words Phytoplasma infections, Phloem tissue, Insect vectors, DNA extraction, Polymerase chain reaction, Diagnosis, Cetyltrimethylammonium bromide, Polysaccharides

1 Introduction

The availability of DNA-based methods into phytoplasmology has greatly improved diagnosis of phytoplasma infections in plant and insect hosts. These methods became available when protocols for isolation and cloning phytoplasma DNA were implemented [1–6]. Currently, polymerase chain reaction (PCR) technology is the method of choice for phytoplasma diagnosis. However, for successful diagnosis of phytoplasma infections by PCR assays the main requirements are that template DNA extracted from diseased plants and insect vectors is of suitable quality and concentration. Since phytoplasmas reside almost exclusively in sieve tubes, the starting material for DNA extraction should include as much phloem tissue as possible. Also, the amount of phytoplasma DNA

Rita Musetti and Laura Pagliari (eds.), *Phytoplasmas: Methods and Protocols*, Methods in Molecular Biology, vol. 1875,
https://doi.org/10.1007/978-1-4939-8837-2_6, © Springer Science+Business Media, LLC, part of Springer Nature 2019

in total DNA extracted from infected plants can be further increased using appropriate phytoplasma-enrichment procedures [2, 4, 7]. This is particularly useful for detecting low-titer phytoplasma infections as is often true for woody plants.

There are a number of different methods for extracting DNA from phytoplasma-infected plant and insect hosts to be used as template in PCR assays [2, 4, 7–14]. The main differences between various procedures lie in the extent of manipulations and hazardous amounts of toxic reagents required, yield and degree of purity of the produced DNA, quantity and physical characteristics of the starting material. Commercially available kits have also been employed for DNA extraction from diseased plants for detection of phytoplasmas by PCR [15].

This chapter presents five different procedures for extracting DNA from phytoplasma-infected plants and insect vectors suitable for PCR assays. The first of these procedures, hereafter referred to as "phytoplasma enrichment procedure," which is largely based on those previously described by Ahrens and Seemüller [4] and Kirkpatrick et al. [2] enriches phytoplasma DNA through differential centrifugation. Most host nuclear and chloroplast DNA is eliminated during a low-speed centrifugation step whereas most polysaccharides, phenolic compounds, and other enzyme-inhibiting contaminants found in plant cells are discarded in the supernatant following a high-speed centrifugation step. The resulting phytoplasma-enriched pellet is then processed using a high-salt cetyltrimethylammonium bromide (CTAB)-based buffer followed by chloroform/isoamyl alcohol extraction prior to isopropanol precipitation. This procedure is effective in producing highly purified DNA from fresh tissues from a wide variety of herbaceous and woody plants. Although the DNA yield is less than those of other known total DNA extraction procedures, a major advantage of the presented phytoplasma-enrichment procedure is that a substantial proportion of the isolated DNA is originating from phytoplasmas.

The other four procedures here described, which are based on the CTAB extraction method described by Doyle and Doyle [16], involve treatments with CTAB-based buffer to lyse cells and purify DNA followed by deproteination and recovery of DNA. These protocols work well for extracting total DNA from fresh, frozen, or lyophilized tissues from a wide variety of plant hosts as well as insect vectors. Because few manipulations are required, they are faster and easier to perform than the phytoplasma enrichment method. In addition, they result in very high yields and provide DNA that is less pure but of suitable quality for use with standard molecular biological techniques including PCR assays.

2 Materials

Use sterile, distilled deionized water to prepare all reagents.

2.1 Phytoplasma Enrichment Procedure

1. Healthy and phytoplasma-infected plants.

2. Phytoplasma grinding buffer: 125 mM potassium phosphate, 10% sucrose, 0.15% bovine serum albumin (BSA), fraction V, 2% polyvinylpyrrolidone (PVP-40), 30 mM ascorbic acid, pH 7.6. To prepare 1 liter of this buffer, dissolve 21.7 g K_2HPO_4-$3H_2O$ or 16.7 g K_2HPO_4, 4.1 g KH_2PO_4, 100 g sucrose, 1.5 g BSA, fraction V, 20 g PVP-40, and 5.3 g ascorbic acid in water to 1 liter. Add ascorbic acid just before use. After adding ascorbic acid, adjust pH to 7.6 with 1 M NaOH. The stock buffer without ascorbic acid can be stored at −20 °C.

3. Extraction buffer: 2% cetyltrimethylammonium bromide (CTAB), 1.4 M sodium chloride, 20 mM EDTA, pH 8, 100 mM Tris–HCl, pH 8, 1% polyvinylpyrrolidone, 0.2% (v/v) 2-mercaptoethanol. To prepare 1 liter of this buffer, mix 40 mL of 0.5 M EDTA stock solution, pH 8 and 100 mL 1 M Tris–HCl stock solution, pH 8 with approximately 600 mL of water and heat to approximately 80 °C. Add 20 g CTAB, 81.8 g NaCl and 10 g PVP-40, stir until dissolved, and then add water to 1 liter. Store up to several months at room temperature. Add 2-mercaptoethanol to the required volume of buffer to give a final concentration of 0.2% (v/v) just before use.

4. Tris/EDTA (TE) buffer: 10 mM Tris–HCl, 1 mM EDTA, pH 8.

5. Chloroform/isoamyl alcohol (24:1, v/v).

6. Isopropanol.

7. 70% (v/v) ethanol.

8. Low-speed refrigerated centrifuge.

9. Bench-top microcentrifuge.

10. 50 mL centrifuge tubes.

11. 2 mL microcentrifuge tubes.

12. Mortar and pestle.

13. Ice bucket.

14. 60 °C water bath.

15. Glass beads or quartz sand.

16. Vacuum desiccator or speedvac evaporator.

2.2 Large-Scale CTAB-Based Extraction Procedure

1. Healthy and phytoplasma-infected plants.

2. Extraction buffer: 2% CTAB, 1.4 M sodium chloride, 20 mM EDTA, pH 8, 100 mM Tris–HCl, pH 8, 1% polyvinylpyrrolidone, 0.2% (v/v) 2-mercaptoethanol. To prepare 1 liter of this buffer, mix 40 mL of 0.5 M EDTA stock solution, pH 8 and 100 mL 1 M Tris–HCl stock solution, pH 8 with approximately 600 mL of water and heat to approximately 80 °C. Add 20 g CTAB, 81.8 g NaCl and 10 g PVP-40, stir until dissolved, and then add water to 1 liter. Store up to several months at room temperature. Add 2-mercaptoethanol to the required volume of buffer to give a final concentration of 0.2% (v/v) just before use.

3. TE buffer: 10 mM Tris–HCl, 1 mM EDTA, pH 8.

4. Chloroform/isoamyl alcohol (24:1, v/v).

5. Isopropanol.

6. 70% (v/v) ethanol.

7. Low-speed refrigerated centrifuge.

8. 50 mL centrifuge tubes.

9. 1.5 mL microcentrifuge tubes.

10. Mortar and pestle.

11. Ice bucket.

12. 60 °C water bath.

13. Liquid nitrogen.

14. Vacuum desiccator or speedvac evaporator.

2.3 Small-Scale CTAB-Based Extraction Procedure

1. Healthy and phytoplasma-infected plants.

2. Extraction buffer: 2% CTAB, 1.4 M sodium chloride, 20 mM EDTA, pH 8, 100 mM Tris–HCl, pH 8, 1% polyvinylpyrrolidone, 0.2% (v/v) 2-mercaptoethanol. To prepare 1 liter of this buffer, mix 40 mL of 0.5 M EDTA stock solution, pH 8 and 100 mL 1 M Tris–HCl stock solution, pH 8 with approximately 600 mL of water and heat to approximately 80 °C. Add 20 g CTAB, 81.8 g NaCl and 10 g PVP-40, stir until dissolved, and then add water to 1 liter. Store up to several months at room temperature. Add 2-mercaptoethanol to the required volume of buffer to give a final concentration of 0.2% (v/v) just before use.

3. TE buffer: 10 mM Tris–HCl, 1 mM EDTA, pH 8.

4. Chloroform/isoamyl alcohol (24:1, v/v).

5. Isopropanol.

6. 70% (v/v) ethanol.

7. Bench-top microcentrifuge.

8. 2 mL microcentrifuge tubes.

9. Micropestles.

10. Ice bucket.

11. 60 °C water bath.

12. Glass beads or quartz sand.

13. Vacuum desiccator or speedvac evaporator.

2.4 CTAB-Based Extraction Procedure Using Polyethylene Grinding Bags

1. Healthy and phytoplasma-infected plants.

2. Extraction buffer: 2% CTAB, 1.4 M sodium chloride, 20 mM EDTA, pH 8, 100 mM Tris–HCl, pH 8, 1% polyvinylpyrrolidone, 0.2% (v/v) 2-mercaptoethanol. To prepare 1 liter of this buffer, mix 40 mL of 0.5 M EDTA stock solution, pH 8 and 100 mL 1 M Tris–HCl stock solution, pH 8 with approximately 600 mL of water and heat to approximately 80 °C. Add 20 g CTAB, 81.8 g NaCl and 10 g PVP-40, stir until dissolved, and then add water to 1 L. Store up to several months at room temperature. Add 2-mercaptoethanol to the required volume of buffer to give a final concentration of 0.2% (v/v) just before use.

3. TE buffer: 10 mM Tris–HCl, 1 mM EDTA, pH 8.

4. Chloroform/isoamyl alcohol (24:1, v/v).

5. Isopropanol.

6. 70% (v/v) ethanol.

7. Bench-top microcentrifuge.

8. 2 mL microcentrifuge tubes.

9. Polyethylene grinding bags.

10. Hand homogenizer.

11. Ice bucket.

12. 60 °C water bath.

13. Vacuum desiccator or speedvac evaporator.

2.5 CTAB-Based Extraction Procedure for Insect Vectors

1. Insect vectors.

2. Extraction buffer: 2% CTAB, 1.4 M sodium chloride, 20 mM EDTA, pH 8, 100 mM Tris–HCl, pH 8, 1% polyvinylpyrrolidone, 0.2% (v/v) 2-mercaptoethanol. To prepare 1 liter of this buffer, mix 40 mL of 0.5 M EDTA stock solution, pH 8 and 100 mL 1 M Tris–HCl stock solution, pH 8 with approximately 600 mL of water and heat to approximately 80 °C. Add 20 g CTAB, 81.8 g NaCl and 10 g PVP-40, stir until dissolved, and then add water to 1 liter. Store up to several months at room temperature. Add 2-mercaptoethanol to the required volume of buffer to give a final concentration of 0.2% (v/v) just before use.

3. TE buffer: 10 mM Tris–HCl, 1 mM EDTA, pH 8.

4. Chloroform/isoamyl alcohol (24:1, v/v).

5. Isopropanol.

6. 70% (v/v) ethanol.

7. Bench-top microcentrifuge.

8. 2 mL microcentrifuge tubes.

9. Micropestles.

10. Ice bucket.

11. 60 °C water bath.

12. Carborundum.

13. Vacuum desiccator or speedvac evaporator.

3   Methods

3.1   Phytoplasma Enrichment Procedure

1. Select approximately 1.0 g of fresh tissue including leaf mid-ribs, petioles, fruit peduncles, young shoots, shoot tips, or phloem tissue from stems and roots (*see* **Note 1**).

2. Cut specimens into small pieces with a scissor and incubate for 10 min in 9 mL of ice-cold phytoplasma grinding buffer in a mortar on ice (*see* **Note 2**).

3. Grind thoroughly with a cold pestle (*see* **Note 3**), add another 10 mL of fresh buffer, and repeat grinding until the tissue is completely broken up. Transfer the homogenate to cold 50 mL centrifuge tubes and keep tubes on ice until all samples are ground.

4. Place the tubes in a cold rotor and centrifuge at $1100 \times g$ for 5 min at 4 °C. Carefully transfer the supernatant to clean, cold 50 mL centrifuge tubes.

5. Centrifuge the supernatant at $14{,}600 \times g$ for 25 min at 4 °C. Carefully discard the supernatant, and drain the tubes thoroughly.

6. Gently resuspend the phytoplasma-enriched pellet in 1 mL of warm extraction buffer (*see* **Note 4**) and transfer the content to 2 mL microcentrifuge tubes.

7. Place the 2 mL microcentrifuge tubes in a 60 °C water bath and incubate for 30 min.

8. Add an equal volume of chloroform/isoamyl alcohol (24:1, v/v), securely cap the tubes, and invert several times to form an emulsion (*see* **Note 5**).

9. Place the tubes to a microcentrifuge rotor and centrifuge at $7600 \times g$ for 10 min at room temperature. Carefully remove

the upper aqueous phase and transfer it to clean 2 mL micro-centrifuge tubes. Discard lower organic phase.

10. Add one volume of ice-cold isopropanol, mix thoroughly, and centrifuge at 15,000 × g for 10 min at room temperature. Discard the supernatant.

11. Wash DNA pellet with 2 mL ice-cold 70% (v/v) ethanol, centrifuge at 15,000 × g for 5 min at room temperature.

12. Carefully pour off the ethanol, invert tubes on a paper towel, and let them dry thoroughly or dry the pellet in a vacuum desiccator or a speedvac evaporator (*see* **Note 6**).

13. Dissolve the pellet in 100 μL of water or TE buffer (*see* **Note 7**).

14. The DNA can be used immediately or stored at −20 °C (*see* **Note 8**).

3.2 Large-Scale CTAB-Based Extraction Procedure

1. Preheat extraction buffer to 60 °C in a water bath.

2. Select 1.5 to 2.5 g of fresh tissue including leaf midribs, petioles, fruit peduncles, young shoots, shoot tips, or phloem tissue from stems and roots. Cut specimens into small pieces with a scissor and grind in liquid nitrogen with mortar and pestle to a fine powder (*see* **Note 9**). Stored frozen or lyophilized tissues can be used as starting material as well (*see* **Note 10**).

3. Transfer the tissue powder to 50 mL centrifuge tubes containing 12 mL of preheated extraction buffer and mix to wet thoroughly. Incubate at 60 °C for 30 min in water bath with occasional gentle mixing.

4. Remove the centrifuge tubes from the water bath, place them in the hood, and let them cool for 2 to 3 min. Add an equal volume of chloroform/isoamyl alcohol (24:1, v/v), securely cap the tubes, and invert several times to form an emulsion (*see* **Note 5**).

5. Place the tubes to a centrifuge rotor and centrifuge at 6700 × g for 10 min at room temperature. Carefully remove the upper aqueous layer and transfer it to clean 50 mL microcentrifuge tubes. Discard lower chloroform layer.

6. Add two-third volume of ice-cold isopropanol, mix thoroughly, leave on ice for 5 to 10 min, and centrifuge at 10,500 × g for 15 min at room temperature. Discard the supernatant.

7. Wash DNA pellet with 2 mL ice-cold 70% (v/v) ethanol, centrifuge at 10500 × g for 5 min at room temperature.

8. Carefully pour off the ethanol, invert tubes on a paper towel, and let them dry thoroughly or dry the pellet under vacuum (*see* **Note 11**).

9. Resuspend DNA in 1 mL of TE buffer and transfer the suspension to 1.5 microcentrifuge tubes.

10. The DNA can be used immediately or stored at −20 °C.

3.3 Small-Scale CTAB-Based Extraction Procedure

1. Preheat extraction buffer to 60 °C in a water bath.

2. Cut 0.5 g of fresh tissue including leaf midribs, petioles, fruit peduncles, young shoots, shoot tips, or phloem tissue into small pieces with a scissor and transfer them to 2 mL microcentrifuge tubes containing 700 μL of preheated extraction buffer.

3. Grind thoroughly with a micropestle (*see* **Note 3**), add another 300 μL of fresh buffer, and repeat grinding until the tissue is completely broken up.

4. Incubate at 60 °C for 30 min in water bath with occasional gentle mixing.

5. Centrifuge samples in a microcentrifuge at 15,000 × g for 30 s at room temperature to pellet debris and transfer the supernatant to clean 2 mL microcentrifuge tubes.

6. Add an equal volume of chloroform/isoamyl alcohol (24:1, v/v), securely cap the tubes, and invert several times to form an emulsion (*see* **Note 5**).

7. Centrifuge at 15,000 × g for 10 min at room temperature. Carefully remove the top aqueous phase and transfer it to clean 2 mL microcentrifuge tubes. Discard organic phase (*see* **Note 12**).

8. Add one volume of ice-cold isopropanol, mix thoroughly, and centrifuge at 15,000 × g for 10 min at room temperature. Discard the supernatant.

9. Wash DNA pellet with 2 mL ice-cold 70% (v/v) ethanol, centrifuge at 15,000 × g for 5 min at room temperature.

10. Carefully pour off the ethanol, invert tubes on a paper towel, and let them dry thoroughly or dry the pellet under vacuum.

11. Dissolve the pellet in 100 μL of water or TE buffer.

12. The DNA can be used immediately or stored at −20 °C.

3.4 CTAB-Based Extraction Procedure Using Polyethylene Grinding Bags

1. Preheat extraction buffer to 60 °C in a water bath.

2. Cut 0.5 g of leaf midribs or other suitable tissues into small pieces with a scissor and transfer them to a polyethylene grinding bag containing 3 mL of preheated extraction buffer.

3. Grind thoroughly with a hand homogeniser, add another 2 mL of fresh buffer, and repeat grinding.

4. Transfer 1 mL of homogenate to 2 mL microcentrifuge tubes and incubate at 60 °C for 30 min in water bath with occasional gentle mixing.

5. Extract the homogenate with an equal volume of chloroform/isoamyl alcohol (24:1, v/v). Mix well by inversion (*see* **Note 5**) and centrifuge at 15,000 × *g* for 5 min at room temperature.

6. Carefully remove the top aqueous phase and transfer it to clean 2 mL microcentrifuge tubes. Discard chloroform phase.

7. Add one volume of ice-cold isopropanol, mix thoroughly and centrifuge at 15,000 × *g* for 30 min at 4 °C. Discard the supernatant.

8. Wash DNA pellet with 2 mL ice-cold 70% (v/v) ethanol, centrifuge at 15,000 × *g* for 5 min at room temperature.

9. Carefully pour off the ethanol, invert tubes on a paper towel, and let them dry thoroughly or dry the pellet under vacuum.

10. Dissolve the pellet in 100 μL of water or TE buffer.

11. The DNA can be used immediately or stored at −20 °C.

3.5 CTAB-Based Extraction Procedure for Insect Vectors

1. Preheat DNA extraction buffer to 60 °C in a water bath.

2. Grind individual insects or batches of 2–5 insects of the same species, either fresh or frozen or stored under 70% ethanol, in 2 mL microcentrifuge tubes containing 500 μL of preheated extraction buffer using a micropestle and sterile carborundum to facilitate grinding. Add another 500 μL of fresh buffer and complete grinding.

3. Incubate at 60 °C for 30 min in water bath with occasional gentle mixing.

4. Centrifuge samples in a microcentrifuge at 15,000 × *g* for 2 min at room temperature to pellet debris and transfer the supernatant to clean 2 mL microcentrifuge tubes.

5. Extract the lysate with an equal volume of chloroform/isoamyl alcohol (24:1, v/v). Mix well by inversion (*see* **Note 5**) and centrifuge at 15,000 × *g* for 5 min at room temperature.

6. Remove the upper aqueous phase and transfer it to clean 2 mL microcentrifuge tubes. Discard lower chloroform phase.

7. Add one volume of ice-cold isopropanol, mix thoroughly, and centrifuge at 15,000 × *g* for 30 min at 4 °C. Discard the supernatant.

8. Wash DNA pellet with 2 mL ice-cold 70% (v/v) ethanol, centrifuge at 15,000 × *g* for 5 min at 4 °C.

9. Carefully pour off the ethanol, invert the tubes on a paper towel, and let them dry thoroughly or dry the pellet under vacuum.

10. Dissolve the pellet in 100 μL of water or TE buffer.

11. The DNA can be used immediately or stored at −20 °C.

4 Notes

1. Phloem tissue is prepared as aseptically as possible from stems and roots, approximately 30 mm in diameter, by removing the outer bark with a knife and excising the layer of conductive tissue with a sterile scalpel [17].

2. It is convenient to process eight samples at a time.

3. The addition of a small amount of glass beads or quartz sand will facilitate grinding.

4. This can be accomplished using a loose-fitting homogenizer such as a plastic transfer pipette bulb.

5. Attention should be paid to avoid that chloroform vapor pressure causes samples to spill. All manipulations with chloroform should be carried out in a well-ventilated fume hood.

6. The DNA pellet will not stick well to walls of the microcentrifuge tube after the 70% ethanol wash. Therefore, care must be taken to avoid aspirating the pellet out of the microcentrifuge tube.

7. Most protocols suggest TE buffer for DNA storage [18]. However, EDTA present in such buffer may chelate the Mg^{2+} in the PCR buffer whose concentration is vital, thereby affecting sensitivity and specificity of the reaction. Therefore, it is suggested to simply storing DNA in distilled water.

8. Yield and quality of the extracted DNA can be assessed by electrophoresing small quantities in a 0.8% agarose gel using known standards or by spectrophotometry at 260 nm or by fluorometry (Fig. 1).

9. Tissue should be kept frozen throughout the grinding operation by replenishing the liquid nitrogen as necessary. Protective clothing, gloves, and goggles should be worn to protect skin and eyes against exposure to liquid nitrogen and the operation should be conducted in a well-ventilated area.

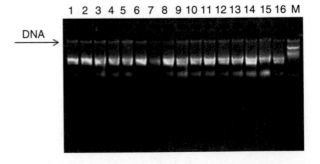

Fig. 1 Agarose gel electrophoresis of DNA (arrow) extracted from phytoplasma-infected woody plants (lanes 1–16) employing the phytoplasma enrichment procedure. M, molecular marker XV (Roche)

10. Stored frozen tissue is ground to a fine powder in liquid nitrogen using a mortar and pestle, whereas lyophilized tissue is ground to a fine powder with a mill.

11. Overdrying will make the pellet difficult to dissolve.

12. If the phases do not resolve well, due for instance to the presence of cellular debris in the aqueous phase, organic extraction should be repeated.

References

1. Kirkpatrick BC, Stenger DC, Morris TJ et al (1987) Cloning and detection of DNA from a nonculturable plant pathogenic mycoplasma-like organism. Science 238:197–200

2. Kirkpatrick BC, Harrison NA, Lee I-M et al (1995) Isolation of mycoplasma-like organism DNA from plant and insect hosts. In: Razin S, Tully JG (eds) Molecular and diagnostic procedures in Mycoplasmology, vol I. Academic Press, San Diego, CA, pp 105–116

3. Kollar A, Seemüller E, Bonnet F et al (1990) Isolation of the DNA of various plant pathogenic mycoplasmalike organisms from infected plants. Phytopathology 80:233–237

4. Ahrens U, Seemüller E (1992) Detection of DNA of plant pathogenic mycoplasmalike organisms by a polymerase chain reaction that amplifies a sequence of the 16S rRNA gene. Phytopathology 82:828–832

5. Lee I-M, Davis RE, Hiruki C (1991) Genetic interrelatedness among clover proliferation mycoplasmalike organisms (MLOs) and other MLOs investigated by nucleic acid hybridization and restriction fragment length polymorphism analyses. Appl Environ Microbiol 57:3565–3569

6. Lee I-M, Davis RE, Sinclair WA et al (1993) Genetic relatedness of mycoplasmalike organisms detected in *Ulmus* spp. in the United States and Italy by means of DNA probes and polymerase chain reactions. Phytopathology 83:829–833

7. Prince JP, Davis RE, Wolf TK et al (1993) Molecular detection of diverse mycoplasmalike organisms (MLOs) associated with grapevine yellows and their classification with aster yellows, X-disease, and elm yellows MLOs. Phytopathology 83:1130–1137

8. Firrao G, Locci R (1993) Rapid preparation of DNA from phytopathogenic mycoplasma-like organisms for polymerase chain reaction analysis. Lett Appl Microbiol 17:280–281

9. Maixner M, Ahrens U, Seemüller E (1995) Detection of the German grapevine yellows (Vergilbungskrankheit) MLO in grapevine, alternative hosts and a vector by a specific PCR procedure. Eur J Plant Pathol 101:241–250

10. Zhang Y-P, Uyemoto JK, Kirkpatrick BC (1998) A small-scale procedure for extracting nucleic acids from woody plants infected with various phytopathogens for PCR assay. J Virol Methods 71:45–50

11. Angelini E, Clair D, Borgo M et al (2001) Flavescence dorée in France and Italy—occurrence of closely related phytoplasma isolates and their near relationships to palatinate grapevine yellows and an alder yellows phytoplasma. Vitis 40:79–86

12. Marzachì C, Palermo S, Boarino A, Boccardo G et al (2001) Optimisation of a one-step PCR assay for the diagnosis of Flavescence dorée-related phytoplasmas in field-grown grapevines and vector populations. Vitis 40:213–217

13. Palmano S (2001) A comparison of different phytoplasma DNA extraction methods using competitive PCR. Phytopathol Mediterr 40:99–107

14. Boudon-Padieu E, Béjat A, Clair D, Angelini E et al (2003) Grapevine yellows: comparison of different procedures for DNA extraction and amplification with PCR for routine diagnosis of phytoplasmas in grapevine. Vitis 42:141–149

15. Green MJ, Thompson DA, MacKenzie DJ (1999) Easy and efficient extraction from woody plants for detection of phytoplasmas by polymerase chain reaction. Plant Dis 83:482–485

16. Doyle JJ, Doyle JL (1990) Isolation of plant DNA from fresh tissue. Focus 12:13–15

17. Ahrens U, Seemüller E (1994) Detection of mycoplasmalike organisms in declining oaks by polymerase chain reaction. Eur J For Pathol 24:55–63

18. Sambrook J, Fritsch EF, Maniatis T (1989) Molecular cloning: a laboratory manual, 2nd edn. Cold Spring Harbor Laboratory Press, Cold Spring Harbor, NY

Chapter 7

Standard Detection Protocol: PCR and RFLP Analyses Based on 16S rRNA Gene

Assunta Bertaccini, Samanta Paltrinieri, and Nicoletta Contaldo

Abstract

Phytoplasma detection and identification is primarily based on PCR followed by restriction fragment length polymorphism analysis. This method detects and differentiates phytoplasmas including those not yet identified. The protocol describes the application of this method for identification of phytoplasmas at 16S rRNA (16Sr) group and 16Sr subgroup levels on amplicons and also in silico on the same sequences.

Key words Phytoplasma, Detection, Identification, Ribosomal group, Ribosomal subgroup, Sequencing

1 Introduction

Following their discovery 50 years ago, phytoplasmas have been difficult to detect due to their low concentration, especially in woody host plants and for their erratic distribution in the sieve tubes of the infected plants [1]. The establishment of electron microscopy (EM)-based techniques represents an alternative approach to the indexing procedure based on graft transmission to healthy indicator plants. Transmission EM [2, 3] and, with less reliability, scanning EM [4] observations can be used as alternative to staining phytoplasmas with DNA-specific dyes [5]. Quite recently cultivation in artificial media has shown great potential in phytoplasma isolation from plant-infected tissues [6, 7]. Protocols for the production of enriched phytoplasma-specific antigens that in the past have been developed but not always effectively applied [8, 9] could also be improved by using culture-purified phytoplasma as antigens. Sensitive and accurate detection of these micro-organisms is a prerequisite for the management of phytoplasma-associated diseases and prevention of severe epidemics with relevant potential economic impact. For this reason, in the last 30 years several detection methods based on a molecular approach,

Rita Musetti and Laura Pagliari (eds.), *Phytoplasmas: Methods and Protocols*, Methods in Molecular Biology, vol. 1875,
https://doi.org/10.1007/978-1-4939-8837-2_7, © Springer Science+Business Media, LLC, part of Springer Nature 2019

combining PCR (*see* **Note 1**) with Restriction Fragment Length Polymorphism (RFLP), were developed. In fact, the collective RFLP pattern characteristic of each phytoplasma is unique [10]. PCR/RFLP analyzes on 16Sr RNA gene and ideally it allows detection and differentiation of all phytoplasmas. This system allocates the up-to-now worldwide detected phytoplasmas in 36 groups and more than 140 subgroups [11–17]. Moreover, this system is more flexible for epidemiological studies than the use of the "*Candidatus* genus" taxa designation [18] formally adopted until now for 43 phytoplasmas (Fig. 1). Several primers designed on the 16Sr RNA sequence universal or group specific were developed (Table 1); they can be used in different combinations in direct, nested, or semi-nested systems for routine detection of phytoplasmas as well as for identification of new phytoplasmas [19–25]. Phytoplasma differentiation is routinely based on 16S rRNA gene sequences, which is carried out by RFLP analysis of PCR amplified DNA sequences using 17 endonuclease restriction enzymes (*see* **Note 9**) [10, 26]. The 16S ribosomal gene is normally in double copy in phytoplasmas, helping its detection; however in a number of cases the samples under analyses may contain phytoplasma (s) with ribosomal RNA interoperon heterogeneity or mixed phytoplasma populations [27–32]. The RFLP patterns will then appear with more restricted fragments of which the sum of total fragment sizes is more than that of the expected amplicon. In the case of mixed infections, the use of specific primers, when available, is suggested (Table 1). Additional tools for strain differentiation using variable single copy genes [33–38], cloning the amplicons [39], or cultivating the phytoplasma if fresh material is available, could be also helpful [6, 7].

The continuous effort to improve the diagnostic procedures aims to develop quicker, more economic, and robust methods. Sensitivity is not an issue per se, as the current nested PCR protocols are extremely sensitive, but the achievement of high levels of sensitivity without the risk of false positive results that can be associated with nested PCR is highly desirable. It is also possible to sequence the PCR or nested-PCR products and then to use the aligned sequences of at least 1200 bp (amplicons obtained with R16F2n/R2 primer combination) for phytoplasma identification in silico [39–45].

2   Materials

1. PCR tubes 200 μL or 500 μL.
2. Primers forward and reverse (20 pmol/μL) Table 1.
3. PCR buffer.
4. *Taq* polymerase (*see* **Note 2**).

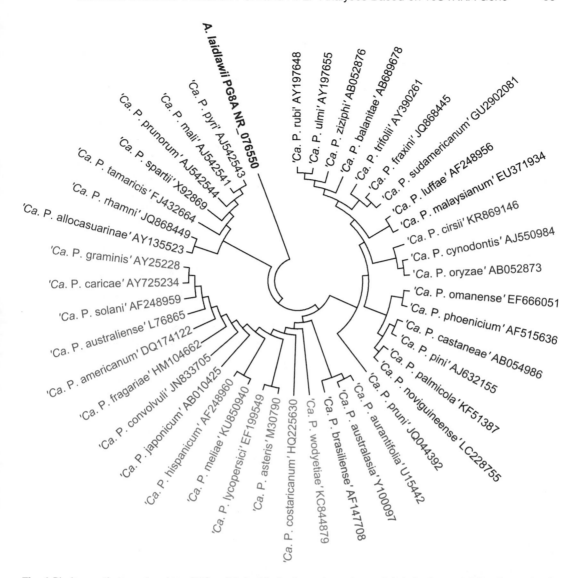

Fig. 1 Phylogenetic tree showing all '*Candidatus* Phytoplasma' species and *Acholeplasma laidlawii* as outroot. A about 1,200 bp fragment of the 16S ribosomal gene was used from phytoplasmas. The evolutionary history was inferred using the Neighbor-Joining method. The evolutionary distances were computed using the Maximum Composite Likelihood method and are in the units of the number of base substitutions per site. Evolutionary analyses were conducted in MEGA6

5. d-NTPs.

6. PCR Thermal cycler.

7. Healthy and infected plant controls preferably of the same species and tissue as samples.

8. Deionized, distilled, sterile water (dd H$_2$O).

9. Agarose for electrophoresis.

10. Acrylamide/bis-acrylamide 29:1 (Sigma-Aldrich 40% solution).

Table 1
Primers for amplification of 16Sr RNA of phytoplasmas

Primer[a]	Specificity 16Sr group	Sequence 5′ – 3′	Position[b]	Reference
P1	Universal	AAGAATTTGATCCTGGCTCAGGATT	16Sr DNA	[46]
P7	Universal	CGTCCTTCATCGGCTCTT	23Sr DNA	[47]
R0	Universal	GAATACCTTGTTACGACTTAACCCC	16Sr DNA	[23]
P3	Universal	GGATGGATCACCTCCTT	16Sr DNA	[47]
P4	Universal	GAAGTCTGCAACTCGACTTC	16Sr DNA	[47]
P5	Universal	CGGCAATGGAGGAAACT	16Sr DNA	[47]
R16F2n	Universal	GAAACGACTGCTAAGACTGG	16Sr DNA	[48]
R16R2	Universal	TGACGGGCGGTGTGTACAAACCCCG	16Sr DNA	[49]
F1	Universal	AAGACGAGGATAACAGTTGG	16Sr DNA	[50]
B6	Universal	TAGTGCCAAGGCATCCACTGTG	IS	[51]
R16mF2	Universal	CATGCAAGTCGAACGGA	16Sr DNA	[48]
R16mR2	Universal	CTTAACCCCAATCATCGA	16Sr DNA	[48]
P1A	Universal	AACGCTGGCGGCGCGCCTAATAC	16Sr DNA	[52]
16Sr-SR	Universal	GGTCTGTCAAAACTGAAGATG	IS	[53]
P7A	Universal	CCTTCATCGGCTCTTAGTGC	23Sr DNA	[52]
Pc399	Universal	AACGCCGCGTGAACGATGAA	16Sr DNA	[54]
Pc1694	Universal	ATCAGGCGTGTGCTCTAACC	IS	[54]
fU5	Universal	CGGCAATGGAGGAAACT	16Sr DNA	[24]
rU3	Universal	TTCAGCTACTCTTTGTAACA	16Sr DNA	[24]
PA2f	Universal	GCCCCGGCTAACTATGTGC	16Sr DNA	[55]
PA2r	Universal	TTGGTGGGCCTAAATGGACTC	IS	[55]
M1(758F)	Universal	GTCTTTACTGACGC	16Sr DNA	[56]
M2(1232R)	Universal	CTTCAGCTACCCTTTGTAAC	16Sr DNA	[56]
SN910601	Universal	GTTTGATCCTGGCTCAGGATT	16Sr DNA	[25]
SN910502	Universal	AACCCCGAGAACGTATTCACC	16Sr DNA	[25]
1F7	Universal	AGTGCTTAACACTGTCCTGCTA	16Sr DNA	[57]
7R3	Universal	TTGTAGCCCAGATCATAAGGGGCA	16Sr DNA	[57]
3Fwd	Universal	ACCTGCCTTTAAGACGAGGA	16Sr DNA	[57]
3rev	Universal	AAAGGAGGTGATCCATCCCCACCT	16Sr DNA	[57]
7R2	Universal	GACAAGGGTTGCGCTCGTTTT	16Sr DNA	[57]
5Rev	Universal	ACCCCGAGAACGTATTCACCGCGA	16Sr DNA	[57]

(continued)

Table 1
(continued)

Primer[a]	Specificity 16Sr group	Sequence 5′ – 3′	Position[b]	Reference
R16(I)F1	I, II, IX, XII, XV	TAAAAGACCTAGCAATAGG	16Sr DNA	[58]
R16(I)R1	I, II, IX, XII, XV	CAATCCGAACTAAGACTCT	16Sr DNA	[58]
R16(III)F2	III	AAGAGTGGAAAAACTCCC	16Sr DNA	[58]
R16(III)R1	III	TTCGAACTGAGATTGA	16Sr DNA	[58]
LY16Sf	IV	CATGCAAGTCGAACGGAAATC	16Sr DNA	[59]
LY16Sr	IV	GCTTACGCAGTTAGGCTGTC,	16Sr DNA	[59]
R16(V)F1	V	TTAAAAGACCTTCTTCGG	16Sr DNA	[58]
R16(V)R1	V	TTCAATCCGTACTGAGACTACC	16Sr DNA	[58]
R16(X)F1	X	GACCCGCAAGTATGCTGAGAGATG	16Sr DNA	[23]
R16(X)R1	X	CAATCCGAACTGAGAGTCT	16Sr DNA	[23]
fO1	X	CGGAAACTTTTAGTTTCAGT	16Sr DNA	[24]
rO1	X	AAGTGCCCAACTAAATGAT	16Sr DNA	[24]
ECA1	X-B	AATAATCAAGAACAAGAAGT	16Sr DNA	[60]
ECA2	X-B	GTTTATAAAAATTAATGACTC	16Sr DNA	[60]

[a]In bold primers for which the protocol is provided here
[b]IS, spacer region between 16Sr and 23Sr DNA

11. Horizontal electrophoresis apparatus.

12. Power supply.

13. Vertical electrophoresis apparatus.

14. Pipettes (20, 100, 200, and 1000 μL).

15. Tips with filter (for 20, 100, 200, and 1000 μL).

16. Tips without filter (for 20 μL).

17. Restriction enzymes (*see* **Note 3**).

18. Water baths or thermo blocks.

19. TAE buffer (*see* **Note 4**).

20. TBE buffer (*see* **Note 5**).

21. Floating racks.

22. Loading dye.

23. DNA ladder (*see* **Note 6**).

24. Ethidium bromide or red gel (*see* **Note 7**).

25. ΦX174 DNA/*BsuRI* (*Hae*III) marker (*see* **Note 10**).

3   Methods

3.1   DNA Extraction

See Chapter 6: a healthy sample of the analyzed species must be always included as negative control.

3.2   Direct Generic PCR Followed by Nested Generic PCR

The P1/P7 primers are recommended [46, 47]: they amplify the near whole length of 16S rRNA, intergenic 16S–23S, and a small part of 23S rRNA gene.

The 25 μL reaction mixture is as follows:
10× PCR buffer 2.5 μL.
10 mM dNTPs 1.75 μL.
primer P1 20 μM 0.5 μL.
primer P7 20 μM 0.5 μL.
Taq polymerase (Sigma-Aldrich Co.) 5 U/μL 1.0 μL (*see* **Note 2**).
DNA extract at 20 ng/μL 1 μL.
dd H_2O to 25 μL.

PCR is then conducted as follows: 1 cycle at 95 °C for 3 min; 35 cycles as follows: 94 °C for 1 min, 55 °C for 2 min; 72 °C for 3 min; final extension 72 °C for 10 min. If the samples are from herbaceous host, an agarose gel can be run as described below. In case of negative results with P1/P7 or of woody or insect samples, nested PCR with R16F2n/R2 primers [48, 49] must be carried out.

Reaction mixture and PCR cycles are as for the direct PCR, substituting the new primers in the same amounts. The DNA is provided as 1.0 μL of the product of the direct PCR (P1/P7), diluted 1: 30. The product is visualized on a 1% agarose gel stained with ethidium bromide (*see* **Notes** 7 and **8**).

3.3   Reaction Mix Preparation

For preparing the PCR reaction mix use the last tube labeled, which will be used as negative control. The PCR reaction mix should be prepared for all tubes together.

1. Label PCR tubes (number of tubes is number of samples + 2, one for negative and one for positive control).

2. Load all reagents without DNA into a last labeled tube and mix well.

3. Aliquot 24 μL of the PCR reaction mix to all tubes.

4. Add 1 μL of DNA samples (diluted to 20 ng/μL or 1 to 100 with ddH_2O), then add mineral oil layer (optional depending on heated lid presence).

5. Load the tubes in PCR thermo cycler and run the program.

3.4   Visualization of PCR Products

To prepare 100 mL of a 1% agarose gel (*see* **Note 8**):

1. Weigh 1 g of agarose.

2. Add 100 mL of 1×TAE buffer (*see* **Note 4**).

3. Carefully dissolve in microwave.

4. Cool to about 38 °C.

5. Pour the agarose into a tray, insert comb, and remove air bubbles.

6. Let it solidify (approximately 15 min).

7. Place the gel into the electrophoresis chamber, cover with 1×TAE buffer.

8. Load 6 μL of each PCR-product (*see* **Note 9**) into the wells.

9. Add 2.5 μL of the DNA ladder in initial or final well (*see* **Note 6**).

10. Connect the electrodes to the power source (5 V/cm) and run the gel avoiding that dye exit the gel (about 30 min).

11. Stain the gel by soaking it in ethidium bromide working solution (0.5 μg/mL) or red gel (*see* **Note 7**) for 10–15 min and wash in water for 10–15 min.

12. Observe the gel on a transilluminator by visualizing DNA under UV light.

13. Take a picture and print and/or store in computer or in laboratory book after proper labeling of the samples.

3.5 RFLP Analyses for Phytoplasma Group/Subgroup Identification from Amplicons

Only the PCR products visible in the agarose gel can be used for RFLP analysis (*see* **Note 3**) following instruction of the enzyme manufacturer. The products are visualized after a 6.7% polyacrylamide gel electrophoresis or after a 3% agarose gel electrophoresis (*see* **Note 5**), stained with ethidium bromide. The size of the products is evaluated using ΦX174 DNA/*BsuRI* (*Hae*III) marker (*see* **Note 10**).

3.6 Reaction Mix Preparation and Performing RFLP

Use 2 to 15 μL of PCR product (*see* **Note 11**) in the selected enzyme mix made according to instructions of enzyme manufacturer and leave at the correct temperature in water bath or thermo blocks for at least 16 h (*see* **Note 12**).

3.7 Visualization of RFLP Products from Amplicons (see Note 13)

To prepare 100 mL of a 3% agarose gel (*see* **Note 8**):

1. Weigh out 3 g of agarose.

2. Add 100 mL of 1×TBE buffer.

3. Carefully dissolve in microwave.

4. Cool it to about 38 °C.

5. Pour the agarose into a tray, insert comb, and remove air bubbles.

6. Let it solidify (approximately 15 min).

7. Place gel in electrophoresis chamber, cover with 1×TBE buffer (*see* **Note 5**).

8. Load each digested mix into each well following the labeled samples order.

9. Add 1.5 μL of ΦX174 DNA/*BsuRI* (*Hae*III) in the first or in the last well.

10. Connect electrodes to power (5 V/cm) and avoid that stain exit the gel (about 30 min).

11. Stain the gel by soaking it in ethidium bromide working solution (0.5 μg/mL) (*see* **Note 7**) for 20–25 min, then wash in water for 20–25 min.

12. Observe the gel on a transilluminator under UV light.

13. Take a picture and print and/or store in computer or in laboratory book after proper labeling of the samples.

To prepare 25 mL of 6.7% polyacrylamide gel (*see* **Note 14**) pour the reagents as follows in a clean beaker and in the following order:

1. ddH$_2$O   17.83 mL,

2. 10× TBE   2.5 mL

3. Acrylamide/bis-acrylamide 29:1 solution   4.37 mL.

4. Ammonium persulfate   310 μL (*see* **Note 15**).

5. TEMED (Sigma Aldrich Co.)   15.6 μL.

6. Gently mix and pour the mix between the glasses.

7. Add the comb and leave 20 min to polymerize.

8. Take out the comb.

9. Fix the glasses containing the gel to vertical electrophoresis apparatus.

10. Cover with 1× TBE.

11. Gently clean the wells with an appropriate pipette.

12. Eliminate air bubbles at the bottom between glasses.

13. Load each digested mix into wells following the labeled samples order.

14. Add 1.5 μL of ΦX174 DNA/*BsuRI* (*Hae*III) marker in the first or in the last well of the gel (*see* **Note 10**).

15. Connect the electrodes to power (7 V/cm) and avoid dye exit the gel (about 30 min).

16. Stain the gel by soaking it in ethidium bromide working solution (0.5 μg/mL) or red gel (*see* **Note 7**) for 10–15 min, then wash in water for 1–2 min.

17. Observe the gel on a transilluminator under UV light.

18. Take a picture and print and/or store in computer or in laboratory book after proper labeling of the samples.

3.8 RFLP Analyses for Phytoplasma Group/Subgroup Identification in Silico

The 16S rDNA amplicons obtained with the listed primers must be sequenced at least in both senses with or without previous cloning. The aligned consensus sequences obtained can be loaded directly in the *i*phyclassifier (https://plantpathology.ba.ars.usda.gov/cgi-bin/resource/iphyclassifier.cgi) or in other systems such as the barcode system of the Qbank (http://www.q-bank.eu/Phytoplasmas/) to generate in silico profiles or provide preliminary phytoplasma classification. Same results can also be obtained using other programs such as pDraw (http://www.acaclone.com/) but remember to use always comparable same size sequences.

The *i*phyclassifier can tentatively (all the data must be confirmed by real RFLP with appropriate enzymes):

1. Provide '*Candidatus* species'.
2. Provide ribosomal group.
3. Provide ribosomal subgroup.
4. Provide identity coefficient.
5. Provide in silico RFLP pictures comparing your sequence to those available in the classifier.

The Barcode system allows you to match your sequence to those of reference available in the Qbank and corresponding to phytoplasma strains maintained in collection in periwinkle (http://www.ipwgnet.org/doc/phyto_collection/collection-august2010.pdf).

4 Notes

1. At all stages while setting up PCR reactions the following precautions should be taken to avoid contamination of samples and reagents:

   (a) Use only dedicated pipettes.

   (b) Use tips with filter and a new tip for each pipetting step.

   (c) Close reagent/sample tubes once aliquot has been removed.

   (d) Wash hands carefully with soap any time they become contaminated.

   (e) Use only clean, sterile plasticware.

   (f) Following steps of analysis must be performed in separated places with separated laboratory equipment:

   Reaction mix preparation (without DNA).
   Adding DNA (UV chamber).
   Amplification of the target sequence.
   Analyzing PCR product.

2. In this protocol, Sigma Aldrich *Taq* DNA polymerase is employed for PCR amplification; however, Promega or other PCR enzymes or ready-to-use master mix may be used. For other *Taq* polymerases usually 25 mM $MgCl_2$ should be added.

3. The 17 restriction enzymes that are used for differentiation of ribosomal groups and subgroups are: *Mse*I (*Tru*1I), *Alu*I, *Rsa*I, *Hha*I, *Hae*III, *Hpa*II, *Taq*I, *Hin*fI, *Sau*3AI, *Kpn*I, *Tha*I, *Bam*HI, *Dra*I, *Eco*RI, *Hpa*I, *Ssp*I, and *Bfa*I. However, some of the most informative restriction enzymes (*Mse*I, *Alu*I, *Rsa*I, *Hha*I, *Hpa*II, *Taq*I) can be used for preliminary classification, since they can distinguish most of the ribosomal groups, and if needed additional enzymes can be employed.

4. For 1 L of TAE buffer 10×.

Trisma base	48.44 g.
Ethylenedianinetetraacetic acid (EDTA)	7.44 g.
dd H_2O	500 mL.
Bring to pH 8.0 with glacial acetic acid.	
dd H_2O	to 1 L.
Keep at 4 °C.	

5. For 1 L of TBE buffer 10×.

Trisma base	108 g.
Boric acid	55 g.
0.5 M Ethylenedianinetetraacetic acid (EDTA) 0.40 mL.	
dd H_2O	to 1 L.
Keep at 4 °C.	

6. In this protocol, Fermentas GeneRuler 1 kb DNA ladder is employed; however, other DNA ladders may be used.

7. In this protocol ethidium bromide solutions or ethidium bromide powder dissolved 10 mg in 1 mL of distilled H_2O is used. Keep mother solution on dark place. Make a 0.5 μg/mL working solution: add one drop (50 μL) of 10 mg/mL ethidium bromide solution to 1000 mL distilled H_2O. As a less toxic alternative other dyes such as red gel could be used following the protocol provided by the manufacturer.

8. For small tray prepare 80 mL of agarose and for larger trays about 120 mL.

9. In this protocol, Sigma Aldrich *Taq* DNA polymerase employed for PCR amplification has loading dye in the 10× PCR buffer. However, if the PCR buffer does not contain a loading dye, add 1.5 μL of loading dye per 6 μL of PCR product before running the gel.

10. In this protocol, Fermentas ΦX174 DNA/*Bsu*RI (*Hae*III) marker, 9 is used; however, other DNA ladders of appropriate size may be used.

11. To determine the amount for enzyme digestion, consider agarose gel band intensity. Very strong bands only need 1–2 μL in RFLP, medium intensity about 3–6 μL, low intensity 10–15 μL, or more. Do not exceed the 20 μL or the well capacity as total volume.

12. Fast restriction enzymes are also available to reduce the digestion time to a few minutes, but remember that the enzyme is inactivate from frequent defrost.

13. For a clear separation of short fragments (below 300 nt) the polycrylamide gel must be used.

14. The amount of the reagents needed for the polyacrylamide gel preparation depends on the size of the glasses and thickness of the gel.

15. Ammonium persulfate preparation: 0.1 g in 1 mL dd H_2O, keep at 4 °C and use fresh (not more than 10 days old solution).

References

1. Berges R, Rott M, Seemüller E (2000) Range of phytoplasma concentration in various plant hosts as determined by competitive polymerase chain reaction. Phytopathology 90:1145–1152

2. Bertaccini A, Marani F (1982) Electron microscopy of two viruses and mycoplasmalike organisms in lilies with deformed flowers. Phytopathol Mediterr 21:8–14

3. Cousin MT, Sharma AK, Isra S (1986) Correlation between light and electron microscopic observations and identification of mycoplasmalikeorganisms using consecutive 350 nm think sections. J Phytopathol 115:368–374

4. Haggis GH, Sinha RC (1978) Scanning electron microscopy of mycoplasmalike organisms after freeze fracture of plant tissues affected with clover phyllody and aster yellows. Phytopathology 68:677–680

5. Seemüller E (1976) Investigation to demonstrate mycoplasmalike organism in diseases plants by fluorescence microscopy. Acta Hortic 67:109–112

6. Contaldo N, Bertaccini A, Paltrinieri S (2012) Axenic culture of plant pathogenic phytoplasmas. Phytopathol Mediterr 51(3):607–617

7. Contaldo N, Satta E, Zambon Y et al (2016) Development and evaluation of different complex media for phytoplasma isolation and growth. J Microbiol Methods 127:105–110

8. Hobbs HA, Reddy DVR, Reddy AS (1987) Detection of a mycoplasma-lke organism in peanut plants with witches' broom using indirect enzyme-linked immunosorbent assay (ELISA). Plant Pathol 36:164–167

9. Bellardi MG, Vibio M, Bertaccini A (1992) Production of a polyclonal antiserum to CY-MLO using infected Catharanthus roseus. Phytopathol Mediterr 31:53–55

10. Lee I-M, Gundersen-Rindal DE, Davis RE et al (1998) Revised classification scheme of phytoplasmas based an RFLP analyses of 16S rRNA and ribosomal protein gene sequences. Int J Syst Bacteriol 48:1153–1169

11. Montano HG, Davis RE, Dally EL et al (2001) 'Candidatus Phytoplasma brasiliense', a new phytoplasma taxon associated with hibiscus witches' broom disease. Int J Syst Evol Microbiol 51:1109–1118

12. Lee I-M, Gundersen-Rindal D, Davis RE et al (2004) 'Candidatus Phytoplasma asteris', a novel taxon associated with aster yellows and related diseases. Int J Syst Bacteriol 54:1037–1048

13. Lee I-M, Martini M, Marcone C et al (2004) Classification of phytoplasma strains in the elm yellows group (16SrV) and proposal of 'Candidatus Phytoplasma ulmi' for the phytoplasma associated with elm yellows. Int J Syst Evol Microbiol 54:337–347

14. Arocha Y, Lopez M, Pinol B et al (2005) 'Candidatus Phytoplasma graminis' and 'Candidatus Phytoplasma caricae', two novel phytoplasmas associated with diseases of sugarcane, weeds and papaya in Cuba. Int J Syst Evol Microbiol 55:2451–2463

15. Al-Saady NA, Khan AJ, Calari A et al (2008) '*Candidatus* Phytoplasma omanense', a phytoplasma associated with witches' broom of *Cassia italica* (Mill.) Lam. in Oman. Int J Syst Evol Microbiol 58:461–466

16. Bertaccini A, Duduk B (2009) Phytoplasma and phytoplasma diseases: a review of recent research. Phytopathol Mediterr 48:355–378

17. Bertaccini A, Duduk B, Paltrinieri S et al (2014) Phytoplasmas and phytoplasma diseases: a severe threat to agriculture. Am J Plant Sci 5:1763–1788

18. IRPCM (2004) '*Candidatus* Phytoplasma', a taxon for the wall-less, non-helical prokaryotes that colonize plant phloem and insects. Int J Syst Evol Microbiol 54:1243–1255

19. Bertaccini A, Davis RE, Hammond RW et al (1992) Sensitive detection of mycoplasmalike organisms in field-collected and *in vitro* propagated plants of *Brassica*, *Hydrangea* and *Chrysanthemum* by polymerase chain reaction. Ann Appl Biol 121:593–599

20. Alvarez E, Mejía JF, Llano GA et al (2009) Characterization of a phytoplasma associated with frogskin disease in cassava. Plant Dis 93:1139–1145

21. Cozza R, Bernardo L, Calari A et al (2008) Molecular identification of '*Candidatus* Phytoplasma asteris' inducing histological anomalies in *Silene nicaeensis*. Phytoparasitica 36:290–293

22. Duduk B, Botti S, Ivanović M et al (2004) Identification of phytoplasmas associated with grapevine yellows in Serbia. J Phytopathol 152:575–579

23. Lee I-M, Bertaccini A, Vibio M et al (1995) Detection of multiple phytoplasmas in perennial fruit trees with decline symptoms in Italy. Phytopathology 85:728–735

24. Lorenz KH, Schneider B, Ahrens U et al (1995) Detection of the apple proliferation and pear decline phytoplasmas by PCR amplification of ribosomal and nonribosomal DNA. Phytopathology 85:771–776

25. Namba S, Kato S, Iwanami S et al (1993) Detection and differentiation of plant-pathogenic mycoplasmalike organisms using polymerase chain reaction. Phytopathology 83:786–791

26. Lee I-M, Gundersen-Rindal DE, Bertaccini A (1998) Phytoplasma: ecology and genomic diversity. Phytopathology 88:1359–1366

27. Schneider B, Seemüller E (1994) Presence of two set of ribosomal genes in phytopatogenic mollicutes. Appl Environ Microbiol 60:3409–3412

28. Liefting LW, Andersen MT, Beever RE et al (1996) Sequence heterogeneity in the two 16S rRNA genes of *Phormium* yellow leaf phytoplasma. Appl Environ Microbiol 62:3133–3139

29. Jomantiene R, Davis RE, Valiunas D et al (2002) New group 16SrIII phytoplasma lineages in Lithuania exhibit rRNA interoperon sequence heterogeneity. Eur J Plant Pathol 108:507–517

30. Davis RE, Jomantiene R, Kalvelyte A et al (2003) Differential amplification of sequence heterogenous ribosomal RNA genes and classification of the '*Fragaria multicipita*' phytoplasma. Microbiol Res 158:229–236

31. Duduk B, Calari A, Paltrinieri S et al (2009) Multigene analysis for differentiation of aster yellows phytoplasmas infecting carrots in Serbia. Ann Appl Biol 154:219–229

32. Montano HG, Contaldo N, David WAT et al (2011) Hibiscus witches' broom disease associated with different phytoplasmas taxa in Brazil. B Insectol 64:S249–S250

33. Schneider B, Gibb KS, Seemüller E (1997) Sequence and RFLP analysis of the elongation factor Tu gene used in differentiation and classification of phytoplasmas. Microbiology 143:3381–3389

34. Marcone C, Lee I-M, Davis RE et al (2000) Classification of aster yellows-group phytoplasmas based on combined analyses of rRNA and tuf gene sequences. Int J Syst Evol Microbiol 50:1703–1713

35. Martini M, Botti S, Marcone C et al (2002) Genetic variability among "flavescence dorée" phytoplasmas from different origins in Italy and France. Mol Cell Probes 16:197–208

36. Martini M, Lee I-M, Bottner KD et al (2007) Ribosomal protein gene-based filogeny for finer differentiation and classification of phytoplasmas. Int J Syst Evol Microbiol 57:2037–2051

37. Lee I-M, Bottner KD, Zhao Y et al (2010) Phylogenetic analysis and delineation of phytoplasmas based on *secY* gene sequences. Int J Syst Evol Microbiol 60:2887–2897

38. Mitrović J, Kakizawa S, Duduk B et al (2011) The groEL gene as an additional marker for finer differentiation of '*Candidatus* Phytoplasma asteris'-related strains. Ann Appl Biol 159:41–48

39. Cai H, Wei W, Davis RE et al (2008) Genetic diversity among phytoplasmas infecting *Opuntia* species: virtual RFLP analysis identifies new subgroups in the peanut witches' broom phytoplasma group. Int J Syst Evol Microbiol 58:1448–1457

40. Duduk B, Bertaccini A (2006) Corn with symptoms of reddening: new host of stolbur phytoplasma. Plant Dis 90:1313–1319

41. Khan AJ, Botti S, Al-Subhi AM et al (2002) Molecular identification of a new phytoplasma associated with alfalfa witches' broom in Oman. Phytopathology 92:1038–1047

42. Tolu G, Botti S, Garau R et al (2006) Identification of 16SrII-E phytoplasmas in *Calendula arvensis* L., *Solanum nigrum* L. and *Chenopodium* spp. Plant Dis 90:325–330

43. Wei W, Davis RE, Lee I-M et al (2007) Computer-simulated RFLP analysis of 16S rRNA genes: identification of ten new phytoplasma groups. Int J Syst Evol Microbiol 57:1855–1867

44. Wei W, Lee I-M, Davis RE et al (2008) Automated RFLP pattern comparison and similarity coefficient calculation for rapid delineation of new and distinct phytoplasma 16Sr subgroup lineages. Int J Syst Evol Microbiol 58:2368–2377

45. Zhao Y, Wei W, Lee I-M et al (2009) Construction of an interactive online phytoplasma classification tool, *i*PhyClassifier, and its application in analysis of the peach X-disease phytoplasma group (16SrIII). Int J Syst Evol Microbiol 59:2582–2593

46. Deng SJ, Hiruki C (1991) Amplification of 16S ribosomal-RNA genes from culturable and nonculturable mollicutes. J Microbiol Methods 14:53–61

47. Schneider B, Seemüller E, Smart CD et al (1995) Phylogenetic classification of plant pathogenic mycoplasmalike organisms or phytoplasmas. In: Razin S, Tully JG (eds) Molecular and diagnostic procedures in Mycoplasmology. Academic press, San Diego, CA, pp 369–380

48. Gundersen DE, Lee I-M (1996) Ultrasensitive detection of phytoplasmas by nested-PCR assays using two universal primer pairs. Phytopathol Mediterr 35:144–151

49. Lee I-M, Hammond RW, Davis RE et al (1993) Universal amplification and analysis of pathogen 16S rDNA for classification and identification of mycoplasmalike organisms. Phytopathology 83:834–842

50. Davis RE, Lee I-M (1993) Cluster-specific polymerase chain reaction amplification of 16S rDNA sequences for detection and identification of mycoplasmalike organisms. Phytopathology 83:1008–1011

51. Padovan AC, Gibb KS, Bertaccini A et al (1995) Molecular detection of the Australian grapevine yellows phytoplasma and comparison with a grapevine yellows phytoplasma from Emilia-Romagna in Italy. Aust J Grape Wine Res 1:2531

52. Lee I-M, Martini M, Bottner KD et al (2003) Ecological implications from a molecular analysis of phytoplasmas involved in an aster yellows epidemic in various crops in Texas. Phytopathology 93:1368–1377

53. Lee I-M, Zhao Y, Bottner KD (2006) SecY gene sequence analysis for finer differentiation of diverse strains in the aster yellows phytoplasma group. Mol Cell Probes 20:87–91

54. Skrzeczkowski LJ, Howell WE, Eastwell KC (2001) Bacterial sequences interfering in detection of phytoplasma by PCR using primers derived from the ribosomal RNA operon. Acta Hortic 550:417–424

55. Heinrich M, Botti S, Caprara L et al (2001) Improved detection methods for fruit tree phytoplasmas. Plant Mol Biol Report 19:169–179

56. Gibb KS, Padovan AC, Mogen BD (1995) Studies on sweet potato little-leaf phytoplasma detected in sweet potato and other plant species growing in northern Australia. Phytopathology 85:169–174

57. Manimekalai R, Soumya VP, Sathish Kumar R et al (2010) Molecular detection of 16SrXI group phytoplasma associated with root (wilt) disease of coconut (*Cocos nucifera*) in India. Plant Dis 94:636

58. Lee I-M, Gundersen DE, Hammond RW et al (1994) Use of mycoplasmalike organism (MLO) group-specific oligonucleotide primers for nested-PCR assays to detect mixed-MLO infections in a single host plant. Phytopathology 84:559–566

59. Harrison NA, Womack M, Carpio ML (2002) Detection and characterization of a lethal yellowing (16SrIV) group phytoplasma in Canary Island date palms affected by lethal decline in Texas. Plant Dis 86:676–681

60. Jarausch W, Lansac M, Saillard C et al (1998) PCR assay for specific detection of European stone fruit yellows phytoplasmas and its use for epidemiological studies in France. Eur J Plant Pathol 104:17–27

Chapter 8

PCR-Based Sequence Analysis on Multiple Genes Other than 16S rRNA Gene for Differentiation of Phytoplasmas

Marta Martini, Kristi D. Bottner-Parker, and Ing-Ming Lee

Abstract

Differentiation and classification of phytoplasmas have been primarily based on the highly conserved 16S rRNA gene, for which "universal" primers are available. To date, 36 ribosomal (16Sr) groups and more than 150 subgroups have been delineated by RFLP analysis of 16S rRNA gene sequences. However, in recent years, the use of moderately conserved genes as additional genetic markers has enhanced the resolving power in delineating distinct phytoplasma strains among members of some 16Sr subgroups.

This chapter describes the methodology of amplification, differentiation, and classification of phytoplasma based on less-conserved non-ribosomal genes, named *rp* and *secY*. Actual and virtual RFLP analyses of amplicons obtained by semi-universal or group-specific *rp* and *secY* gene-based primers are used for finer differentiation of phytoplasma strains within a given group. The *rp* and *secY* gene-based classification not only readily resolves 16Sr subgroups within a given 16Sr group, but also provides finer differentiation of closely related phytoplasma strains within a given 16Sr subgroup.

Key words *rplV* (*rpl22*), *rpsC* (*rps3*), *secY*, Semi-universal primers, 16Sr group-specific primers, Restriction enzymes, Virtual RFLP, Phytoplasma strain differentiation

---

1 Introduction

The highly conserved 16S rRNA gene sequence for which "universal" primers are available, has been used as the primary genetic marker for phylogenetic studies and classification of phytoplasmas [1]. Thus far, 36 16Sr groups and more than 150 16Sr subgroups have been identified [2, 3]. At present, species designation is primarily based on an arbitrary threshold of 2.5% dissimilarity of 16S rDNA sequences among phytoplasmas [4]. This guideline may exclude many ecologically or biologically distinct phytoplasma strains with a sequence similarity higher than 97.5%, some of which may warrant designation as a new taxon [5]. It is not uncommon that closely related phytoplasma strains on the basis of the 16S rRNA gene sequences, have unique ecological niches encompassing both plant host range and insect vectors. Therefore, additional

Rita Musetti and Laura Pagliari (eds.), *Phytoplasmas: Methods and Protocols*, Methods in Molecular Biology, vol. 1875, https://doi.org/10.1007/978-1-4939-8837-2_8, © Springer Science+Business Media, LLC, part of Springer Nature 2019

unique biological properties, as well as other molecular criteria, need to be included for speciation. Several "*Candidatus* Phytoplasma" species in the 16SrX group have been designated that way. Concerning the significance of including additional molecular criteria besides unique biological properties for phytoplasma speciation, in recent years, interest has focused on less conserved non-ribosomal genes including ribosomal protein (*rp*), *secY*, *secA*, and *tuf* genes that permit finer differentiation of closely related strains [5]. For epidemiological studies and quarantine purposes, it is relevant and essential to use multiple biomarkers to differentiate and identify these closely related but ecologically distinct strains. Besides 16S rRNA gene, only the more variable genes *rp* and *secY* genes have been extensively characterized among the majority of phytoplasma groups [6, 7] and include a quite comprehensive database.

Ribosomal protein (*rp*) genes are part of the *rp* operon, which contains at least 21 genes in phytoplasma genomes [8]; the *secY* gene encodes for an essential component of the Sec protein translocation system [9]. The *rp* and *secY* genes are more variable than 16S rRNA genes and have more phylogenetically informative characters, which substantially enhance the resolving power for classifying genetic closely related but distinct phytoplasma strains within a given 16Sr group or "*Candidatus* Phytoplasma" species. They have been used as molecular markers especially for finer differentiation of phytoplasma strains of the groups 16SrI [10, 11], 16SrIII [12], 16SrV [13, 14], 16SrIX [15]. These studies indicated that analysis of *rp* and *secY* gene sequences not only readily delineated subgroups that are consistent with 16Sr subgroups but also identified, within some subgroups, additional distinct strains that could not be resolved by analysis of highly conserved 16S rRNA gene sequences. For example, ten and twelve RFLP subgroups were differentiated on the basis of ribosomal protein gene sequences of 16SrI and 16SrV phytoplasma strains, respectively [10, 13]. Most of the additional strains identified have distinct biological and ecological properties; e.g., subgroup 16SrV-C can be further differentiated into several rp subgroups [13, 16]. The *secY* gene sequence variability is similar to that of *rp* genes; *secY* subgroups delineated by RFLP analyses of *secY* gene sequences from phytoplasma strains in taxonomic groups 16SrI and 16SrV generally coincided with those delineated with *rp* gene sequences [10, 11, 13]. However, the resolving power of *secY* is slightly better than *rp* gene sequences.

Recently, Martini et al. [6] and Lee et al. [7] constructed two comprehensive phylogenetic trees based on the analysis of two ribosomal protein genes [*rplV* (*rpl22*), *rpsC* (*rps3*)] and *secY* gene using semi-universal primers. These *rp* and *secY* gene-based phylogenetic trees, which were congruent with that inferred from the 16S rRNA gene, yielded more clearly defined phylogenetic interrelationships among phytoplasma strains and delineated more distinct phytoplasma subclades and distinct lineages than those resolved by the 16S rRNA gene-based tree. Primers for these genes have also been designed that are 16Sr group-specific [6, 7].

This chapter describes the protocols for PCR amplification of phytoplasma *rp* and *secY* gene sequences using semi-universal or selective (16Sr-group-specific) primers followed by actual and/or virtual RFLP analyses for finer differentiation and classification of phytoplasma strains. The restriction fragment length polymorphism (RFLP) analysis of PCR-derived amplicons has become increasingly important in phytoplasma identification as well as in typing for epidemiological studies.

2   Materials

2.1   PCR Components

All PCR reagents and template DNAs are stored at -20 °C unless otherwise stated.

1. Nuclease-free water.
2. Appropriate DNA polymerase buffer and $MgCl_2$.
3. dNTPs solution containing all four dNTPs 2.5 mM (*see* **Note 1**).
4. Forward and reverse primers (20 µM) (*see* **Note 2**).
5. A high-fidelity thermostable DNA polymerase such as GeneAmp High Fidelity polymerase (Life Technologies, Gaithersburg, MD, USA) for amplification of *rp* genes; a high-fidelity PCR Enzyme for Long Range PCR such as TaKaRa LA Taq DNA Polymerase (Takara Mirus Bio, Madison, WI, USA) for amplification of *secY* gene.
6. Template DNA: total plant genomic DNA 20 ng/µL, obtained from test plant samples and phytoplasma reference strains maintained in periwinkle.
7. Micropipettes and barrier or filter tips (*see* **Note 3**).
8. Microfuge tubes (0.2–0.5 mL thin-walled for PCR reactions, sterile 1.5 mL).
9. Thermal cycler, programmed with desired amplification protocol.

2.2   Agarose Gel Electrophoresis Components

1. Standard agarose.
2. 50× TAE electrophoresis buffer: 242 g of Tris base, 57.1 mL of glacial acetic acid, 100 mL of 0.5 M EDTA (pH 8.0). Use diluted to 1× TAE in ddH$_2$O for agarose gel electrophoresis of PCR products.
3. 10× TBE electrophoresis buffer: 108 g of Tris base, 55 g of boric acid, 40 mL of 0.5 M EDTA (pH 8.0). Use diluted to 1×

TBE in ddH$_2$O for agarose gel electrophoresis of RFLP products.

4. 6× gel-loading buffer type I: 0.25% bromophenol blue, 0.25% xylene cyanol FF, 40% (w/v) sucrose in H$_2$O.

5. Staining solution: ethidium bromide 0.5 μg/mL stock solution, GelRed™ 10,000× (Biotium, Hayward, CA, USA) or SYBR®Safe DNA Gel stain 10,000× stock solution in DMSO (Invitrogen, Carlsbad, CA, USA) (*see* **Note 4**).

6. DNA size standard.

7. PCR products (*see* **Note 5**) or RFLP products.

8. Micropipettes and tips.

9. Gel sealing tape.

10. Horizontal electrophoresis apparatus with chamber and comb.

11. Power supply device for electrophoresis apparatus.

12. Microwave oven.

13. Magnetic heating stirrer.

14. Gel documentation system, e.g., DigiDoc-It (UVP, Upland, CA, USA).

2.3 Actual RFLP Analysis of PCR Products Components

1. Nuclease-free water.

2. Selected key restriction enzyme stored at −20 °C.

3. 10× buffer supplied with the restriction enzyme stored at −20 °C.

4. PCR products obtained from test samples and phytoplasma reference strains.

5. Mineral oil.

6. Micropipettes and tips.

7. Microfuge tubes (0.5 mL).

8. Floating racks.

9. Water bath.

2.4 Purification of PCR Products

1. Wizard® SV Gel and PCR Clean-Up System Kit (Promega, Madison, WI, USA) or other column-based purification kit may be used.

2. PCR products obtained from test samples and phytoplasma reference strains.

3. Micropipettes and tips.

4. Microfuge tubes (1.5 mL).

5. Microfuge.

Table 1
Semi-universal and16Sr group-specific primers designed for amplification of *rpl22-rps3* and *secY* genes

Primer pair	Sequence (5′–3′)	Expected size of PCR product (bp)	Specificity	Reference
Rp gene				
rpF1/ rpR1	ggacataagttaggtgaattt/ acgatatttagttctttttgg	1245–1389[a]	16SrI, III, IV, V, VII, VIII, IX, XIII	[18]
rpL2F3/ rp(I)R1A	wccttggggyaaaaaagctc/ gttcttttggcattaacat	1600	16SrI, III, IV, V, VI, VII, IX, X, XII, XIII, XVIII	[6]
rpF1C/ rp(I)R1A	atggtdggdcayaarttagg/ gttcttttggcattaacat	1212–1386[a]	16SrI, II, III, IV, V, VI, VII, IX, X, XII, XIII, XVIII	[6]
rp(I)F1A/ rp(I)R1A	tttcccctacacgtactta/ gttcttttggcattaacat	1200	16SrI	[19]
rp(II)F1/ rp(II)R1	gctcttactcgtaaayatgtagt/ ttacttgattttctggttttga	1200	16SrII	[6]
rp(II)F/ rp(I)R1A	acttattctcgtgatactag/ gttcttttggcattaacat	1390	16SrII	[20]
rp(II)F2/ rp(I)R1A	atggtaggttataaattagg/ gttcttttggcattaacat	1290	16SrII	[20]
rp(III)F1/ rp(III)R1	ttagagaaggcattaaac/ ctctttccccatctaggacg	1200	16SrIII	[6]
rp(III)-FN/ rp(I)R1A	ggtgaattttctccaactcg/ gttcttttggcattaacat	1400	16SrIII	[12]
rp(V)F1/ rpR1	tcgcggtcatgcaaaaggcg/ acgatatttagttctttttgg	1200	16SrV	[1, 18]
rp(V)F2/ rpR1	ttgcctcgtttatttccgagagcta/ acgatatttagttctttttgg	950	16SrV	[1, 18]
rp(V)F1A/ rp(V)R1A	aggcgataaaaaagtttcaaaa/ ggcattaacataatatattatg	1200	16SrV	[13]
rp(VI)F2/ rp(VI)R2	ggttgttgatttaattcgtggtc/ ccagatattcgtctagtatcagaa	1000	16SrVI	[6]
rp(VIII)F2/ rp(VIII)R2	agttgtcgatttaattcgtggca/ cagcagatatttgtctagtatctgcg	1000	16SrVII, VIII	[6]
rp(IX)F2/ rp(IX)R2	gcacaagctattttaatgtttacaccc/ caaagggactaaacctaaag	800	16SrIX	[6]
rpAP15f2/ rp(I)R1A	ctcctaaatcagcttcaagt/ ttcttttggcattaacat	1036	16SrX-A	[21]
rpAP15f/ rpAP15r	agtgctgaagctaatttgg/ tgctttttatagcaaaaggtt	920	16SrX-A	[21]

(continued)

Table 1
(continued)

Primer pair	Sequence (5′–3′)	Expected size of PCR product (bp)	Specificity	Reference
rpStolF/ rpStolR	cgtacaaaataatcgggaga/ cgaaacaaaaggtttacgag	1372	16SrXII-A	[6]
rpStolF2/ rpStolR	aaacttggtcacgtagttcc/ cgaaacaaaaggtttacgag	1253	16SrXII-A	[6]
Secϒ gene				
L15F1/ MapR1	cctggtagtggyamtggwaaaac/ attarraatatarggytcttcrtg	2800 2200	16SrI, 16SrXII, 16SrXIII, 16SrXVIII, 16SrII, 16SrIII, 16SrIV, 16SrV, 16SrVI, 16SrVII, 16SrX	[7]
L15F1A-a/ MapR1A-a	tggwaaaactkcbggwaargg/ aagmtkyaccratdccatg	2800	16SrI, 16SrXII, 16SrXIII	[7]
L15F1A-b/ MapR1A-b	ggwaaaacytshggymrvgghcataaagg/ ccwatmccrtgwccdgwaaaa	2200	16SrII, 16SrIII, 16SrV, 16SrVI	[7]
L15F1A (I)/ MapR1A (I)	cttctggtaaaggacataaagg/ gttcttcgtgcaaagatgtacc	2800	16SrI	[7]
AYsecYF1/ AYsecYR1	cagccattttagcagttggtgg/ cagaagcttgagtgcctttacc	1400	16SrI	[11]
L15F1A (II)/ MapR1A (II)	cttgcggtcgcggccataaagg/ ggttcttcgtgtaaagatttacc	2200	16SrII	[7]
SecYF1 (II)/ SecYR1(II)	cgcgtataggttttgaaggtg/ cctgccattttcattatagcg	1700–1850	16SrII	[7]
SecYF2 (II)/ SecYR1(II)	tgaaggtggtcaaactcct/ cctgccattttcattatagcg	1700–1850	16SrII	[7]
L15F1A (III)/ MapR1A (III)	cttctggtaaaggacataaagg/ ggttcttcgtgcaattgcaaacc	2200	16SrIII	[7]
SecYF1 (III)/ SecYR1 (III)	ctagaccaggttttgaagg/ gacctgcttttctcattatagc	1700–1850	16SrIII	[7]

(continued)

Table 1
(continued)

Primer pair	Sequence (5′–3′)	Expected size of PCR product (bp)	Specificity	Reference
SecYF2 (III)/ SecYR1 (III)	tgaaggyggacaaatccct/ gacctgcttttctcattatagc	1700–1850	16SrIII	[7]
FD9f/ FD9r	gaattagaactgtttgaagac/ tttgctttcacatcttgtatcg	1400	16SrV	[22]
FD9f2L/ FD9r	gttttagctaaaggtgatttaac/ tttgctttcatatcttgtrtcg	1343	16SrV	[14]
FD9f3L/ FD9r2L	aataaggtagttttatatgacaag/ taaaagactagtcccrccaaaag	1174	16SrV	[14]
L15F1A (VI)/ MapR1A (VI)	cttcaggyaarggtcataaagg/ ggtttcttcatcaagtctagtacc	2200	16SrVI	[7]
SecYF1 (VI)/ SecYR1 (VI)	ctagattaggattygaggg/ gaccrccaaaaccttgataatc	1700–1850	16SrVI	[7]
SecYF2 (VI)/ SecYR1 (VI)	attygagggygggycaaacac/ gaccrccaaaaccttgataatc	1700–1850	16SrVI	[7]
L15F2 (IX)/ MapR2 (IX)	ttcaaagaattcctaaaagagg/ gtacaactgcttcgtttacaga	1900	16SrIX	[15]
L15F4 (IX)/ MapR7 (IX)	ggaaatgttgaaataataaggta/ tgtaaccaaaacaaaatagaacc	1500	16SrIX	[15]
SecYF1 (X)/ SecYR1(X)	ggtggtgttagaccaggttt/ ggaataccytgaacaactac	1700–1850	16SrX	[7]
SecYF1a/ SecYR1 (XII)	ggacaattagcwcgttcagg/ caggaactaacttcccttga	1700–1850	16SrXII	[7]
SecYF2a/ SecYR1 (XII)	ctcttcgmccyggttttgaagg/ caggaactaacttcccttgag	1700–1850	16SrXII	[7]

[a]Product size is group-dependent

3 Methods

3.1 PCR Analysis Using Semi-Universal or 16Sr Group-Specific Primers

1. Chose the *rp* and *secY* primer pair according to your experiment. For amplification of *rp* and *secY* genes of phytoplasmas several semi-universal and 16Sr group-specific primer pairs have been designed (Table 1). The majority of them can also be used in nested or seminested-PCR (*see* **Note 6**).

2. Thaw the PCR reagents on ice. For *rp*-based PCR in a sterile 1.5 mL microfuge tube, make a master mix using the reagents in the following order (volumes are given for one reaction, multiply these as necessary for the number of samples):

Component	Final volume (μL)	Final concentration
Nuclease-free H_2O	16.75	–
10× PCR Buffer	2.5	1×
$MgCl_2$ solution, 25 mM	1.5	1.5 mM
dNTPs 2.5 mM	2	200 μM
Forward primer 20 μM	0.5	0.4 μM
Reverse primer 20 μM	0.5	0.4 μM
GeneAmp High Fidelity polymerase (5 U/μL) (*see* **Note 7**)	0.25	1.25 U
Total volume	**24**	

For *secY*-based PCR in a sterile 1.5 mL microfuge tube, make a master mix as previously described using the reagents in the following order:

Component	Final volume (μL)	Final concentration
Nuclease-free H_2O	27.5	–
10× LA Buffer II (Mg^{2+} free)	5.0	1×
$MgCl_2$ solution, 25 mM	3.0	1.5 mM
dNTPs 2.5 mM	8	400 μM
Forward primer 20 μM	2.0	0.8 μM
Reverse primer 20 μM	2.0	0.8 μM
Takara LA Taq™ DNA Polymerase (5 U/μL) (*see* **Note 8**)	0.5	2.5 U
Total volume	**48**	

3. Mix thoroughly and briefly spin down the reaction mixture, then aliquot 24 or 48 μL (*see* **Note 9**) of the reaction mixture in each PCR tube.

4. Add 1–2 μL of template DNA (≥20 ng/μL) into the reaction mixture. Include PCR positive and negative controls in each set (*see* **Note 10**).

5. Mix thoroughly, spin down, and place the tubes in the thermal cycler to amplify the nucleic acids using number of cycles, denaturation, annealing and polymerization times and temperatures listed (Table 2) for the given primer pair.

6. To perform nested-PCR, make a 1:30 dilution (in ddH$_2$O) of the direct-PCR product in 0.5 mL tubes using 1 μL of the diluted product as the template DNA in the nested-PCR.

3.2 Agarose Gel Electrophoresis

1. Seal the open ends of the plastic tray supplied with the electrophoretic apparatus with tape to form a mold. Place the mold on a horizontal section of the bench. Choose an appropriate comb and position it 0.5–1.0 mm above the plate so that a complete well is formed when the agarose is added to the mold.

2. Prepare sufficient 1× TAE electrophoresis buffer to fill the tank and cast the gel (*see* **Note 11**).

3. Prepare agarose 1% (w/v) solution in 1× TAE in a flask or a glass bottle (*see* **Note 12**).

4. Heat the slurry in a microwave oven or with a magnetic heating stirrer until the agarose dissolves (*see* **Note 13**).

5. Add a magnetic stirring bar to the flask or bottle and then transfer it to a stirrer to let the melted agarose solution to cool to about 60 °C.

6. (Optional) Add 1 μL of GelRed™ (10,000× stock solution) or SYBR®Safe DNA Gel stain (10,000× stock solution) to 100 mL of agarose solution and mix the gel solution thoroughly by gentle swirling.

7. Pour the warm agarose solution into the mold (*see* **Note 14**). Allow the gel to set completely for 20–30 min at room temperature, and then pour a small amount of 1× TAE buffer on the top of the gel, and carefully remove the comb. Pour the electrophoresis buffer off and carefully remove the tape. Mount the gel in the electrophoresis tank and add enough 1× TAE buffer to cover the gel completely.

8. Withdraw 5 μL from the sample and the control PCR reaction mixtures. Add 1 μL of 6× gel-loading buffer (*see* **Note 15**) and slowly load the sample mixture into the wells of the gel using a P10 or P20 micropipette. Load 5 μL of a DNA size standard such as GeneRuler™ 1 kb DNA Ladder (Fermentas, Vilnius, Lituania) into wells on both the right and left sides of the gel.

Table 2
PCR cycling conditions (number of cycles, denaturation, annealing and polymerization times and temperatures) used with *rp* and *secY* gene-based PCR primers

Primer pair	Initial denaturation	Cycles	Denaturation	Anneal	Extension	Final extension
Rp gene						
rpF1/rpR1 rpL2F3/rp(I)R1A rpF1c/rp(I)R1A rp(II)F1/rp(II)R1 rp(II)F/rp(I)R1A rp(II)F2/rp(I)R1A rp(III)F1/rp(III)R1 rp(III)-FN/rp(I)R1A rp(V)F1/rp(V)R1 rp(V)F1A/rp(V)R1A rp(VI)F2/rp(VI)R2 rp(VIII)F2/rp(VIII)R2 rp(IX)F2/rp(IX)R2	94 °C, 2 min	38	94 °C, 1 min	50 °C, 2 min	72 °C, 3 min	72 °C, 7 min
rp(I)F1A/rp(I)R1A	94 °C, 2 min	38	94 °C, 1 min	55 °C, 2 min	72 °C, 3 min	72 °C, 7 min
rpAP15f2/rp(I)R1A	94 °C, 2 min	35 40[a]	94 °C, 1 min	54 °C, 45 s 52 °C, 80 s[a]	72 °C, 1.5 min 72 °C, 2 min[a]	72 °C, 7 min
rpAP15f/rpAP15r	94 °C, 2 min	35/ 40[b] 36[a]	94 °C, 1 min	55 °C, 45 s 53 °C, 1 min[a]	72 °C, 1.5 min 72 °C, 2 min[a]	72 °C, 7 min
rpStolF/rpStolR rpStolF2/rpStolR	94 °C, 2 min	40	94 °C, 1 min	53 °C, 45 s	72 °C, 1.5 min	72 °C, 7 min
SecY gene						
L15F1/MapR1 SecYF1(III)/SecYR1(III) SecYF1(VI)/SecYR1(VI) L15F1A(VI)/MapR1A(VI) L15F2(IX)/MapR2(IX) SecYF1(X)/SecYR1(X)	94 °C, 1 min	35	94 °C, 30 s	50 °C, 1 min	68 °C, 5 min	72 °C, 10 min
L15F1A(I)/MapR1A(I)	94 °C, 1 min	35	94 °C, 30 s	52 °C, 1 min	68 °C, 5 min	72 °C, 10 min
L15F1A-a/MapR1A-a SecYF1(II)/SecYR1(II) SecYF2(II)/SecYR1(II) SecYF2(III)/SecYR1(III) L15F1A(III)/MapR1A(III) SecYF2(VI)/SecYR1(VI) SecYF1a/SecYR1(XII) SecYF2a/SecYR1(XII)	94 °C, 1 min	35	94 °C, 30 s	55 °C, 1 min	68 °C, 5 min	72 °C, 10 min

(continued)

Table 2
(continued)

Primer pair	Initial denaturation	Cycles	Denaturation	Anneal	Extension	Final extension
L15F1A(II)/MapR1A(II)	94 °C, 1 min	35	94 °C, 30 s	60 °C, 1 min	68 °C, 5 min	72 °C, 10 min
L15F1A-b/MapR1A-b	94 °C, 1 min	35	94 °C, 30 s	64 °C, 1 min	68 °C, 5 min	72 °C, 10 min
FD9f/FD9r	92 °C, 90 s	40	92 °C, 30 s	54 °C, 30 s	72 °C, 80 s	66 °C, 5 min
FD9f2L/FD9r FD9f3L/FD9r2L	92 °C, 1 min	40	92 °C, 1 min	55 °C, 1 min	66 °C, 30 s	66 °C, 5 min
AYsecYF1/AYsecYR1	94 °C, 2 min	38	94 °C, 1 min	55 °C, 2 min	72 °C, 3 min	72 °C, 7 min

[a]PCR conditions used with insect total DNA template
[b]n° of cycles used in direct PCR with primers rpAP15f/ rpAP15r

9. Close the lid of the gel tank and apply a voltage of 1–5 V/cm (*see* **Note 16**). Run the gel until the bromophenol blue and xylene cyanol FF have migrated an appropriate distance through the gel.

10. If GelRed™ or SYBR®Safe DNA Gel stain is present in the gel (*see* **Note 17**), examine the gel by UV light on a transilluminator and photograph the gel by a fixed-focus digital camera using a gel documentation system. Otherwise, stain the gel by immersion in ethidium bromide (0.5 μg/mL in water) for 10 min at room temperature followed by destaining for 10 min in water.

3.3   RFLP Analysis of PCR Products

1. Choose the appropriate key restriction enzymes to use according to the 16Sr group phytoplasma strains you are studying. Different sets of selected key restriction enzymes have been used in actual RFLP analysis or putative restriction sites analysis to differentiate phytoplasma strains of different 16Sr groups (Table 3).

2. Set the water bath(s) to the appropriate temperature as indicated by the enzyme manufacturer.

3. Thaw the enzyme reagents. In a sterile 0.5 mL microfuge tube, combine the reagents in the following order (volumes are given for one sample reaction):

Component	Final volume (µL)	Final concentration
Nuclease-free H_2O	9.5–11.5	–
10× restriction buffer	2	1×
Appropriate restriction enzyme	0.5	0.5–1 U
PCR product	6–8	300–500 ng/20 µL
Total volume	20	

4. Mix thoroughly, spin down, and add 1 drop of mineral oil only to restriction enzyme reactions which need incubation at 60–65 °C.

5. Place the tubes in the water bath(s) at the temperature indicated by the manufacturer and leave overnight.

6. Prepare the agarose gel as explained at paragraph Subheading 3.2 with the following modifications: (a) prepare sufficient 1× TBE electrophoresis buffer to fill the tank and cast the gel (*see* **Note 18**); (b) prepare agarose 3% (w/v) solution in 1× TBE in a flask or a glass bottle (*see* **Note 19**).

7. Mix the digested DNA sample (20 µL) with 4 µL of 6× gel-loading buffer. Load about 12 µL of the mixture with a P20 micropipette. Load 10 µL of a DNA size standard such as *Hae*III-digested ΦX174 DNA into wells on both the right and left sides of the gel.

8. Connect the electrodes to a power supply and apply a voltage of 1–8 V/cm. Run the gel until the bromophenol blue and xylene cyanol FF have migrated an appropriate distance through the gel.

9. Stain the gel in ethidium bromide (0.5 µg/mL in water) for 10 min at room temperature then destain for 10 min in water.

10. Examine the gel by UV light on the transilluminator and photograph the gel by a fixed–focus digital camera using a gel documentation system.

11. Compare the obtained RFLP patterns with those of phytoplasma reference strains or previously published patterns (Fig. 1a, c) [10, 11, 13].

3.4 Purification of PCR Products for Sequencing and Further Analysis

1. We recommend that PCR products are cleaned using commercially available products based on column purification which usually give good results.

2. Add the PCR products (about 50 µL) to the column according to the manufacturer's instructions (*see* **Note 20**).

3. Sequence the PCR-derived amplicon in both directions using Sanger sequencing, inspect the raw sequence chromatograms, and then assemble and edit your sequences using any available software such as BioEdit or MEGA, which are free online (*see* **Note 21**).

Table 3
Set of key restriction enzymes which are useful in differentiating phytoplasma strains within different 16Sr groups or subgroups on the bases of *rp* and *secY* genes

16Sr group	Restriction enzymes for RFLP on *rp* gene	Reference	Restriction enzymes for RFLP on *secY* gene	Reference
I	*Alu*I, *Mse*I, *Tsp*509I	[1, 10, 19]	*Alu*I, *Mse*I, *Tsp*509I	[11]
II	*Alu*I, *Mse*I, *Tsp*509I	[20]	*Alu*I, *Bfa*I, *Mse*I, *Sau*3AI, *Tsp*509I	[7]
III	*Alu*I, *Dra*I, *Mse*I	[1, 23]	*Alu*I, *Bfa*I, *Mse*I, *Tsp*509I	[7]
V	*Alu*I, *Hha*I, *Mse*I, *Tsp*509I	[13, 16]	*Alu*I, *Bfa*I, *Mse*I, *Rsa*I, *Taq*I	[13]
VI	*Alu*I, *Dra*I, *Taq*I, *Tsp*509I	[6]	*Alu*I, *Bfa*I, *Mse*I, *Taq*I, *Tsp*509I	[7]
VII	*Mse*I	[24]		
IX	*Alu*I, *Hha*I, *Rsa*I	[15]	*Alu*I, *Hha*I, *Mse*I	[15]
X	*Alu*I, *Dra*I	[6, 21]		
XII-A	*Mse*I	[25]		
XII-B	*Alu*I, *Dra*I, *Mse*I	[26]		

4. The edited sequences can be subjected to in silico RFLP using pDRAW32 software (*see* **Note 22**) using key restriction enzymes (Table 3; *see* **Note 23**) and the resulting RFLP pattern can be compared to those derived from *rp* and *secY* gene sequences of phytoplasma reference strains available in GenBank or previously published patterns (Fig. 1b, d) [7, 10, 11, 13, 15].

4   Notes

1. Combine equal volumes of 100 mM stock solutions of each dATP, dCTP, dGTP, and dTTP in a 1.5 mL microfuge tube, mix thoroughly, and then dilute the mixture 1:10 with nuclease-free water. Store at −20 °C.

2. Dilute 200 μM primer stock solutions 1:10 with nuclease-free water.

3. A complete set of micropipettes is normally dedicated for PCR use only (most commonly used micropipettes: P10, P20, P100, P200, and P1000).

4. In recent years, there has been increasing interest in the use of alternative DNA intercalating dyes such as GelRed™ (Biotium, Hayward, CA) or SYBR®Safe DNA Gel Stain (Invitrogen, Carlsbad, CA) as replacements for staining with the hazardous ethidium bromide.

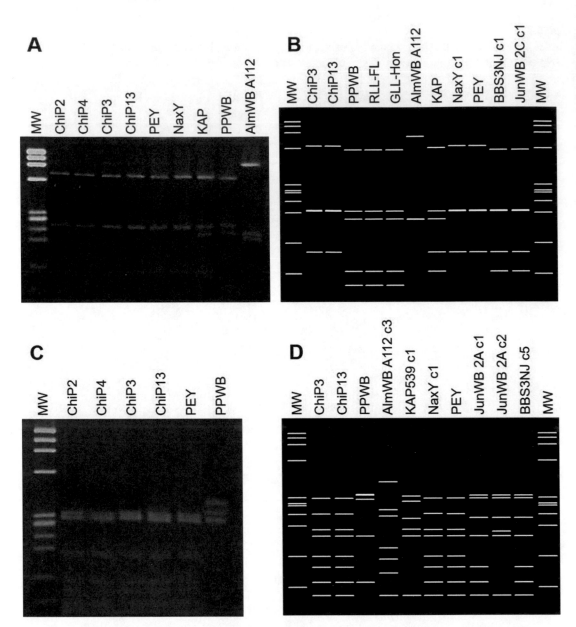

Fig. 1 (a) Actual RFLP patterns of *rp* operon sequence (1.2–1.3 kbp, containing *rpl22* and *rps3* genes) amplified with primer pair rpF1C/rp(l)R1A, (b) computer-simulated virtual RFLP patterns derived from in silico digestions of *rp* operon sequence (1.2 kbp containing *rpl22* and *rps3* genes), (c) actual RFLP patterns of *secY* gene fragment (about 1.5 kbp), amplified with primer pair L15F4/MAPR7 and (d) computer-simulated virtual RFLP patterns derived from in silico digestions of *secY* gene fragment (about 1.2–1.3 kbp) from chicory phyllody (ChiP) phytoplasma strains (16SrIX-C) [17] and some representative phytoplasma strains in the PPWB phytoplasma group (16SrIX) [15] with key restriction enzyme *AluI*. Lanes MW, ΦX174 DNA-*Hae*III digest; fragment sizes (bp) from top to bottom: 1353, 1078, 872, 603, 310, 281, 271, 234, 194, 118, 72

5. Normally only nested-PCR products are analyzed by agarose gel electrophoresis.

6. The semi-universal forward primers rpL2F3 and rpF1C are degenerated primers and are used together with the reverse primer rp(I)R1A. Primers rpF1C/rp(I)R1A can also be used in semi-nested PCR after amplification with rpL2F3/rp(I)R1A for a more sensitive semi-universal PCR for phytoplasmas. Primer pairs rp(I)F1A/rp(I)R1A (16SrI group-specific) can be used in nested-PCR following direct-PCR with rpF1/rpR1 primer pair. The group-specific primer pairs rp(II)F1/rp(II)R1 and rp(III)-FN/rp(I)R1A can be used in nested-PCR following direct-PCR primed by rpF1C/rp(I)R1A and rpL2F3/rp(I)R1A, respectively; whereas group-specific primer pairs rp(III)F1/rp(III)R1, rp(VI)F2/rp(VI)R2, rp(VIII)F2/rp(VIII)R2, rp(IX)F2/rp(IX)R2 can be used in nested-PCR following direct-PCR with either rpL2F3/rp(I)R1A or rpF1C/rp(I)R1A. For amplification of partial rp operon from phytoplasma strains of 16SrII group seminested-PCR with primer pairs rp(II)F/rp(I)R1A followed by rp(II)F2/rp(I)R1A can also be used. Partial rp operon from phytoplasma strains of 16SrV group can be amplified in direct-PCR with primer pair rp(V)F1/rpR1 followed by semi-nestedPCR with rp(V)F2/rpR1 or nested-PCR with rp(V)F1A/rp(V)R1A. The latter primer pair amplifies a longer and therefore more informative *rp* operon fragment. For amplification of partial *rp* operon from apple proliferation phytoplasma strains of 16SrX-A subgroup primer pair rpAP15f/rpAP15r can be used in direct PCR or in nested-PCR following direct-PCR with rpAP15f2/rp(I)R1A. Finally, partial *rp* operon of stolbur phytoplasmas (subgroup 16SrXII-A) can be amplified by semi-nested PCR with primers rpStolF/rpStolR followed by rpStolF2/rpStolR.

The semi-universal degenerate primer pair L15F1/MapR1 can be used to amplify the partial *spc* operon from groups 16SrI-VIII, X, XII, XIII, and XVIII. Groups 16SrI, XII, and XIII have the *adk* gene in the *spc* operon, however the gene is lacking in groups 16SrII, III, VI, IX, and X. The primer pair L15F1A-a/MapR1A-a is used in nested PCR following L15F1/MapR1 for groups 16SrI, XII, and XIII. The primer pair L15F1A-b/MapR1A-b is used in nested PCR following L15F1/MapR1 for groups 16SrII, III, V, and, VI. Also, primer pair L15F1/MapR1 can be used in nested-PCR followed by L15F1A(I)/MapR1A(I), L15F1A(II)/MapR1A(II), L15F1A(III)/MapR1A(III), L15F1A(VI)/MapR1A(VI), L15F2(IX)/MapR2(IX), depending on the group being amplified. All these nested PCRs yield a large portion of the *spc* operon, whereas the following more group-specific primers [e.g.,

SecYF1(II)/SecYR1(II)] yield the complete *secY* gene and a small portion of the flanking genes. The group-specific primers SecYF1(II)/SecYR1(II), SecYF1(III)/SecYR1(III), SecYF1 (VI)/SecYR1(VI), L15F2(IX)/MapR2(IX), SecYF1(X)/ SecYR1(X), and SecYF1a(XII)/SecYR1(XII) can be used in nested-PCR following L15F1/MapR1. They can also be used, except primers SecYF1(X)/SecYR1(X), as direct primers in nested-PCR followed by primer pairs SecYF2(II)/SecYR1 (II), SecYF2(III)/SecYR1(III), SecYF2(VI)/SecYR1(VI), L15F4(IX)/MapR7(IX), SecYF2a(XII)/SecYR1(XII), respectively. We used the primer pair AYsecYF1/AYsecYR1 only in direct-PCR to amplify the *secY* gene in 16SrI strains. However, the pair should work as nested primers following L15F1/ MapR1, L15F1A-a/MapR1A-a, or L15F1A(I)/MapR1A(I). The group-specific primer pair FD9f3L/FD9r2L can be used in nested-PCR following direct-PCR with either FD9f/FD9r or FD9f2L/FD9r.

7. A high-fidelity thermostable DNA polymerase, e.g., GeneAmp High Fidelity polymerase (Life Technologies, Gaithersburg, MD, USA), is especially recommended for downstream applications such as DNA sequencing; standard Taq polymerases from different manufacturers can be also used in PCR reactions followed by actual RFLP analyses.

8. A high-fidelity PCR Enzyme for Long Range PCR, e.g., TaKaRa LA Taq DNA Polymerase (Takara Mirus Bio, Madison, WI, USA), is especially recommended for amplification of long DNA fragments containing *secY* gene and for downstream applications such as DNA sequencing.

9. A master mix final volume of 24 or 48 μL depends on the PCR reagents used, however 48 μL are recommended for PCR products to be sequenced.

10. Positive controls (phytoplasma reference strains can be used as positive controls, but they must be handled with great care to avoid any risk of contamination) are required to monitor the efficiency of the PCR, whereas negative controls are required to detect contamination with the DNAs that contain the target sequence.

11. It is important to use the same batch of electrophoresis buffer in both the electrophoresis tank and the gel, since small differences can greatly distort the DNA migration. To prepare 1× TAE solution, add 20 mL of 50× TAE to 980 mL of distilled water in a 1 L graduated cylinder, seal the cylinder with Parafilm, and mix the solution inverting the cylinder.

12. In a flask or a glass bottle add 1 g of standard agarose to 100 mL of 1× TAE buffer.

13. Heat the slurry for the minimum time required allowing agarose to dissolve; undissolved agarose appears as translucent chips floating in the solution. Carefully swirl the flask or bottle from time to time to make sure that any grains of undissolved agarose do not stick to the walls but enter the solution.

14. The thickness of the gel should be between 3 and 5 mm. No air bubbles should be present under or between the teeth of the comb.

15. 1 µL of 6× gel-loading buffer can be deposited on a piece of Parafilm forming a drop and then 5 µL of the sample can be added and mixed on the Parafilm before loading the sample.

16. Distance (cm) is measured between the positive and negative electrodes.

17. The presence of GelRed™ or SYBR®Safe DNA Gel stain into the gel allows you to examine the gel by UV illumination at any stage during electrophoresis. The gel tray may be removed and placed directly on the transilluminator.

18. It is important to use the same batch of electrophoresis buffer in both the electrophoresis tank and the gel, since small differences can greatly distort the DNA migration. To prepare 1× TBE solution, add 100 mL of 10× TBE to 900 mL of distilled water in a 1 L graduated cylinder, seal the cylinder with Parafilm, and mix the solution inverting the cylinder.

19. In a flask or a glass bottle add 3 g of standard agarose to 100 mL of 1× TBE buffer. For RFLP analyses of *rp* and *secY* gene-based PCR products a 3% agarose gel in 1× TBE buffer can be used with all restriction enzymes. However, for frequent cutter enzymes such as *Mse*I and *Tsp*509I, the use of a 7–12% polyacrylamide gel is recommended for improved resolution (*see* Chapter 7).

20. This DNA cleanup procedure is intended for at least 50 µL volume PCR samples.

21. The BioEdit software can be downloaded free from http://www.mbio.ncsu.edu/BioEdit/bioedit.html; the MEGA software can be downloaded free from http://www.megasoftware.net/.

22. The pDRAW32 software (AcaClone Software) can be downloaded free from http://www.acaclone.com).

23. For the molecular characterization of a new phytoplasma strain we suggest performing a multiple sequence alignment of the edited sequences with *rp* and/or *secY* gene sequences of closely related phytoplasma reference strains available in GenBank using BioEdit or MEGA software; then the multiple alignment can be used to scan for sequence variability and to further select key restriction enzymes that will allow phytoplasma strain differentiation and classification.

References

1. Lee I-M, Gundersen-Rindal DE, Davis RE et al (1998) Revised classification scheme of phytoplasmas based on RFLP analyses of 16SrRNA and ribosomal protein gene sequences. Int J Syst Bacteriol 48:1153–1169

2. Zhao Y, Wei W, Davis RE et al (2010) Recent advances in 16S rRNA gene-based phytoplasma differentiation, classification and taxonomy. In: Weintraub PG, Jones P (eds) Phytoplasmas: genomes, plant hosts and vectors. CAB International, Wallingford, Oxfordshire, pp 64–92

3. Pérez-López E, Wei W, Wang J et al (2017) Novel phytoplasma strains of X-disease group unveil genetic markers that distinguish north American and south American geographic lineages within subgroups 16SrIII-J and 16SrIII-U. Ann Appl Biol 171:405–416

4. The IRPCM Phytoplasma/Spiroplasma Working Team—Phytoplasma taxonomy group (2004) '*Candidatus* Phytoplasma', a taxon for the wall-less, non-helical prokaryotes that colonize plant phloem and insects. Int J Syst Evol Microbiol 54:1243–1255

5. Lee I-M, Zhao Y, Davis RE (2010) Prospects of multiple gene-based systems for differentiation and classification of phytoplasmas. In: Weintraub PG, Jones P (eds) Phytoplasmas: genomes, plant hosts and vectors. CAB International, Wallingford, Oxfordshire, pp 51–63

6. Martini M, Lee I-M, Bottner KD et al (2007) Ribosomal protein gene-based phylogeny for finer differentiation and classification of phytoplasmas. Int J Syst Evol Microbiol 57:2037–2051

7. Lee I-M, Bottner-Parker KD, Zhao Y et al (2010) Phylogenetic analysis and delineation of phytoplasmas based on *secY* gene sequences. Int J Syst Evol Microbiol 60:2887–2897

8. Hodgetts J, Dickinson M (2010) Phytoplasma phylogeny and detection based on genes other than 16S rRNA. In: Weintraub PG, Jones P (eds) Phytoplasmas: genomes, plant hosts and vectors. CAB International, Wallingford, Oxfordshire, pp 93–113

9. Kakizawa S, Oshima K, Kuboyama T et al (2001) Cloning and expression analysis of *Phytoplasma* protein translocation genes. Mol Plant-Microbe Interact 14:1043–1050

10. Lee I-M, Gundersen DE, Davis RE et al (2004) '*Candidatus* Phytoplasma asteris', a novel phytoplasma taxon associated with aster yellows and related diseases. Int J Syst Evol Microbiol 54:1037–1048

11. Lee I-M, Zhao Y, Bottner KD (2006) *SecY* gene sequence analysis for finer differentiation of diverse strains in the aster yellows phytoplasma group. Mol Cell Probes 20:87–91

12. Davis RE, Zhao Y, Dally EL et al (2013) '*Candidatus* Phytoplasma pruni', a novel taxon associated with X-disease of stone fruits, *Prunus* spp.: multilocus characterization based on 16S rRNA, *secY*, and ribosomal protein genes. Int J Syst Evol Microbiol 63:766–776

13. Lee I-M, Martini M, Marcone C et al (2004) Classification of phytoplasma strains in the elm yellows group (16SrV) and proposition of '*Candidatus* Phytoplasma ulmi' for the phytoplasma associated with elm yellows. Int J Syst Evol Microbiol 54:337–347

14. Arnaud G, Malembic-Maher S, Salar P et al (2007) Multilocus sequence typing confirms the close genetic interrelatedness of three distinct flavescence dorée phytoplasma strain clusters and group 16SrV phytoplasmas infecting grapevine and alder in Europe. Appl Environ Microbiol 73:4001–4010

15. Lee I-M, Bottner-Parker KD, Zhao Y et al (2012) Differentiation and classification of phytoplasmas in the pigeon pea witches'--broom group (16SrIX): an update based on multiple gene sequence analysis. Int J Syst Evol Microbiol 62:2279–2285

16. Martini M, Botti S, Marcone C et al (2002) Genetic variability among Flavescence dorée phytoplasmas from different origins in Italy and France. Mol Cell Probes 16:197–208

17. Martini M, Ermacora P, Moruzzi S et al (2012) Molecular characterization of phytoplasma strains associated with epidemics of chicory phyllody. J Plant Pathol 94:S4.50

18. Lim PO, Sears BB (1992) Evolutionary relationships of a plant-pathogenic mycoplasmalike organism and *Acholeplasma laidlawii* deduced from two ribosomal protein gene sequences. J Bacteriol 174:2606–2611

19. Lee I-M, Martini M, Bottner KD et al (2003) Ecological implications from a molecular analysis of phytoplasmas involved in an aster yellows epidemic in various crops in Texas. Phytopathology 93:1368–1377

20. Martini M (2004) Ribosomal protein gene-based phylogeny: a basis for phytoplasma classification. PhD Dissertation, University of Udine, Udine, Italy, p 106

21. Martini M, Ermacora P, Falginella L et al (2008) Molecular differentiation of '*Candidatus* Phytoplasma mali' and its spreading in

Friuli Venezia Giulia region (north-east Italy). Acta Hortic 781:395–402

22. Daire X, Clair D, Larrue J et al (1997) Survey for grapevine yellows phytoplasmas in diverse European countries and Israel. Vitis 36:53–54

23. Gundersen DE, Lee I-M, Schaff DA et al (1996) Genomic diversity and differentiation among phytoplasma strains in 16S rRNA group I (aster yellows and related phytoplasmas) and III (X-disease and related phytoplasmas). Int J Syst Bacteriol 46:64–75

24. Griffiths HM, Sinclair WA, Smart CD et al (1999) The phytoplasma associated with ash yellows and lilac witches'-broom: '*Candidatus* Phytoplasma fraxini'. Int J Syst Bacteriol 49:1605–1614

25. Durante G, Casati P, Quaglino F et al (2008) Bois noir in Lombardy (northern Italy): identification of molecular markers for diagnosis and characterization of 16SrXII-A phytoplasmas. In: Proceedings of the 4th national meeting on phytoplasma diseases, Rome, Italy, 28–30 May 2008

26. Streten C, Gibb K (2005) Genetic variation in '*Candidatus* Phytoplasma australiense'. Plant Pathol 54:8–14

Chapter 9

Real-Time PCR Protocol for Phytoplasma Detection and Quantification

Yusuf Abou-Jawdah, Vicken Aknadibossian, Maan Jawhari, Patil Tawidian, and Peter Abrahamian

Abstract

Phytoplasmas are mollicutes restricted to plant phloem tissue and are normally present at very low concentrations. Real-time polymerase chain reaction (qPCR) offers several advantages over conventional PCR. It is a fast, sensitive, and reliable detection technique amenable to high throughput. Two fluorescent chemistries are available, intercalating dyes or hybridization probes. Intercalating dyes are relatively less expensive than TaqMan® hybridization probes but the latter chemistry is the most commonly used for phytoplasma detection. qPCR may be designed for universal detection of phytoplasma, group or subgroup specific detection, or for simultaneous detection of up to three or four phytoplasmas (multiplexing). qPCR may be used for relative or absolute quantification in host plants and in insect vectors. Therefore, qPCR plays an important role in phytoplasma detection as well as in host-pathogen interaction and in epidemiological studies. This chapter outlines the protocols followed in qPCR assay for phytoplasma detection and quantification, focusing mainly on the use of TaqMan® probes.

Key words Mollicutes, qPCR, Diagnosis, SYBR green, TaqMan® probes, Phytoplasma, Quantitation, Internal control genes

1 Introduction

New diseases associated with "phytoplasmas are being discovered at an increasingly rapid pace" [1]. Phytoplasmas are tiny cell wall-less microorganisms that live only in the plant phloem tissue and are normally present at extremely low concentrations. It was generally agreed that phytoplasmas cannot be grown in axenic cultures. However, only very recently, a culturing method was reported which requires special culture media and experience [2]. Therefore, efficient methods are required for their detection/diagnosis. Polymerase chain reaction (PCR)-based detection techniques are the most commonly used methods. The developed PCR methods allow detection of the phytoplasma in the host plant as well as in the insect vectors. Due to low phytoplasma titer in plant tissue, a nested

Rita Musetti and Laura Pagliari (eds.), *Phytoplasmas: Methods and Protocols*, Methods in Molecular Biology, vol. 1875,
https://doi.org/10.1007/978-1-4939-8837-2_9, © Springer Science+Business Media, LLC, part of Springer Nature 2019

PCR assay is followed in order to improve detection efficiency. This is typically followed by restriction fragment length polymorphism (RFLP) or cleaved amplified polymorphic sequences (CAPS) for identification of phytoplasma groups or subgroups. Alternatively, amplicons are sequenced and subjected to in silico RFLP (Chapter 8). However, advances in fluorescence technology allowed the development of qPCR with several advantages over conventional PCR [3–5]. qPCR is faster, more sensitive, highly specific and allows quantification of phytoplasma titer. Furthermore, the throughput is highly increased since no post-amplification steps are needed, no gels to run, stain, and photograph, thus leaving less room for operator error compared to PCR.

Two qPCR fluorescence chemistries are commonly used: intercalating binding dyes (SYBR® Green, EvaGreen®, PicoGreen®) and hybridization probes (TaqMan® probes, Locked Nucleic Acid probes, molecular beacons). Fluorescence occurs during the elongation step of the qPCR when using the intercalating dyes. The intercalating dye binds to the double-stranded DNA leading to 1000-fold increase in fluorescence, but fluorescence decreases significantly during denaturation [6]. A melt-curve analysis is typically performed after the qPCR run to verify the specificity of the test and to check for the presence of primer dimers or unspecific amplicons based on the melting temperature difference of amplicons. Each amplicon has an expected melting temperature (T_m), the temperature at which 50% dissociations of dsDNA to ssDNA occur. The melting temperature curve may be transformed by the thermocycler software into melting peaks. Slight differences in amplicon sequence, sometimes of even one base pair, may be reflected by a change in the T_m peak. This allows multiplexing, i.e., detection of more than one phytoplasma in a single run, when their respective amplicon melting temperatures have distinct peaks in the melt-curve analysis [7].

On the other hand, hybridization probes typically possess a fluorophore linked to the 5' end and a quencher linked to the 3' end. The presence of the quencher in close proximity to the fluorophore prevents its fluorescence. TaqMan® probes are the most commonly used probes. During qPCR, upon primer annealing and extension, the 5'–3' exonuclease activity of the Taq polymerase cleaves the fluorophore from the TaqMan® probe. The cleaved fluorophore is excited by light at a specific spectrum and the fluorescent signal is detected through special optical devices. The availability of different fluorophores allows simultaneous detection of up to three or four phytoplasmas [8]. The TaqMan® probes and some of the related modifications such as minor groove binding (MGB™) [9, 10] or BHQplus® [11] may considerably improve assay specificity and especially for AT-rich regions in phytoplasma genomes, thus significantly reducing false positives. Internal controls are used to evaluate the quality of DNA extracts, which may contain polyphenolics, polysaccharides, or other inhibitors [12]. Internal control genes, such as plant or vector DNA, are typically targeted along with the phytoplasma target gene either in singleplex or multiplex reactions. Several internal control/reference genes have been reported (Table 1), for example the 18S rDNA gene [4, 13, 14] has been used as internal control of plant host DNA gene for several plant species including some fruit trees. Likewise, the 18S rDNA gene has been reported as internal control target gene in DNA extractions from insect vectors [13] (Table 2).

Depending on the objectives, the primers used in qPCR may be universal primers, that allow detection of all or several phytoplasma groups [4, 15] (Table 3) or may be specific to one phytoplasma or a phytoplasma group or subgroup [16].

In addition to diagnostics, qPCR is useful for quantification of phytoplasma titer in plants and vectors, and has a wide range of applications in studies pertaining to host–pathogen interaction, insect transmission, and varietal reactions. qPCR quantification can be either absolute or relative [17]. Relative quantification provides relative amounts of phytoplasma titer when two infected samples are compared and normalized to a reference gene. The selected reference gene should be moderately or highly expressed and stable under biotic and abiotic stresses. Several other reference genes have been reported in the literature (Tables 1 and 2). On the other hand, absolute quantification requires the development of a standard curve using a known amount of plasmid DNA containing the phytoplasma target gene [11].

2   Materials

If only detection of phytoplasma is desired without quantification, omit materials Subheading 2.1, **items 5** and **6** and Subheading 2.3.

2.1 Samples and Controls

1. Extracted host (plant or vector) DNA to be tested (*see* **Note 1**).

2. Extracted DNA of a negative control (NC): same host but known to be uninfected (*see* **Note 2**).

3. No template control (NTC): use sterile deionized water instead of DNA template (*see* **Note 3**).

4. Extracted DNA of a positive control (PC): same host but known to be infected (*see* **Note 3**).

5. Series of tenfold dilutions of a positive DNA cloned in a plasmid to be used in quantification analysis (*see* **Note 4**).

6. Series of tenfold dilutions of host DNA to be used in quantification analysis (*see* **Note 5**).

Table 1
List of primers and probes designed for some internal control plant genes

Host	Target gene	Forward primer 5'–3'	Reverse primer 5'–3'	Probe 5'–3'	Size	Reference
Plants	18S rDNA	C18S-F2 (F) CAGCTCGGGTTGACTACGTC	D-C18Sr6 (R) GATCCGAACACTTCACCGG AIIIICAATCGGTA	C18S-Pt (P) TAMRA-ACACACGGCCCGTCGCTCC-BHQ2a	84 bp	[8]
Plants	18S rDNA	GACTACGTCCCTGCCCTTTG	AACACTTCACCGGACCA TTCA	VIC- ACACACGGCCCGTCGCTCC -TAMRA	67 bp	[4]
Plants (Fruit Trees)	18S rDNA (Apple)	AGAGGGAGCCTGAGAAACGG	CAGACTCATAGAGGCCGGTA TTG	CCACA TCCAAGGAAGGCAGCAGGCG	115 bp	[18]
Periwinkle, Poinsettia, Prunus spp.	18S rDNA	GACTACGTCCCTGCCCTTTG	AACACTTCACCGGACCA TTCA	ACACACGGCCCGTCGCTCC	67 bp	[4, 14]
Grapevine, potato	Cytochrome oxidase	COX-F CGTCGCATTCCAGATT ATCCA	COX-R CAACTACGGATATATAAGA GCCAAAACTG	COXP TGCTTACGCTGG ATGGAATGCCCT	78 bp	[19]
Grapevine	Chaperonin	Chaperonin grapevine gene Forward GGTCCTTTGGATGAGG ATGG	Chaperonin grapevine gene Reverse GAAGTCAITTCCCTGCAT ACTTGG	Chaperonin grapevine gene Probe GAAACCACTGTCT GTGAGCCAGGA	89 bp	[20]
Grapevine	18S rDNA	CCGTTGCTCTGATGAITTCA TGA	CGTCGCGGCACGAT	FAM -AACTCGACGGATCGCACGGC -BHQ1	58 bp	[21]
Apple	tRNA leucine	qMd-cpLeu-F CCTTCATCCTTTCTGAAG TTTCG	qMd-cpLeu-R AACAAATGGAGTTG GCTGCAT	qMd-cpLeu VIC-TGGAAGGATTCCTTTACTAAC	68 bp	[9]
Marguerite	18S rDNA (ITS1)	ChrysFw AAGGAAAACTAAACTTAAGA AGCTT–GTT	ChrysRv GTGGCTTCTTTATAATCAC	Chrys Probe CCCGTTCGCGGT GTGCTCATG	64 bp	[13]
Sesame	18S rDNA	CGCGGAAGTTTGAGGCAATA	CTGTCGGCCAAGGCTA TAGACT	HEX-TAGATGTTCTGGGCCGCACGCG-TAMRA	103 bp	[22]

Table 2
List of primers and probes designed for some internal control genes of insect vectors

Insect	Target gene	Forward primer 5'-3'	Reverse primer 5'-3'	Probe 5'-3'	Size	Reference
Leafhopper species	18S rDNA	MqFw AACGGCTACCACA TCCAAGG	MqRv GCCTCGGATGAG TCCCG	Mq Probe AGGCAGCAGGCA CGCAAATTACCC	100 bp	[13]
Hyalesthes obsoletus	–	TCTGTCTGCTGCG TGAACAT	ACTGTAGC TCGCCCAGGTT	TR -TAACAAACTCCGGAAGAACATAACTCA -DDQ2	128 bp	[21]
Macrosteles quadrilineatus	CP6	GGGCAAGAAGGGCAAG TA	AGGCTCCAGATACAC TAGGTC	EvaGreen®	90 bp	[23]

Table 3
List of primers and probes designed for universal detection of phytoplasma DNA by real-time PCR

Gene targeted	Forward primer 5'-3'	Reverse primer 5'-3'	Probe 5'-3'	Size	Reference
16S rDNA	CGTACGCAAGTATGAA ACTTAAAGGA	TCTTCGAATTAAA CAACATGATCCA	FAM- TGACGGGACTCCGCACAAGCG-TAMRA	75 bp	[4]
16S rDNA	CYS2Fw AGGTTGAACGGCCACA TTG	CYS2Rv TTGCTCGGTCAGA GTTTCCTC	CYS2 Probe ACACGGCCCAAACTCCTACGGGA	98 bp	[16]
16S rDNA	UniRNA Forward AAATATAGTGGAGGT TATCAGGGATACAG	UniRNA Reverse AACCTAACATCTCA CGACACGAACT	UniRNA Probe FAM-ACGACAAACCATGCACCA-NFQ	73 bp	[19]
16S rDNA	UPH-F (F) CGTACGCAAGTA TGAAAC TTAAAGGA	D-UPHr2 (R) CGACAACCA TGCACCACCTG IIIIICTGATAACC	UPH-P (P) FAM-TGACGGGACTCCGCACA- MGB	169 bp	[8]
23S rDNA	JH-F 1 (F) GGTCTCCGAA TGGGAAAACC + JH-F all (F) ATTTCCGAA TGGGGCAACC (2 primers mixed in equal amounts)	JH-R (R) CTGGTCACTACTACC RGAA TCGTTATTAC	JH-P uni (P) FAM-AACTGAAATATCTAAGTAAC- MGB	138- 149 bp	[15]
16S rDNA (Mycoplasma and phytoplasma primers)	GPO3F TGGGGAGCAAAACAGGA TTAGATACC	MGSO TGCACCATCTGTCAC TCTGTTAACCTC	SYBR® Green I	274 bp	[24]

2.2 qPCR Reagents	1. qPCR Taq polymerase Master Mix or SYBR® Green Master Mix (*see* **Note 6**).

2. Primers and probes (*see* **Note 7** and Table 1 for universal detection of phytoplasma). For the detection of specific phytoplasma, please refer to the literature and to the following two references [15, 16].

3. Sterile DNase-free water (*see* **Note 8**).

2.3   Cloning of PCR Product into Plasmid

1. PCR product purification Kit (e.g., Illustra™ GFX PCR DNA, Sigma-Aldrich, or QIAquick® PCR Purification Kit, Qiagen).

2. Suitable restriction digestion enzymes along with their reaction buffers.

3. Cloning vector kit (e.g., pGEM®-T Easy Vector System II, Promega, or CloneJET™ PCR Cloning Kit, Thermofisher).

4. *E. coli* competent cells (e.g., strain DH5α).

5. *E. coli* growth media (e.g. S1797—SOC Medium, Sigma-Aldrich).

6. Plasmid purification kit (e.g. QIAprep® Spin Miniprep Kit, Qiagen, or ChargeSwitch®-Pro Plasmid Miniprep Kit, Thermofisher).

2.4   qPCR Equipment and Accessories

1. qPCR plates: 96 or 384-well hard reaction plates.

2. Optical adhesive plate covers.

3. Real-time thermal cycler.

4. Software for the process of plate setup, data collection, and analysis of real-time PCR results.

5. Autoclaved tubes, pipettes, and aerosol-barrier pipette tips.

6. UV chamber.

7. Microcentrifuge and a plate centrifuge.

8. UV-Vis Spectrophotometer for the quantification and purity of DNA templates (e.g., Nanodrop 2000, Thermofisher).

3   Methods

Detection of phytoplasma is simpler than quantification. If the purpose is to merely detect the presence of phytoplasma or survey for the presence of a phytoplasma in a given region, then all steps relating to the preparation of standard curves of phytoplasma or host DNA may be omitted (Subheadings 3.2, 3.3, **step 3**, and 3.6, **steps 3** and **4**).

3.1   DNA Extraction   Refer to Chapters 6 and 14.

3.2 Cloning PCR Product into Plasmid

1. Perform a PCR with primers amplifying a region (>100 bp) of the phytoplasma genome which contains the sequence to be amplified in qPCR such as 16S rDNA, or 23S rDNA. An amplicon of the expected size should be verified by gel electrophoresis.

2. Purify PCR product with a purification kit according to the kit manufacturer's instructions.

3. Insert purified PCR product into a cloning vector by T4 DNA ligase either directly by ligation of PCR product with a vector such as pGEM®-T Easy, or after restriction enzyme digestion of the PCR product and the vector at appropriate sites found on the vector and designed into the PCR product with the primers. Insert to vector molar ratios 3:1 up to 10:1 have been successful. Blunt-end ligation is also possible, but is more difficult and less likely to succeed.

4. Transform competent *E. coli* cells with the cloned vector following instructions of the competent cell of choice (e.g. XL1 Blue) and incubate on LB-agar+antibiotic plates overnight. If blue-white screening is desired, add X-Gal/S-Gal and IPTG to the plates. Plate non-transformed cells as negative control to verify antibiotic selection success. Plate competent cells transformed with control plasmid (e.g., pUC19) as positive control to verify transformation efficiency.

5. Prepare cultures of transformed cells in LB + antibiotic broth by inoculating with single colonies with a sterile loop.

6. Preform miniprep from the cultures to purify recombinant vector (plasmid) according to the instructions of the manufacturer of the plasmid purification kit.

7. Sequence the cloned insert either with PCR primers or preferably with vector sequencing primers to verify presence of target sequence.

3.3 Plate Preparation

1. Arrange the samples using the plate layout manually or according to the qPCR software.

2. Assign two or three wells for each NTC, NC, and PC.

3. Each plate should contain at least 12 wells (for duplicates) or 18 wells (for triplicates) for standard curves.

4. *See* Fig. 1 for a typical plate layout.

3.4 Sample Preparation

1. Remove all reagents from the freezer and thaw on ice and away from light. Mix all thawed tubes using a vortex and spin them briefly (~5 s).

2. Prepare two separate reaction mixes: one reaction mix for the phytoplasma primers and the other for the internal control gene (host primers). The final volume of the master mix

	1	2	3	4	5	6	7	8	9	10	11	12
A	Std-1 FAM phytoplasma DNA	Std-1 FAM phytoplasma DNA	Std-2 FAM phytoplasma DNA	Std-2 FAM phytoplasma DNA	Std-3 FAM phytoplasma DNA	Std-3 FAM phytoplasma DNA	Std-4 FAM phytoplasma DNA	Std-4 FAM phytoplasma DNA	Std-5 FAM phytoplasma DNA	Std-5 FAM phytoplasma DNA	Std-6 FAM phytoplasma DNA	Std-6 FAM phytoplasma DNA
B												
C	Std-7 FAM Host DNA	Std-7 FAM Host DNA	Std-8 FAM Host DNA	Std-8 FAM Host DNA	Std-9 FAM Host DNA	Std-9 FAM Host DNA	Std-10 FAM Host DNA	Std-10 FAM Host DNA	Std-11 FAM Host DNA	Std-11 FAM Host DNA	Std-12 FAM Host DNA	Std-12 FAM Host DNA
D												
E												
F	Unk-1 FAM	Unk-1 FAM	Unk-2 FAM	Unk-2 FAM	Unk-3 FAM	Unk-3 FAM	Unk-4 FAM	Unk-4 FAM	Unk-5 FAM	Unk-5 FAM	Unk-6 FAM	Unk-6 FAM
G	Unk-7 FAM	Unk-7 FAM	Unk-8 FAM	Unk-8 FAM	Unk-9 FAM	Unk-9 FAM	Unk-10 FAM	Unk-10 FAM	Pos FAM	Pos FAM	Neg FAM	Neg FAM
H											NTC FAM	NTC FAM

Fig. 1 Typical 96-well plate layout for a real-time PCR assay. The plate layout represents an assay in which ten host samples or unknowns (Unk) are checked for the presence and/or quantification of phytoplasma. Standard wells (Std) are to be used in constructing two standard curves. Wells Std1 to Std6 pertain to the serial dilutions of a plasmid carrying a phytoplasma target gene in order to get the efficiency of the assay and phytoplasma titer. Wells Std7 to Std12 pertain to the serial dilutions of the host (plant/vector) DNA for standardization of input DNA. A positive control (Pos), a negative control (Neg), and a no template control (NTC) are also included. Each sample is run in two replicates. FAM is the reporter dye in this layout, it should be changed to read the reporter dye used in your probe. NB: in case no quantification is needed, the Std-1 to 12 may be omitted

depends on the number of reactions. Use the qPCR machine software (e.g., Biorad CFX manager 3.1 -> Tools -> Mastermix Calculator) to calculate the volumes of each of the components of the master mix. Increase the number of samples by 10% when preparing the master mix to factor for pipetting error (*see* **Note 9**).

3. Fill the appropriate wells with 18 μL of the Master Mix (*see* **Note 10**).

4. Use 300 nM and 100 nM of each primer and probe, respectively, for each reaction (*see* **Note 11**).

5. Add 2 μL of DNA extract (*see* **Note 1**), cover the qPCR with an optical adhesive cover, and then centrifuge for 30 s at $2000 \times g$ to collect the DNA and the mix at the bottom of the plate wells (*see* **Note 12**).

3.5 Perform qPCR Run

1. Turn the thermocycler on.

2. Load the plate into the thermocycler.

3. Launch the thermocycler's software.

4. Select the cycling protocol with 20 s at 95 °C for initial denaturation step followed by 40 cycles of 3 s denaturation at 95 °C and 30 s annealing and elongation at 60 °C (standard protocol to be optimized depending on your working conditions) and import the plate data then click Start Run (*see* **Note 13**).

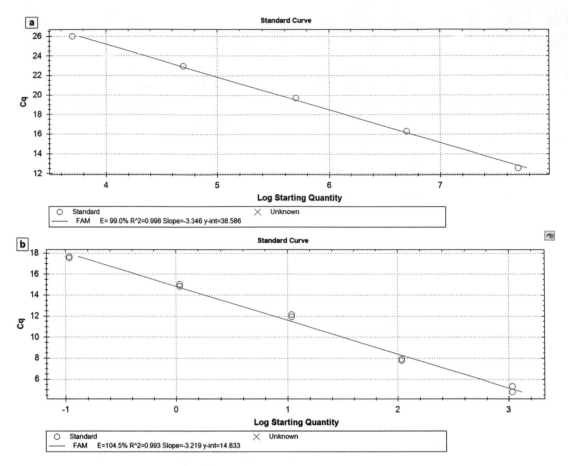

Fig. 2 Standard curves of two real-time PCR assays showing the linear regression line of the Cq against the log dilution of plasmid containing an insert of a phytoplasma target gene (**a**) and from 18S rDNA of total nucleic acids extract from a healthy plant (**b**). Reproduced from Jawhari et al. with permission [11]

3.6 Analysis of qPCR Results

1. When the reaction is complete, save the results of the run and remove the plate (*see* **Note 14**).

2. The software will automatically determine the quantification cycle (Cq) values of all samples; but you have the choice to modify slightly the threshold line. The NTC and NC samples should get no or $\geq$38 Cq.

3. Examine the linear regression line of the Cq on the log dilution of the plasmid containing the target DNA and that of the 18S rDNA of the plant (*see* Fig. 2). The slope of the standard curve should be -3.3 and the R^2 values should be close to 1.00 and efficiency values close to 100% (*see* **Note 15**).

4. For phytoplasma titer quantification, the software calculates the phytoplasma copy number in each sample using the Cq value of the sample and subsequently extrapolated on the standard curve. Since phytoplasmas carry two copies of 16S

rDNA gene, the copy number should be divided by two. In case the standard curves are run separately on a different plate than the samples, then calculations can be done manually (*see* **Note 16**). To get the number of genomic units (copy number)/ng DNA, the phytoplasma copy number obtained for each sample is then divided by the quantity (ng) of DNA of the same sample obtained from the standard curve for plant/vector DNA [11].

4   Notes

1. The amount of template DNA added to each reaction should be diluted to less than 100 ng in sterile DNase-free water. Mix the DNA tubes and their dilutions thoroughly, then spin them briefly to collect the DNA at the bottom of the tube.

2. In every qPCR assay a healthy sample of the same host (plant or vector) should be included to check for possible cross reactions of the primers with the sample material (false positives).

3. A non-template control (NTC) and a positive control (PC) should be used in every qPCR run. NTC uses purified deionized water as template instead of sample DNA. The positive control is a sample that contains the target phytoplasma DNA.

4. For quantitative analysis, at least five to six tenfold dilutions of a positive sample with known target phytoplasma DNA copy number are used to generate a standard curve. Normally, a plasmid cloned with the target phytoplasma DNA sequence is used. The number of plasmid copies is calculated based on molecular weight using the formula: number of copies = plasmid concentration/[(plasmid size + insert (bp) × 660)/(Avogadro's number)]. Avogadro's number = 6.023×10^{23}. During plate set up on the software, the copy number is entered for each standard, thus, the software will automatically calculate the copy numbers at the end of the reaction.

5. Quantity (ng) of DNA of the sample can be quantified by UV-Vis Spectrophotometer (e.g., Nanodrop 2000, Thermofisher).

6. Master Mix can be used from any supplier, it is advised to compare products from at least two suppliers to check for reproducibility.

7. Purchased lyophilized primers are to be suspended to 100 µM in sterile DNase-free water. Dilute each stock till 10 µM in aliquots and store at −20 °C. Cover the tubes containing probes to protect from light exposure.

8. Use sterile DNase-free water in all recipes and protocol steps.

9. Take maximum precautions at all stages of setting up qPCR reactions. To avoid contamination, the preparation of the reaction mixes and the addition of DNA have to be performed in a UV chamber.

10. Take into consideration each sample and standard are run in duplicates. Reactions can be scaled down to 10 μL depending on the master mix used.

11. To determine the optimum primer/probe combination for the target gene, an equimolar (or possibly different ratios) of primers (100, 200, 300, 400 nM) and probe (100, 200, 300 nM) concentrations can be used in order to screen for the highest relative fluorescence unit (RFU), the lowest quantitation cycle (Cq), and the highest reaction efficiency. The reaction efficiency can be determined using a standard curve for each primer/probe set.

12. The plate and the optical adhesive cover should be handled by the sides. Run the plate directly after spinning; otherwise keep on ice in the dark for few hours only.

13. Using the thermocycler software: (1) adjust the reaction volume and the samples location on the plate; (2) assign the correct filters of the probes used.

14. When using SYBR® Green, a melting curve analysis is required to ensure the specificity of the reaction. After the qPCR run is finished, the temperature is steadily increased normally by 0.5 °C in each step, and the fluorescence emission due to dsDNA denaturation is monitored by the thermocycler and recorded as a function of the temperature. A unique, well-defined peak on the melt curve indicates one product (*see* Fig. 3).

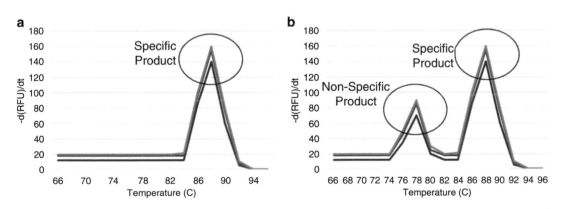

Fig. 3 Diagrammatic representation of melt curves showing the presence of specific products (**a**) and nonspecific products (**b**)

15. Efficiency of each standard curve was calculated by using the following equation: $E = [10^{(-1/\text{slope})} - 1]*[100]$. Acceptable efficiency should range between 90 to 110%.

16. For manual calculation of phytoplasma titer values can be extrapolated on a standard curve using the following method. The estimated number of phytoplasmas is derived from the equation: $NT = 10^{(Cq - b)/a}$, where NT is the target copy number, Cq for any sample, a and b the slope and intercept of the regression line, respectively. The copy numbers obtained should be normalized for input amount of DNA using the qPCR assay for the internal control plant 18S rRNA. The standard curves are constructed by plotting the $\log_{10}$ of the target copy number against their respective Cq. Phytoplasma titer in plants or vectors may be expressed as genomic units (GU) of phytoplasma per nanogram (ng) of plant or vector DNA.

References

1. Zhao Y, Davis R (2016) Criteria for phytoplasma 16Sr group/subgroup delineation and the need of a platform for proper registration of new groups and subgroups. Int J Syst Evol Microbiol 66(5):2121–2123

2. Contaldo N, Satta E, Zambon Y, Paltrinieri S, Bertaccini A (2017) Development and evaluation of different complex media for phytoplasma isolation and growth. J Microbiol Methods 127:105–110

3. Bianco PA, Casati P, Marziliano N (2004) Detection of phytoplasmas associated with grapevine flavescence dorée disease using real-time PCR. J Plant Pathol 86:257–261

4. Christensen NM, Nicolaisen M, Hansen M, Schulz A (2004) Distribution of phytoplasmas in infected plants as revealed by real-time PCR and bioimaging. Mol Plant-Microbe Interact 17:1175–1184

5. Wei W, Kakizawa S, Suzuki S, Jung HY, Nishigawa H, Miyata S, Oshima K, Ugaki M, Hibi T, Namba S (2004) In planta dynamic analysis of onion yellows phytoplasma using localized inoculation by insect transmission. Phytopathology 94:244–250

6. Dragan AI, Pavlovic R, McGivney JB, Casas-Finet JR, Bishop ES, Strouse RJ et al (2012) SYBR green I: fluorescence properties and interaction with DNA. J Fluoresc 4:1189–1199

7. Anniballi F, Auricchio B, Delibato E, Antonacci M, De Medici D, Fenicia L (2012) Multiplex real-time PCR SYBR green for detection and typing of group III clostridium botulinum. Vet Microbiol 154(3–4):332

8. Ito T, Suzaki K (2017) Universal detection of phytoplasmas and xylella spp. by TaqMan singleplex and multiplex real-time PCR with dual priming oligonucleotides. PLoS One 12(9): e0185427. https://doi.org/10.1371/journal.pone.0185427

9. Baric S, Dalla-Via J (2004) A new approach to apple proliferation detection: a highly sensitive real-time PCR assay. J Microbiol Methods 57:135–145

10. Kostina EV, Ryabinin VA, Maksakova GA, Sinyakov AN (2007) TaqMan probes based on oligonucleotide–hairpin minor groove binder conjugates. Russ J Bio Organichemistry 33:614–616

11. Jawhari M, Abrahamian P, Abdel Sater A, Sobh H, Tawidian P, Abou-Jawdah Y (2015) Specific PCR and real-time PCR assays for detection and quantitation of 'Candidatus Phytoplasma phoenicium'. Mol Cell Probes 29 (1):63–70

12. Rezadoost MH, Kordrostami M, Kumleh HH (2016) An efficient protocol for isolation of inhibitor-free nucleic acids even from recalcitrant plants. 3. Biotech 6:61

13. Marzachì C, Bosco D (2005) Relative quantification of chrysanthemum yellows (16Sr I) phytoplasma in its plant and insect host using real-time polymerase chain reaction. Mol Biotechnol 30:117–127

14. Martini M, Loi N, Ermacora P, Carraro L, Pastore M (2007) A real-time PCR method for

detection and quantification of 'Candidatus Phytoplasma prunorum' in its natural hosts. B Insectol 60(2):251–252

15. Hodgetts J, Boonham N, Mumford R, Dickinson M (2009) Panel of 23S rRNA gene-based real-time PCR assays for improved universal and group-specific detection of phytoplasmas. Appl Environ Microbiol 75:2945–2950

16. Galetto L, Bosco D, Marzachì C (2005) Universal and group-specific real-time PCR diagnosis of flavescence dorée (16Sr-V), bois noir (16Sr-XII) and apple proliferation (16Sr-X) phytoplasmas from field-collected plant hosts and insect vectors. Ann Appl Biol 147:191–201

17. Pfaffl MW (2012) Quantification strategies in real-time polymerase chain reaction. In: Filion M (ed) Quantitative real-time PCR. Applied microbiology. Horizon Scientific Press, Norfolk, USA

18. Oberhänsli T, Altenbach D, Bitterlin W (2011) Development of a duplex TaqMan real-time PCR for the general detection of phytoplasmas and 18S rRNA host genes in fruit trees and other plants. B Insectol 64(Supplement): S37–S38

19. Hren M, Boben J, Rotter A, Kralj P, Gruden K, Ravnikar M (2007) Real-time PCR detection systems for Flavescence dorée and bois noir phytoplasmas in grapevine: comparison with conventional PCR detection and application in diagnostics. Plant Pathol 56:785–796

20. Angelini E, Bianchi GL, Filippin L, Morassutti C, Borgo M (2007) A new TaqMan method for the identification of phytoplasmas associated with grapevine yellows by real-time PCR assay. J Microbiol Methods 68:613–622

21. Fahrentrapp J, Michl G, Breuer M (2013) Quantitative PCR assay for detection of bois noir phytoplasmas in grape and insect tissue. Vitis 52(2):85–89

22. Ikten C, Ustun R, Catal M, Yol E, Uzun B (2016) Multiplex real-time qPCR assay for simultaneous and sensitive detection of phytoplasmas in sesame plants and insect vectors. PLoS One 11(5)

23. Frost K, Willis D, Groves R (2011) Detection and variability of aster yellows phytoplasma titer in its insect vector, *Macrosteles quadrilineatus* (Hemiptera: Cicadellidae). J Econ Entomol 104(6):1800–1815

24. Satta E, Nanni I, Contaldo N, Collina M, Poveda J, Ramírez A, Bertaccini A (2017) General phytoplasma detection by a q-PCR method using mycoplasma primers. Mol Cell Probes 35(2017):1–7

Chapter 10

Duplex TaqMan Real-Time PCR for Rapid Quantitative Analysis of a Phytoplasma in Its Host Plant without External Standard Curves

Sanja Baric

Abstract

The chapter describes a simple quantitative approach to assess phytoplasma load in samples obtained from "*Candidatus* Phytoplasma mali"-infected apple plants without the use of external standard curves. The assay is based on the simultaneous detection of a gene of the pathogen and a gene of the host plant in a duplex single-tube real-time PCR reaction using TaqMan chemistry. The quantity of the phytoplasma, relative to its host plant, is determined as the difference between the C_T values of the two target genes (ΔC_T). A critical data analysis step, affecting the inter-assay reproducibility between different amplification runs, is the setting of the threshold level, which is achieved by the recurrent analysis of a calibrator sample. The relative quantification procedure allows analyzing 45 DNA samples in duplicates on a 96-well reaction plate, in addition to the control and calibrator samples, and thus contributes to a substantial increase of analysis throughput and decrease of reagent/consumable costs per sample.

Key words Quantitative real-time PCR, Relative quantification, Pathogen load, "*Candidatus* Phytoplasma mali", *Malus domestica*

1 Introduction

Real-time PCR has developed into a widely applied diagnostic tool, which has been used in many plant pathology laboratories due to its high sensitivity, specificity and robustness, the unnecessity for post-PCR manipulations, and its suitability for high-throughput testing applications [1]. The technology represents an advanced variant of the polymerase chain reaction, in which the accumulation of amplified fragments of nucleic acid is measured as the increase of fluorescent signal during or after each reaction cycle. Two types of detection chemistry are available: (1) intercalating dyes binding nonspecifically to double-stranded DNA that is generated during PCR and (2) sequence-specific fluorogenic hybridization probes, which employ the principle of fluorescence resonance energy transfer (FRET) [2]. Independent of the reaction chemistry, real-time

Rita Musetti and Laura Pagliari (eds.), *Phytoplasmas: Methods and Protocols*, Methods in Molecular Biology, vol. 1875,
https://doi.org/10.1007/978-1-4939-8837-2_10, © Springer Science+Business Media, LLC, part of Springer Nature 2019

PCR operates with threshold cycle (C_T) values, which are defined as the cycle of the PCR at which the increasing fluorescent signal crosses a threshold line, and are thus also referred to as the crossing points (or C_P) [3]. Since C_T values are inversely proportional to the amplicon amount in the reaction, real-time PCR is well suited for quantitative analyses, such as the determination of gene expression levels or the estimation of pathogen load.

Numerous real-time PCR protocols have so far been described for phytoplasmas. These include both, qualitative assays for the detection of the presence or absence of pathogens in host plants and/or vectors (e.g., [4–7]) and quantitative approaches for the determination of pathogen titer (e.g., [8–14]). The most commonly applied procedure for assessing plant pathogen load is the absolute quantitative real-time PCR, which depends on external standard curves to which the amplification signal of the target nucleic acid is related. Absolute quantification provides the advantage of assessing the exact number of pathogen copies in infected host tissues. However, it must be considered that the accuracy of quantification depends on the quality of standards. Imprecisions can occur during the determination of initial standard concentrations or during the preparation of serial dilutions, while the stability of standard curves can decline over time [3, 15]. In addition, the analysis of serially diluted standards in each real-time PCR experiment can considerably reduce the sample throughput and consequently increase the overall analysis time and reagent/consumable costs [16].

As an alternative approach, relative quantification can be applied, which is frequently used in quantitative gene expression assays [17, 18], but is less commonly used for quantification of plant pathogens [16]. This approach does not provide the exact number of pathogen cells per unit, but relates the C_T value of the target nucleic acid to that of an internal control or reference gene. Since the analysis of external standard curves is not required, the quantification procedure is much simpler and contributes to a substantial increase of analysis throughput and decrease of reagent/consumable costs per sample.

This chapter describes a relative quantitative real-time PCR assay for "*Candidatus* Phytoplasma mali" in its host plant, the apple tree (*Malus* × *domestica* Borkh.) [16]. Phytoplasmas are obligate endoparasites [19] and nucleic acid isolates from phytoplasma-infected plant tissue thus contain a mixture of pathogen and plant DNA. The DNA isolates are analyzed by real-time PCR, where the pathogen and the plant target genes are detected simultaneously in a single-tube duplex reaction using TaqMan probes. Both target genes (16S rRNA gene of "*Ca.* P. mali" and the gene for 1-aminocyclopropane-1-carboxylate oxidase of *M. domestica*) are present in two copies on the phytoplasma chromosome and in the diploid host plant genome, respectively (discussed in [12]). Following real-time PCR analysis, the C_T values of the target genes of the pathogen and the host plant are subtracted and relative quantities are obtained.

There are several application possibilities of the comparative quantitation method, such as the assessment of the "*Ca.* P. mali" pathogen load in plant tissue in dependence of plant organ, symptom expression, sampling season, endophytic colonization, or genotype of the host plant. In addition, methodological questions, such as the most suitable sampling procedure, the best tissue preparation technique, or the optimization of DNA isolation protocols, could be addressed [16]. Although the described protocol is specific for the detection of "*Ca.* P. mali" in apple plant tissue, indications are given, how the protocol could be adapted to the analysis of other obligate host-pathogen systems.

2 Materials

DNA samples from phloem tissue of phytoplasma-infected samples of apple trees are obtained by following a procedure described in Chapter 6 of this book (*see* **Note 1**). Prior to quantitative real-time PCR analysis, the quality and quantity of DNA is assessed by a NanoDrop Spectrophotometer ND-1000 (Thermo Scientific) or equivalent instrument. The DNA samples need to be free of contaminants that might inhibit DNA polymerase and show absorbance A260/A280 ratios between 1.7 and 1.9. Based on the spectrophotometric measurements, the concentration of each DNA sample is adjusted to 10 ng/µL in DNase-free molecular biology grade water (*see* **Note 2**). The DNA isolates to be analyzed by quantitative real-time PCR were previously tested positive for the presence of "*Ca.* P. mali." A total of 45 DNA samples can be analyzed in duplicate on a 96-well reaction plate, in addition to the control and calibrator samples.

All pre-PCR steps, including the preparation and aliquotation of reagents, are carried out under a laminar flow hood or PCR workstation. A clean laboratory coat and clean gloves, which were never in contact with PCR products, are worn. All tubes containing DNA or reagents are gently mixed and shortly centrifuged before opening. The reagent mixes and reaction plates are prepared by the same experienced operator until the conclusion of the study.

2.1 Real-Time PCR

The following materials are needed:

1. Sterile molecular biology grade water (*see* **Note 3**).

2. Low-retention filtered pipette tips and calibrated pipettes (that are exclusively dedicated for real-time PCR use) covering a volume from 0.1 µL to 1000 µL.

3. Primers and probe targeting the 16S rRNA gene of "*Ca.* P. mali" (AP-16S):

   Forward primer qAP-16S-F: 5'-CGAACGGGTGAGTAACAC GTAA-3'.
   Reverse primer qAP-16S-R: 5'-CCAGTCTTAGCAGT CGTTTCCA-3'.
   Probe qAP-16S: 5'-FAM–TAACCTGCCTCTTAGACG–MGB-NFQ-3' (*see* **Note 4**).

4. Primers and probe targeting the *M. domestica* gene for 1-aminocyclopropane-1-carboxylate oxidase (md-ACO1):

   Forward primer qMd-ACO-F: 5'-CCAGAATGTCGATAG CCTCGTT-3'.
   Reverse primer qMd-ACO-R: 5'-GGTGCTGGGCTGAT GAATG-3'.
   Probe qMd-ACO: 5'-VIC–TACAACCCAGGCAACG–MGB-NFQ-3' (*see* **Note 4**).

5. The working concentration of all the primers is 9 μM and of the probes 10 μM (*see* **Note 5**).

6. TaqMan Universal PCR Master Mix (2×) (Applied Biosystems) (*see* **Note 6**).

7. Optical 96-well reaction plate and optical adhesive film (*see* **Note 7**).

8. Two positive control samples, one of which is employed as a calibrator sample to set the threshold line (*see* **Note 8**).

9. Real-Time PCR instrument with filters calibrated for FAM, VIC and ROX dyes (e.g., 7500 Fast Real-Time PCR System).

3   Methods

3.1   Quantitative Real-Time PCR

1. Prepare a reaction mix for the respective number of samples to be analyzed as well as for the control/calibrator samples (*see* **Note 9**). Each sample is analyzed in duplicate in a 20 μL reaction volume, which contains 10 μL TaqMan Universal PCR Master Mix [2×], 2 μL of each primer qAP-16S-F [9 μM] and qAP-16S-R [9 μM], 0.44 μL of each primer qMd-ACO-F [9 μM] and qMd-ACO-R [9 μM], 0.4 μL of each TaqMan MGB probe qAP-16S [10 μM] and qMd-ACO [10 μM], and 2.32 μL of sterile molecular biology grade water. Vortex the reaction mix thoroughly and spin it down briefly in a centrifuge.

2. Transfer 18 μL of the reaction mix to the bottom of each well of a 96-well reaction plate (*see* **Note 10**).

3. Add 2 μL of template DNA (normalized to 10 ng/μL) directly into the reaction mix by pipetting up and down at least five

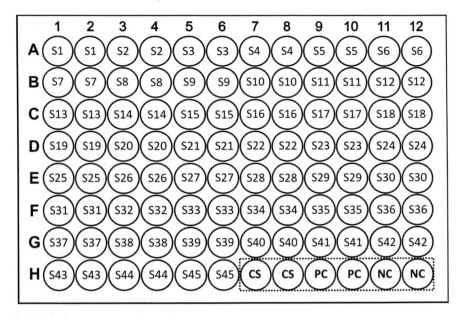

Fig. 1 96-well reaction plate showing the positions of the samples (S1 to S45), the calibrator sample (CS), the positive control (PC), and the negative control (NC)

times to mix well. Ninety positions on the reaction plate are destined for the 45 DNA samples to be quantified in duplicates (*see* **Note 11**) (Fig. 1).

4. Add 2 μL of each of the two positive control samples in duplicates and 2 μL of sterile molecular biology grade water (also in duplicate) as the negative control to the remaining positions of the 96-well reaction plate (Fig. 1).

5. Seal the 96-well reaction plate with the optical adhesive film (*see* **Note 12**).

6. Place the 96-well reaction plate into a centrifuge equipped with an adequate rotor and spin shortly to collect all the content on the bottom of each well. Control each well for the absence of air bubbles.

7. Place the 96-well reaction plate in the correct orientation into the real-time PCR instrument.

8. Programme the real-time PCR instrument for the run under the following conditions: 2 min at 50 °C, 10 min at 95 °C and 40 cycles of 15 s at 95 °C, and 1 min at 60 °C. The reaction volume per well is set to 20 μL (*see* **Note 13**).

3.2 Data Analysis

Data is analyzed using the 7500 Software Version 2.0.1 (Applied Biosystems) after the termination of the amplification reaction:

1. Apply the default automatic baseline setting for both targets, qAP-16S and qMd-ACO.

2. Adjust the threshold level manually for each target gene, qAP-16S and qMd-ACO (*see* **Note 14**) (Fig. 2). Display the amplification plots of one of the two positive controls that were defined as the calibrator sample for all the runs being part of the experiment. Drag the threshold line for each target gene and bring it in a position to cross the amplification plot of the calibrator sample at a specified C_T value. Make sure that the threshold line is positioned in the exponential phase of the amplification curve. Initially, the C_T value to set the threshold line can be chosen arbitrarily; however, it has to remain identical over different runs (e.g., in all the runs of this protocol the threshold line was set to cross the amplification plot of the calibrator sample at cycle number 20.0 and 23.2 for the pathogen target gene qAP-16S and for the host plant target gene qMd-ACO, respectively). Reanalyze the data by using the calibrator sample-adjusted threshold values for each of the two target genes to obtain the C_T values for the remaining samples analyzed on the 96-well reaction plate. Check the C_T values of the second positive control to monitor the reproducibility of data between different runs.

3. Control the C_T values of each replicate sample for both target genes. If the C_T values of the replicate samples differ by more than 0.5 for any target gene, the results need to be discarded and the analysis repeated.

4. Export the results into an Excel file.

5. Sort the data in Excel and calculate the average C_T value from the replicate data for each sample and each target gene.

6. Subtract the average C_T values of the two target genes for each sample ($\Delta C_T = C_T$ AP-16S – C_T Md-ACO1) [20]. The relative quantity data is expressed as the difference between the C_T of the pathogen AP-16S rRNA gene and the plant Md-ACO1 gene, where negative values indicate higher and positive values indicate lower phytoplasma loads per host plant cell.

7. Perform statistical analysis appropriate to the hypothesis being tested and interpret the results.

The here-described protocol is specific for quantitative analysis of "*Ca*. P. mali" in host plant tissues. However, the principle of relative quantification could be applied to other phytoplasma species or uncultivable plant pathogens if suitable amplification targets for simultaneous analysis of pathogen and host DNA were selected (*see* **Note 15**).

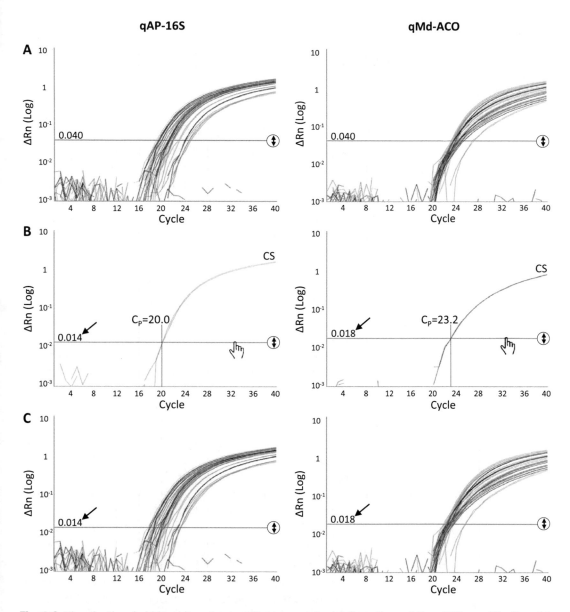

Fig. 2 Setting the threshold level by using a calibrator sample. (**a**) Open the real-time PCR amplification data for each target gene, qAP-16S and qMd-ACO, separately. (**b**) Display only the amplification plot of the calibrator sample (CS) analyzed in duplicate of each target gene. Drag the threshold line to cross the amplification plot at a predefined cycle in the exponential phase of the reaction that will be kept constant for all different amplification runs (in this case, cycle 20 was selected for target qAP-16S and cycle 23.2 for target qMd-ACO as the crossing point [C_P]). Record the threshold value for each target gene. (**c**) Display the amplification plot of all the samples analyzed on the 96-well reaction plate and reanalyze the amplification data by using the threshold values from step B for each target gene to obtain the C_T values that can be used for relative quantification

4   Notes

1. Phloem-enriched tissue should be used for DNA isolation. Depending on the plant organ, a sterile blade is used to obtain phloem shavings from branches or roots—after the removal of the outer bark—or to prepare midribs from leaves. Furthermore, leaf petioles can be subjected to DNA extraction.

2. DNA samples are adjusted to the concentration of 10 ng/μL shortly before the quantitative real-time PCR analyses are initiated, and kept at 4 °C until the termination of the analyses, which should be completed over a span of a few weeks. If diluted DNA samples are maintained for longer periods in the fridge or if they are stored at −20 °C, they should be re-quantified by spectrophotometry prior to use. Before performing spectrophotometric quantification or transferring DNA to the reaction mix, sample tubes need to be gently mixed by tipping and centrifuged by a short spin.

3. Commercial molecular biology grade water is recommended to be used for the dilution of primers and for preparing the quantitative real-time PCR reaction mix. In order to prevent contamination of larger stocks, the water is aliquoted under the laminar flow hood in smaller portions into sterile, individually wrapped microtubes (e.g., Eppendorf Biopur) and stored frozen until use.

4. Both probes were designed as TaqMan MGB probes, which are conjugated with a minor groove binder (MGB) and a non-fluorescent quencher (NFQ) at the 3′-end. These probes are much shorter than standard probes while highly specific and are characterized by lower background signal and higher sensitivity. Ordering another type of a dual-labeled fluorescent probe may have a negative effect on the melting temperature and the performance of the real-time PCR.

5. The primers and probes are diluted with DNase-free molecular biology grade water to obtain the working concentration of 9 μM and 10 μM, respectively. Primers and probes are aliquoted as single-use volumes into sterile, individually wrapped 0.5 mL microtubes and stored at −20 °C. Aliquots of TaqMan probes are transferred to sterile amber microtubes for light-sensitive liquids to prevent photobleaching. In order to maintain the quality of the probes and the primers, repeated freezing and thawing should be avoided.

6. Prepare single-use aliquots of the TaqMan Universal PCR Master Mix in 2 mL sterile, individually wrapped microtubes and store them at 4 °C. If the exact volume of TaqMan Universal PCR Master Mix (i.e., 1060 μL) required for a 96-well reaction plate is transferred to the microtube, all the other

reagents can just be added to the same tube when preparing the reaction mix for the real-time PCR analysis.

7. The 96-well reaction plates need to be suitable for real-time PCR and compatible with the specific real-time PCR instrument. Optical adhesive film or optical cap strips are required to ensure reproducible quantitative data.

8. The volume of the positive control samples has to be sufficient for repeated analysis on successive reaction plates, in order to analyze all the samples that are part of the quantification study with the same threshold setting conditions. For example, if 450 samples are to be analyzed in duplicate, at least ten 96-well reaction plates are required. For each duplicate analysis on a reaction plate, 2 µL of the positive control are needed or 4 µL per plate. Thus, at least 50 µL of positive control samples normalized to 10 ng/µL have to be prepared to analyze all the remaining samples and, if necessary, to repeat the analysis for up to 20% of the samples on two additional reaction plates. If larger volumes of positive control samples are normalized to the concentration of 10 ng/µL and stored in aliquots at −20 °C for forthcoming analyses, they should be re-quantified by spectrophotometry prior to use.

9. The reaction mix should be prepared for 110 percent of the samples in order to compensate for pipetting error. For example, if 96 samples/controls are analyzed on a reaction plate, the reaction mix is prepared for a total of 106 samples.

10. Since the accuracy of real-time PCR can be affected by imprecise pipetting, the most reproducible results are obtained when the reaction volume is aliquoted with a calibrated single-channel pipette to each well of the reaction plate and not with a dispenser.

11. The most reproducible results are obtained if the 2 µL volumes are transferred with a calibrated single-channel pipette and low-retention tips in the volume range of 0.1–2.5 µL. Multi-channel pipettes can compromise the reproducibility of replicate samples.

12. Use a film-sealing paddle or a plate roller to assure that all the wells of the reaction plate are closed. Pay particular attention to seal the wells on the borders and corners of the reaction plate.

13. The here-described protocol runs on a 7500 Fast Real-Time PCR System: The programme is set up as a *Standard Curve Experiment*, and *Quantitation* is selected as the experiment type. In the Methods & Materials screen, *Standard Curve* is selected as the quantitation method, *TaqMan Reagents* for the reagents, *Standard (~2 hours to complete a run)* for the ramp speed, and *gDNA (genomic DNA)* for the template type. The two targets, qAP-16S and qMd-ACO, are selected for each

sample. For qAP-16S, FAM is selected as the reporter and NFQ-NGB as the quencher. For qMd-ACO, VIC is selected as the reporter and NFQ-NGB as the quencher. All the DNA samples as well as the control/calibrator samples are indicated as *unknown samples*.

14. In my experience, the best performance and inter-assay reproducibility of relative quantification is achieved with a calibrator sample to set the threshold value in different amplification runs. Alternatively, the threshold could also be fixed manually at a defined level (e.g., 0.05) for both target genes and all amplification runs, if the amount of DNA of the calibrator sample is limited or samples are analyzed in different experiments over longer time periods. Also in this situation, the threshold level can be selected subjectively, but it has to be in the exponential phase of the amplification plot and it has to be kept constant over all different runs. However, it is not recommended to apply the default automatic threshold option implemented in the 7500 Fast Real-Time PCR analysis software, as it leads to higher degrees of variation between different runs compared to the two manual threshold-setting procedures [16, 21].

15. If the relative quantification assay without standard curves is to be adapted to a different pathogen-host system, new primers and probes need to be designed. The selection of target genes of the pathogen and the host organism, which are to be related to each other, is a critical step, as they should allow specific detection and occur in a defined copy number. Ribosomal RNA genes of the host plant may not be appropriate as a reference to relate the pathogen titer, as these genes are present in varying copy numbers within populations and, in dependence of the tissue type, the developmental stage or environmental factors, they can also vary within individual organisms (discussed in [12]). For this reason, the quantitative assay for "*Ca.* P. mali" employs Md-ACO1 as the reference gene, which was mapped as a single-copy gene to chromosome 10 in the haploid genome of *M. domestica*. After the design of primers and probe systems, the duplex real-time PCR has to be optimized and validated. It has to be confirmed that both target genes have comparable amplification efficiencies and that the performance of the duplex reaction does not differ from singleplex reactions.

Acknowledgments

The author is grateful to J. Dalla Via for critically reading and discussing the manuscript.

References

1. Ravnikar M, Mehle N, Gruden K, Dreo T (2016) Real-time PCR. In: Boonham N, Tomlinson J, Mumford R (eds) Molecular methods in plant disease diagnostics: principles and protocols. CABI Wallingford, Oxfordshire, Boston, pp 28–58

2. Bustin SA, Nolan T (2004) Chemistries. In: Bustin SA (ed) A–Z of quantitative PCR. International University Line, La Jolla, pp 215–278

3. Pfaffl MW (2004) Quantification strategies in real-time PCR. In: Bustin SA (ed) A–Z of quantitative PCR. International University Line, La Jolla, pp 87–120

4. Baric S, Dalla Via J (2004) A new approach to apple proliferation detection: a highly sensitive real-time PCR assay. J Microbiol Methods 57:135–145

5. Galetto L, Bosco D, Marzachì C (2005) Universal and group-specific real-time PCR diagnosis of flavescence dorée (16Sr-V), bois noir (16Sr-XII) and apple proliferation (16Sr-X) phytoplasmas from field-collected plant hosts and insect vectors. Ann Appl Biol 147:191–201

6. Pelletier C, Salar P, Gillet J, Cloquemin G, Very P, Foissac X, Malembic-Maher S (2009) Triplex real-time PCR assay for sensitive and simultaneous detection of grapevine phytoplasmas of the 16SrV and 16SrXII-A groups with an endogenous analytical control. Vitis 48:87–95

7. Linck H, Krüger E, Reineke A (2017) A multiplex TaqMan qPCR assay for sensitive and rapid detection of phytoplasmas infecting *Rubus* species. PLoS One 12:e0177808

8. Christensen NM, Nicolaisen M, Hansen M, Schulz A (2004) Distribution of phytoplasmas in infected plants as revealed by real-time PCR and bioimaging. Mol Plant-Microbe Interact 17:1175–1184

9. Torres E, Bertolini E, Cambra M, Montón C, Martín MP (2005) Real-time PCR for simultaneous and quantitative detection of quarantine phytoplasmas from apple proliferation (16SrX) group. Mol Cell Probes 19:334–340

10. Saracco P, Bosco D, Veratti F, Marzachì C (2006) Quantification over time of chrysanthemum yellows phytoplasma (16Sr-I) in leaves and roots of the host plant *Chrysanthemum carinatum* (Schousboe) following inoculation with its insect vector. Physiol Mol Plant Pathol 67:212–219

11. Bisognin C, Schneider B, Salm H, Grando MS, Jarausch W, Moll E, Seemüller E (2008) Apple proliferation resistance in apomictic rootstocks and its relationship to phytoplasma concentration and simple sequence repeat genotypes. Phytopathology 98:153–158

12. Baric S, Berger J, Cainelli C, Kerschbamer C, Letschka T, Dalla Via J (2011) Seasonal colonisation of apple trees by 'Candidatus Phytoplasma mali' revealed by a new quantitative TaqMan real-time PCR approach. Eur J Plant Pathol 129:455–467

13. Jawhari M, Abrahamian P, Sater AA, Sobh H, Tawidian P, Abou-Jawdah Y (2015) Specific PCR and real-time PCR assays for detection and quantitation of 'Candidatus Phytoplasma phoenicium'. Mol Cell Probes 29:63–70

14. Arratia-Castro AA, Santos-Cervantes ME, Arce-Leal ÁP, Espinoza-Mancillas MG, Rodríguez Negrete EA, Méndez-Lozano J, Arocha-Rosete Y, Leyva-López NE (2016) Detection and quantification of 'Candidatus Phytoplasma asteris' and 'Candidatus Liberibacter asiaticus' at early and late stages of Huanglongbing disease development. Can J Plant Pathol 38:411–421

15. Rutledge RG, Côté C (2003) Mathematics of quantitative kinetic PCR and the application of standard curves. Nucleic Acids Res 31:e93

16. Baric S (2012) Quantitative real-time PCR analysis of 'Candidatus Phytoplasma mali' without external standard curves. Erwerbs-Obstbau 54:147–153

17. Schmittgen TD, Livak KJ (2008) Analyzing real-time PCR data by the comparative C_T method. Nat Protoc 3:1101–1108

18. Nolan T, Hands RE, Bustin SH (2006) Quantification of mRNA using real-time RT-PCR. Nat Protoc 1:1559–1582

19. Hogenhout SA, Oshima K, Ammar ED, Kakizawa S, Kingdom HN, Namba S (2008) Phytoplasmas: bacteria that manipulate plants and insects. Mol Plant Pathol 9:403–423

20. Gachon CM, Strittmatter M, Müller DG, Kleinteich J, Küpper FC (2009) Detection of differential host susceptibility to the marine oomycete pathogen *Eurychasma dicksonii* by real-time PCR: not all algae are equal. Appl Environ Microbiol 75:322–328

21. Liu ZL, Palmquist DE, Ma MG, Liu J, Alexander NJ (2009) Application of a master equation for quantitative mRNA analysis using qRT-PCR. J Biotechnol 143:10–16

Chapter 11

A Multiplex-PCR Method for Diagnosis of AY-Group Phytoplasmas

Shigeyuki Kakizawa

Abstract

Polymerase chain reaction (PCR) methods using phytoplasma-specific primers are widely used to detect phytoplasmas from infected plants and insects. Here, I describe a method of multiplex-PCR to amplify nine gene fragments in PCR reactions from AY-group phytoplasmas. Strain-identification was possible after electrophoresis and direct sequencing was also possible after PCR. The combinations of primers can be easily modified, so this method could be applied to other phytoplasma strains.

Key words PCR, Multiplex-PCR, AY-group phytoplasma, Detection, Strain-identification, Direct sequencing

1 Introduction

Polymerase chain reaction (PCR) is a powerful tool to detect phytoplasmas from infected hosts and identify the phytoplasma strains because it is highly sensitive and easily handled [1–4]. In addition, several secondary experiments such as restriction fragment length polymorphism (RFLP), direct sequencing, phylogenetic analyses, etc., could be possible after PCR amplification. Many phytoplasma genes were reported as PCR targets, including genes encoding 16S rDNA [5, 6], ribosomal proteins [3, 7–10], the protein secretion machinery SecY [9–11], the elongation factor TufB [4, 12, 13], the molecular chaperon GroEL [14] (*see* Chapters 8 and 9). Further target genes would be required for the specific detection of many phytoplasma strains.

Multiplex-PCR allows the amplification of multiple DNA fragments in a single reaction [15, 16] and for this reason it has been used to detect several pathogens or genes [15, 17–20]. More than 15 fragments can be successfully amplified in a reaction [17] and fragments as large as 2.0 kbp can also be amplified. In addition, the combinations of primers can be easily modified and secondary experiments are also possible after multiplex-PCR [21, 22].

Rita Musetti and Laura Pagliari (eds.), *Phytoplasmas: Methods and Protocols*, Methods in Molecular Biology, vol. 1875, https://doi.org/10.1007/978-1-4939-8837-2_11, © Springer Science+Business Media, LLC, part of Springer Nature 2019

Previously we showed a method of multiplex-PCR to amplify nine gene fragments in PCR reactions from 16SrI group phytoplasmas [23]. Here I show detailed methods for the multiplex-PCRs for AY-group phytoplasma.

2 Materials

2.1 Multiplex-PCR

1. TE buffer: 5 mM Tris–HCl pH 8.0, 1 mM EDTA.
2. Primers: 100 µM solutions of each primer (*see* **Note 1**).
3. Primer-mixture solution: 2 µM of all primers in a multiplex-PCR reaction (*see* **Note 2**).
4. Multiplex PCR kit (*see* **Note 3**).
5. 200 µL PCR tubes.
6. Thermal cycler.

2.2 Electrophoresis

1. TAE buffer: 40 mM Tris, 40 mM acetate, and 1 mM EDTA, pH 8.3.
2. 2% agarose gel with TAE buffer (*see* **Note 4**).
3. LoadingQuick 100 bp DNA Ladder.
4. Loading dye.
5. GelRed or EtBr.
6. Trans-illuminator.

2.3 Sequencing

1. ExoSAP-IT (*see* **Note 5**).
2. Cycle Sequencing Kit.
3. DNA purification column after the cycle sequencing reaction.
4. An ABI 3130×l Genetic Analyzer or other capillary sequencers.
5. CLC Main workbench, a genetic analysis software.

3 Methods

3.1 Primer Design

1. Annealing temperatures of all primers should be almost same, around 50–60 °C. Since Phytoplasma genomes are AT-rich, primers tend to be long, e.g., 20–30 bp (*see* **Notes 6–8**). It is easy to use software like Primer3 [24] to design primers.
2. Amplified DNA fragments by PCR should be approximately 200–1500 bp.
3. Amplified fragments were designed to have 80–100 bp differences between each other so that they could be distinguished by electrophoresis (*see* **Note 9**).

3.2 Multiplex-PCR, Electrophoresis

1. Make multiplex-PCR reaction mixture in 200 μL PCR tubes. Mix 7.5 μL of 2× Master Mix of Qiagen multiplex PCR kit (*see* **Note 10**), 1.5 μL of 2 μM Primer Mix (*see* **Note 11**), 1.0 μL of template DNA solution (*see* **Note 12**), and 6.0 μL of water, for a total volume of 15 μL.

2. Put PCR tubes on a thermal cycler and heat the lid.

3. Run PCR reaction as follows:

   94 °C for 15:00.
   35 times of 94 °C for 0:30, 52 °C for 1:30, 68 °C for 2:00 (*see* **Note 13**).
   68 °C for 3:00.
   4 or 10 °C forever

4. After PCR, mix each PCR product with a loading dye according to the manufacturer's instructions.

5. Thaw LoadingQuick 100 bp DNA Ladder, a DNA marker of electrophoresis according to the manufacturer's instructions.

6. Load the samples and the DNA ladder in 2% agarose gels and run an electrophoresis, at 100 V for 45 min at room temperature.

7. Stain the gel with a dye, GelRed, or EtBr for 20–30 min (*see* **Note 14**).

8. Examine by UV trans-illuminator.

3.3 Direct Sequencing of Multiplex-PCR Products

1. Thaw ExoSAP solution and dilute ExoSAP-IT solutions by 16-fold with water (*see* **Notes 5** and **15**).

2. Mix 5 μL of PCR products and 2 μL of ExoSAP-IT solution in PCR tubes.

3. Set the PCR tubes on a thermal cycler and run it as follows:

   37 ° C for 30:00 (treatment step) (*see* **Note 16**).
   80 ° C for 15:00 (inactivation of enzymes).
   4 or 10 °C forever

4. The solution will be directly used for sequencing reaction. Mix DNA, primer, BigDye Terminator v3.1 Cycle Sequencing Kit and buffer as follows: 0.5 μL of BigDye Ready Reaction Mix, 1.75 μL of BigDye Sequencing Buffer (*see* **Note 17**), 0.32 μL of 5 μM primer solution, 10–40 ng of purified PCR product (up to 7.75 μL) and add water until total amount is 10 μL (the amount of water is 7.75 μL minus amount of PCR product).

5. Run a cycle in a thermal cycler as follows:

   96 °C for 1:00.
   25 times of 96 °C for 0:10, 50 °C for 0:05, 60 °C for 4:00.
   4 or 10 °C forever.

6. Purify the products with DNA purification column according to the manufacturer's instructions.

7. Run the samples on an ABI 3130 × L Genetic Analyzer or other capillary sequencers according to the manufacturer's instructions (*see* **Note 18**).

8. Analyze resulted files with CLC main workbench or other software.

4 Notes

1. All primers were synthesized and dissolved as 100 μM with TE or water.

2. When nine fragments were aimed to be amplified in a multiplex reaction, 18 primers were mixed together to make the primer-mixture solution. 5 μL of each 100 μM primer stock solution was mixed together, then TE buffer added up to the total volume as 250 μL. When 18 primers were mixed, 90 μL of primers (5 μL × 18) plus 160 μL of TE were mixed. The concentration of each primer was 2 μM. In single PCR analyses, only 1 primer set was added.

3. Alternatively normal PCR with rTaq or exTaq would be also possible for multiplex-PCRs.

4. The 2% agarose solution is difficult to resolve. At first, add TAE buffer in a Flask, rotate a magnetic stirrer in it, then add agarose powder. Several minutes later, heat the solution by microwave. To dissolve the agarose completely, run an autoclave at 121 °C for 10 min is recommended.

5. ExoSAP-IT solution could be diluted. We tested several dilutions and found that up to 16-fold dilution makes no differences in sequencing results (data not shown). Dilution could be done with just water. The diluted solution must be used in the same day; it is not recommended to keep it in freezer or refrigerator.

6. It is better to include a number of G and C nucleotides in primer-annealing regions. It is better to check primer dimers or hairpin structures with primer analysis software by lots of oligo venders, but it is not necessary.

7. Previously there have been a lot of phytoplasma-specific primes to amplify 16S rDNA, SecY, ribosomal proteins, etc. These primers could be also used for the multiplex-PCRs.

8. We designed all primers on single-copy-genes in "*Candidatus* Phytoplasma asteris" genome [23]. We thought that if primers could anneal with multiple loci of genome, it is difficult to judge whether the target gene was correctly amplified or not. However in most cases, it is not necessary to design primers on single-copy genes.

9. DNA fragment with 80–100 bp differences could be distinguished by 2% agarose gel electrophoresis, but not by normal 0.7% or 1% gels. If there are only 2 or 3 amplified fragments, it is better to have ca. 500 bp differences between each fragment, so that it would be easily distinguished by normal gels.

10. Normal PCR conditions with rTaq or exTaq would be also possible for multiplex-PCRs. In this case, it is better to pre-examine the combinations of primer sets in a reaction.

11. In single PCR analyses, only 1 primer set was added and other conditions were completely similar to those of the multiplex-PCR.

12. The purity of extracted DNA is very important in multiplex-PCRs (and even in normal PCRs). In many cases, extracted DNAs from infected plants and insects contain inhibitory materials, such as polyphenols, proteins, lipids, etc. Using less amount of template DNA in PCR reactions is often good way to obtain bands. As a template, various DNA concentrations could be acceptable for successful PCR amplification (e.g., 0.1–100 ng/μL DNA), since most DNAs were derived from plant genome and organelle. It is difficult to know the amount of phytoplasma DNA since it must be varied in each extraction experiment; therefore, DNA concentration of the template DNAs might not be directly related to the successful PCR results.

13. Several annealing temperature should be examined since the best annealing temperature would be varied and it depends on each primer set.

14. EtBr or other dyes are also fine.

15. PCR-amplified fragments were purified using ExoSAP-IT (GE Healthcare), which inactivates primers and nucleotides by exonuclease and shrimp alkaline phosphatase. Alternatively, fragments were purified by other methods, e.g., column purification method, gel-extraction method, phenol-chloroform extraction, magnetic bead purification, etc. However, the ExoSAP-IT treatment is good because it is easy and DNA amount will not be decreased during procedures.

16. The treatment time of ExoSAP-IT is 15 min according to the manufacturer's instructions, but I recommend extending it to 30 min when the reagent was diluted.

17. This is a diluted protocol: the sequencing reaction solution could be diluted and total amount of reaction solution could be half (10 μL) of original protocol (20 μL). Using less amount of BigDye Kit makes sequencing signals weaker, but it is usually enough to read 600–900 bp. This protocol is also good because it can decrease possibilities to let capillaries dirty with

lots of amount of fluorescent dyes when excess amount of template DNA was used in the cycle-sequencing reactions. Thus it could save the effort to strictly control of DNA amount in the cycle-sequencing reactions.

18. The manufacturer's instructions recommend replacing the solvent of DNA from water to the Hi-Di Formamide, but DNA with water could be directly applied to the ABI Genetic Analyzer. There was almost no significant difference between water and formamide in sequencing results. It might be because phytoplasma genes are usually highly AT-rich.

References

1. Win N, Lee Y, Kim Y, Back C, Chung H, Jung H (2012) Reclassification of aster yellows group phytoplasmas in Korea. J Gen Plant Pathol 78(4):264–268

2. Namba S, Kato S, Iwanami S, Oyaizu H, Shiozawa H, Tsuchizaki T (1993) Detection and differentiation of plant-pathogenic mycoplasmalike organisms using polymerase chain-reaction. Phytopathology 83(7):786–791

3. Lee IM, Gundersen-Rindal DE, Davis RE, Bartoszyk IM (1998) Revised classification scheme of phytoplasmas based an RFLP analyses of 16S rRNA and ribosomal protein gene sequences. Int J Syst Bacteriol 48:1153–1169

4. Marcone C, Lee IM, Davis RE, Ragozzino A, Seemuller E (2000) Classification of aster yellows-group phytoplasmas based on combined analyses of rRNA and tuf gene sequences. Int J Syst Evol Microbiol 50 (Pt 5):1703–1713

5. Gundersen DE, Lee IM, Rehner SA, Davis RE, Kingsbury DT (1994) Phylogeny of mycoplasmalike organisms (phytoplasmas): a basis for their classification. J Bacteriol 176 (17):5244–5254

6. IRPCM (2004) 'Candidatus Phytoplasma', a taxon for the wall-less, non-helical prokaryotes that colonize plant phloem and insects. Int J Syst Evol Microbiol 54(Pt 4):1243–1255

7. Toth KF, Harrison N, Sears BB (1994) Phylogenetic relationships among members of the class *Mollicutes* deduced from rps3 gene sequences. Int J Syst Bacteriol 44(1):119–124

8. Jomantiene R, Davis RE, Maas J, Dally EL (1998) Classification of new phytoplasmas associated with diseases of strawberry in Florida, based on analysis of 16S rRNA and ribosomal protein gene operon sequences. Int J Syst Bacteriol 48(Pt 1):269–277

9. Lee IM, Bottner-Parker KD, Zhao Y, Bertaccini A, Davis RE (2012) Differentiation and classification of phytoplasmas in the pigeon pea witches'-broom group (16SrIX): an update based on multiple gene sequence analysis. Int J Syst Evol Microbiol 62(Pt 9):2279–2285

10. Davis RE, Zhao Y, Dally EL, Lee IM, Jomantiene R, Douglas SM (2012) 'Candidatus Phytoplasma pruni', a novel taxon associated with X-disease of stone fruits, Prunus spp.: multilocus characterization based on 16S rRNA, secY, and ribosomal protein genes. Int J Syst Evol Microbiol

11. Lee IM, Zhao Y, Bottner KD (2006) SecY gene sequence analysis for finer differentiation of diverse strains in the aster yellows phytoplasma group. Mol Cell Probes 20(2):87–91

12. Lee IM, Gundersen-Rindal DE, Davis RE, Bottner KD, Marcone C, Seemuller E (2004) 'Candidatus Phytoplasma asteris', a novel phytoplasma taxon associated with aster yellows and related diseases. Int J Syst Evol Microbiol 54(Pt 4):1037–1048

13. Malembic-Maher S, Salar P, Filippin L, Carle P, Angelini E, Foissac X (2011) Genetic diversity of European phytoplasmas of the 16SrV taxonomic group and proposal of 'Candidatus Phytoplasma rubi'. Int J Syst Evol Microbiol 61 (Pt 9):2129–2134

14. Mitrovic J, Kakizawa S, Duduk B, Oshima K, Namba S, Bertaccini A (2011) The groEL gene as an additional marker for finer differentiation of 'Candidatus Phytoplasma asteris'-related strains. Ann Appl Biol 159(1):41–48

15. Edwards MC, Gibbs RA (1994) Multiplex PCR: advantages, development, and applications. PCR Methods Appl 3(4):S65–S75

16. Chamberlain JS, Gibbs RA, Ranier JE, Nguyen PN, Caskey CT (1988) Deletion screening of the Duchenne muscular dystrophy locus via multiplex DNA amplification. Nucleic Acids Res 16(23):11141–11156

17. Caliendo AM (2011) Multiplex PCR and emerging technologies for the detection of respiratory pathogens. Clin Infect Dis 52 (Suppl 4):S326–S330

18. Kim Y, Win N, Back C, Yea M, Yim K, Jung H (2011) Multiplex PCR assay for simultaneous detection of Korean quarantine Phytoplasmas. Plant Pathol 27(4):367–371

19. Gibson DG, Benders GA, Axelrod KC, Zaveri J, Algire MA, Moodie M, Montague MG, Venter JC, Smith HO, Hutchison CA 3rd (2008) One-step assembly in yeast of 25 overlapping DNA fragments to form a complete synthetic *Mycoplasma genitalium* genome. Proc Natl Acad Sci U S A 105 (51):20404–20409

20. Tao Y, Man J, Wu Y (2012) Development of a multiplex polymerase chain reaction for simultaneous detection of wheat viruses and a phytoplasma in China. Arch Virol 157 (7):1261–1267

21. Manam S, Nichols WW (1991) Multiplex polymerase chain reaction amplification and direct sequencing of homologous sequences: point mutation analysis of the ras genes. Anal Biochem 199(1):106–111

22. Lo KW, Mok CH, Chung G, Huang DP, Wong F, Chan M, Lee JC, Tsao SW (1992) Presence of p53 mutation in human cervical carcinomas associated with HPV-33 infection. Anticancer Res 12(6B):1989–1994

23. Kakizawa S, Kamagata Y (2014) A multiplex-PCR method for strain identification and detailed phylogenetic analysis of AY-group phytoplasmas. Plant Dis 98(3):299–305. https://doi.org/10.1094/PDIS-03-13-0216-RE

24. Rozen S, Skaletsky H (2000) Primer3 on the WWW for general users and for biologist programmers. In: Krawetz S, Misener S (eds) Bioinformatics methods and protocols in the series methods in molecular biology. Humana Press, Totowa, NJ, pp 365–386

Chapter 12

One-Step Multiplex Quantitative RT-PCR for the Simultaneous Detection of Viroids and Phytoplasmas

Ioanna Malandraki, Christina Varveri, and Nikon Vassilakos

Abstract

A one-step multiplex quantitative reverse transcription polymerase chain reaction protocol is described, for the detection in pome trees of *Pear blister canker viroid* and *Apple scar skin viroid*, together with universal detection of phytoplasmas. Total nucleic acids extraction is performed according to a modified CTAB protocol and TaqMan MGB probes are used to surpass high genetic variability of viroids. The multiplex real-time assay is at least ten times more sensitive than conventional protocols and its features make it suitable for rapid and massive screening of pome fruit trees phytoplasmas and viroids in certification schemes and surveys.

Key words Real-time PCR, Universal detection, Multiplex detection, MGB probes, "*Candidatus* phytoplasma mali", "*Candidatus* phytoplasma pyri", Viroids

1 Introduction

Large surveys and certification schemes for propagation material require methods capable of a fast and reliable detection of plant pathogens. Typically, a large number of harmful organisms need to be screened in each occasion. Hence, the availability of multiplex assays capable of the simultaneous detection of more than one target-pathogen per sample provides rapidity and cost-effectivity of the diagnosis. Novel chemistries and improved instrumentation platforms have made multiplexing possible for quantitative PCR (qPCR) [1]. Generally, qPCR has been proved to be more sensitive than conventional PCR, less laborious (no post-PCR processing required), and with reduced risk of carry-over contaminations [2].

Viroids and phytoplasmas infecting pome fruit trees are important pathogens, recommended for regulation as quarantine pests in Europe in the EPPO A2 list [3]. This protocol is based on the assay developed by Malandraki et al. [4] and describes the simultaneous detection of *Pear blister canker viroid* (PBCVd), *Apple scar skin viroid* (ASSVd), and phytoplasmas in pome fruit trees, using a

Rita Musetti and Laura Pagliari (eds.), *Phytoplasmas: Methods and Protocols*, Methods in Molecular Biology, vol. 1875, https://doi.org/10.1007/978-1-4939-8837-2_12, © Springer Science+Business Media, LLC, part of Springer Nature 2019

multiplex RT-qPCR. The RNA of viroids was effectively co-extracted with phytoplasmic DNA from shoots using a modified CTAB protocol described by Ahrens and Seemüller [5]. The usage of Minor Groove Binder (MGB) quenchers allowed the design of shorter TaqMan probes and made it possible to overcome the obstacle of the high genetic variability of viroids. Universal detection of phytoplasmas was achieved using primers and TaqMan probe as described by Christensen et al. [6] with the difference that the probe was MGB modified.

2   Materials

1. Type 1 Ultrapure water, pyrogen-, nuclease-, protease-, and bacteria-free.

2. Grinding buffer: 125 mM potassium phosphate, 30 mM ascorbic acid, 10% sucrose, 0.15% BSA, 2% PVP-10, adjust to pH 7.6, filter sterilize, store at 4 °C.

3. CTAB buffer: 2% CTAB, 1.4 M NaCl, 20 mM EDTA pH 8.0, 100 mM Tris–HCl pH 8.0, autoclave, store at 4 °C. Before use, heat CTAB buffer at 60 °C until it becomes transparent, transfer in new glass bottle the amount needed for the current extraction, add 2-Mercaptoethanol (0.2%) and use immediately (*see* **Note 1**).

4. Chloroform:isoamyl alcohol (24:1 v/v).

5. 70% ice-cold absolute ethanol.

6. One Step PrimeScript™ RT-PCR Kit (Perfect Real Time, TAKARA) (*see* **Note 2**).

7. 50 mM $MgCl_2$ provided with any commercial PCR kit. It is required in the setup of the RT-qPCR reaction (*see* **Note 2**).

8. Oligonucleotide primers (Table 1).

9. TaqMan MGB probes labeled with distinct reporter dyes (Table 1).

10. Flake ice (laboratory flaker ice machine).

11. Set of micro-pipettes (preferentially autoclavable) covering a range of volumes from 0.1 μL to 1 mL and their respective filter tips.

12. Lancet and sterile blades.

13. Extraction bags for ELISA tissue extraction (BIOREBA art. no. 430100 or equivalent, preferentially from heavy duty plastic with synthetic intermediate layer for optimal filtration) (*see* **Note 3**).

Table 1
Sequence of primers and probes designed for simultaneous detection of *Pear blister canker viroid* (PBCVd), *Apple scar skin viroid* (ASSVd), and universal detection of phytoplasmas in a single-tube RT-qPCR reaction. Probes incorporate a 5′ reporter dye and a 3′ minor groove binder (MGB) moiety along with a nonfluorescent quencher (NFQ)

Target	Primer/Probe name	Sequence (5′–3′)	Reference
PBCVd	PB-F	CGCGCGGCTGTGAGTAAT	[4]
	PB-R	GGCTCAGGCAGGAAGCAA	
	PB-P	**VIC**-TGGAGAAGAAAACCAGC-MGB NFQ	
ASSVd	ASS-F	CCCCTGTTCTCTCACGCTCTT	
	ASS-R	TTTACCGGGAAACACCTATTGTGT	
	ASS-P	**NED**-TGACGCAGCGGCG-MGB NFQ	
Phytoplasma (universal)	UPhy-F	CGTACGCAAGTATGAAACTTAAAGGA	[6]
	UPhy-R	TCTTCGAATTAAACAACATGATCCA	
	UPhy-P	**FAM**-TGACGGGACTCCGCACA-MGB NFQ	

14. Tissue homogenizer for ELISA extraction bags (BIOREBA HOMEX 6 with standard rack, Art. No. 400014, or equivalent) (*see* **Note 3**).

15. 5 mL centrifuge tubes.

16. Fume hood.

17. Refrigerated benchtop centrifuge, set at 4 °C.

18. Heat block or water bath, set at 60 °C.

19. Vortex mixer.

20. Pathogen tested-negative plant tissues and viroid/phytoplasma infected positive controls of the same species of the tested samples.

21. Tube- and cap-strips (or 96-well plates with cover films) suitable for qPCR.

22. Mini-centrifuge for tube strips.

23. Microvolume spectrophotometer.

24. Quantitative PCR apparatus (STEP-ONE PLUS Applied Biosystems).

3   Methods

3.1   Nucleic Acids Extraction

1. Debark shoots using a lancet with sterile blade and place pieces of vascular tissue (130 mg) into ELISA extraction bags. Keep bags on ice. Change blade between samples.

2. Add 4 mL ice-cold grinding buffer.

3. Homogenize once shortly and keep on ice for 10 min (or more, until all samples are homogenized).

4. Homogenize again thoroughly and keep bags on ice.

5. Transfer homogenate in ice-cold 5 mL centrifuge tubes.

6. Centrifuge at $1,100 \times g$ for 5 min at 4 °C.

7. Transfer supernatant in new ice-cold tubes (*see* **Note 4**).

8. Centrifuge at $14,000 \times g$ for 25 min at 4 °C (meanwhile add 2-Mercaptoethanol to heated CTAB buffer as described in Subheading 2).

9. Keep pellet.

10. Work under fume hood from this step onward.

11. Add 400 µL heated CTAB buffer with added 0.2% 2-Mercaptoethanol, pipet gently, close well lids, and then vortex to resuspend the pellet.

12. Place at 60 °C (heatblock or water bath) for 30 min, vortex every 10 min (*see* **Note 5**).

13. Centrifuge shortly to clear the cap of the tube.

14. Add 400 µL chloroform:isoamyl alcohol (24:1 v/v) and vortex to obtain an emulsion.

15. Centrifuge at $11,000 \times g$ for 10 min at 4 °C.

16. Transfer the supernatant in a new tube.

17. Add 2/3 V ice-cold isopropanol and vortex.

18. Centrifuge at $16,000 \times g$ for 20 min at 4 °C.

19. Keep pellet. Fume hood is not necessary from this step onward.

20. Add 900 µL 70% ice-cold ethanol and invert gently several times to wash the pellet.

21. Centrifuge at $16,000 \times g$ for 15 min at 4 °C.

22. Remove ethanol and dry pellet at room temperature for about 6 min or until it becomes transparent.

23. Resuspend pellet by gently pipetting in 50 µL Type 1, RNase/ DNase-free water (*see* **Note 6**).

24. Store extracted nucleic acids at −30 °C for short periods (up to a month) or −80 °C for longer periods, until use.

3.2 Reverse Transcription Quantitative PCR (RT-qPCR)

1. Let all reagents and nucleic acids thaw on ice. Centrifuge briefly tubes before use. Probes should be protected from light and thawed just before use.

2. Mix all reagents well by flicking and centrifuge briefly. Use sterile filter tips. Work on ice (*see* **Note 7**).

Table 2
Final concentration of RT-qPCR reaction components in the master mix

Reaction component	Stock concentration	Final concentration
One Step RT-PCR Buffer III (includes dNTP Mixture, Mg2+)	2×	1×
ROX Reference Dye (optional)	50×	1×
MgCl₂ (*see* **Note 3**)		2 mM
Primer PB-F		200 nM
Primer PB-R		200 nM
Primer ASS-F		200 nM
Primer ASS-R		200 nM
Primer UPhy-F		300 nM
Primer UPhy-R		300 nM
Probe PB-P		100 nM
Probe ASS-P		100 nM
Probe UPhy-P		150 nM
TaKaRa Ex Taq HS	5 µ/µL	0.1 µ/µL
PrimeScript RT enzyme Mix II (includes RNase inhibitor)		0.4 µL
Total nucleic acids	150–400 ng/µL	1 µL
Type 1, RNase/DNase-free water		Add to 20 µL final reaction volume

3. Prepare a master mix to a final volume according to the number of reactions performed, following the RT-qPCR kit manual instructions. Estimate excess of master mix depending on the number of reactions and the accuracy of used pipettes, to compensate pipetting errors. Final concentrations of reaction components are summarized in Table 2. Reaction volume is set at 20 µL. Each assay should include non-template, negative, and positive controls as well as the test samples, all in duplicates or triplicates.

4. Prepare master mix by gentle pipetting. Spin down all droplets and dissolve bubbles by centrifugation at low speed. Load 19 µL of the master mix into each tube. Add 1 µL of the total nucleic acids (*see* **Note 6**) into the reaction mix, pipette very gently twice up and down, taking precautions to avoid carry-over contaminations. Close caps carefully. Mix by flicking the

tube strip. Centrifuge strips at low speed and make sure that caps are clear and there are no bubbles remaining (*see* **Note 8**).

5. Load tube strips onto the real-time PCR apparatus.

6. Assign targets to the reporter dyes and include ROX dye as a passive reference for normalization by background subtraction (optional, depending on the instrument). Set the steps of the amplification profile as follows: reverse transcription incubation time at 42 °C for 30 min, 95 °C for 2 min, 40 cycles of 95 °C for 15 s, and 60 °C for 1 min (collection of data at this point).

7. Collect and analyze data according to qPCR apparatus manual instructions (*see* **Note 9**).

4   Notes

1. Handle 2-Mercaptoethanol under fume hood.

2. This protocol has been developed using this specific kit. In case of using a different kit new optimization of the method might be required. The addition of 2 mM $MgCl_2$ improves the simultaneous detection of the three targets and it is recommended.

3. Another system that could be used for tissue homogenization is mortar and pestle. Mortars and pestles should be kept at 4 °C before use.

4. At this step, to avoid tip obstruction by tissue particles, it is often useful to use sterile, already cut at the extreme end with clean scissors pipette tips.

5. It is advised to secure that caps are tightly closed before performing vortex.

6. Our experience shows that with this extraction protocol the yield of total nucleic acids is usually 10–20 μg (200–400 ng/μL). This yield depends on the species of the sampled tissue (e.g., pear or apple), on the age of the shoot, on the condition of the plant etc. Using 1 μL within the above-mentioned range of concentrations for the RT-qPCR, has always been proven effective in our hands. Therefore, it is recommended to verify the quality and quantity of some randomly selected total nucleic acid extracts using a microvolume spectrophotometer before proceeding to the RT-qPCR.

7. Fine mixing all reagents in the master mix is essential. Probes tend to float while enzymes gather in the bottom of the tube. To avoid bubbling, gentle pipetting is required.

8. To avoid cross-contamination, it is useful to keep nucleic acids in a separate ice box. A stand for the tube strips placed firmly on ice is also helpful. Finally, closing the caps of the tubes should be done gently.

9. This protocol has been proven to detect PBCVd down to 10^3 copies, ASSVd to 10^4 copies, and "*Ca*. P. mali" to 10^4 copies. Depending on the target is 10–100 times more sensitive than conventional RT-PCR/PCR protocols [4].

Acknowledgment

This work was supported by the Hellenic GSRT Project "09SYN-22-638" and partially by FP7- REGPOT-2008-1 project "BPI-PlantHeal 230010."

References

1. Shipley G (2006) An introduction to real-time PCR. In: Dorak MT (ed) Real-time PCR. Taylor & Francis Group, New York, pp 1–37

2. Heid CA, Stevens J, Livak KJ, Williams PM (1996) Real time quantitative PCR. Genome Res 6:986–994

3. EPPO Standards (2017) EPPO A1 and A2 lists of pests recommended for regulation as quarantine pests, PM 1/2(26) English, pp 8–9

4. Malandraki I, Varveri C, Olmos A, Vassilakos N (2015) One-step multiplex quantitative RT-PCR for the simultaneous detection of viroids and phytoplasmas of pome fruit trees. J Virol Methods 213:12–17

5. Ahrens U, Seemüller E (1992) Detection of DNA of plant pathogenic mycoplasmalike organisms by a polymerase chain reaction that amplifies a sequence of the 16S rRNA gene. Phytopathology 82:828–832

6. Christensen NM, Nyskjold H, Nicolaisen M (2013) Real-time PCR for universal phytoplasma detection and quantification. Methods Mol Biol 938:245–252

Chapter 13

A Rapid Protocol of Crude RNA/DNA Extraction for RT-qPCR Detection and Quantification

Claudio Ratti, Stefano Minguzzi, and Massimo Turina

Abstract

Most of the molecular diagnostic protocols used for phytoplasmas detection are based on the purification of total nucleic acids and on the use of genomic DNA of the pathogen as the target of amplification. Here we describe a diagnostic approach that, avoiding the purification of nucleic acids and exploiting the amplification of the abundant phytoplasma ribosomal RNA molecules produced during the infectious process, allows reducing the time and the costs necessary for the analysis, without affecting sensitivity and specificity. This is useful in particular when high numbers of analyses are required, as in certification programs, to monitor phytoplasmas classified as quarantine or quality pathogens. The protocol here described can be used for the detection and quantification of *Candidatus* Phytoplasma mali, Ca. P. pyri, Ca. P. prunorum, Ca. P. vitis, and Ca. P. solani by qPCR, RT-qPCR, ddPCR, and ddRT-PCR techniques based on TaqMan chemistry.

Key words Crude extract, DNA RNA extraction, Phytoplasmas, qPCR

1 Introduction

Several molecular assays have been published based on nested-PCR methods for phytoplasma detection by two step amplification of the 16S/23S rRNA gene [1, 2]. Although the introduction of a second round of amplification allows a more specific detection of phytoplasmas, it dramatically increases the risk of false positives due to cross contamination.

Quantitative or qualitative Real-Time PCR methods were developed to detect a wide range of phytoplasma strains using TaqMan probe or SYBR Green chemistry. These methods allow the detection of several quarantine phytoplasmas of the 16SrX group affecting pear, apple, and stone fruit species [3–5] as well as of *Candidatus* P. vitis and Ca. P. solani in grapevine [5–8] or Ca. P. mali [9].

To evaluate a detection protocol based on PCR amplification, the nucleic acid extraction method must be carefully selected.

Rita Musetti and Laura Pagliari (eds.), *Phytoplasmas: Methods and Protocols*, Methods in Molecular Biology, vol. 1875,
https://doi.org/10.1007/978-1-4939-8837-2_13, © Springer Science+Business Media, LLC, part of Springer Nature 2019

Several commercial RNA and DNA purification kits are available [10]. These kits are quite rapid and efficient but they strongly increase final detection costs and are not suitable when hundreds of samples need to be tested in limited time. Protocols developed by using cetyltrimethyl ammonium bromide (CTAB) [11, 12] have been shown to be as efficient as commercial kits, but they have the main disadvantage of being time-consuming. To solve this problem many protocols were published showing different efficient ways to perform PCR amplification after rapid crude sap preparation [13–16]. As a general concept, the efficiency of the method depends on the specific pathogen-plant host combination, presenting wide variability even between different phytoplasma strains within the same plant host species [17].

All previously reported protocols for phytoplasmas detection are based on amplification of a DNA target, although previous works on Flavescence dorée phytoplasma detection showed a higher sensitivity by adding a reverse transcriptase (RT) step therefore adding ribosomal RNA as target of amplification [15, 18]. Recently, it has been demonstrated by RT-qPCR protocol that combining RNA and DNA detection provides the best sensitivity in the Ca. P. prunorum detection. Moreover, using the same protocol for relative quantification, we calculated that the number of 16S rRNA copies in active cells is at least ten times higher than the corresponding 16S rRNA gene [19].

More recently, a precise absolute quantification by Droplet Digital RT-PCR (ddRT-PCR) of RNA and DNA in samples infected by different phytoplasmas, confirmed the relative higher abundance of RNA compared to the corresponding DNA. In particular, the ddRT-PCR analyses indicate that in both CTAB and crude extract preparations the amount of Ca. P. mali, pyri, prunorum, vitis, or solani rRNA range between 92% and 99% of the total nucleic acids target of the phytoplasmas and, as a consequence, phytoplasmatic DNA only range between 1% and 8% (Ratti, unpublished).

Here, we describe a rapid and inexpensive crude sap extraction method, which exploits Nylon membrane discs and can be applied to a new TaqMan® based RT-qPCR protocol for specific 16S rRNA amplification of Ca. Phytoplasma mali (associated with Apple Proliferation, AP), Ca. P. pyri (Pear Decline, PD), Ca. P. prunorum (European stone fruit yellows, ESFY), Ca. P. vitis (Flavescence dorée, FD), and Ca. P. solani (Bois noir, BN). By simultaneously using internal control primers and probe for plant 18S rRNA, the highly sensitive protocol is not only suitable for large-scale analysis, but also for relative quantification of phytoplasma in the host tissues. Finally, an additional advantage of RNA based detection protocols is the possibility of using the same extract for analysis of viruses with RNA genome [15, 19].

2   Material

Prepare all solutions using ultrapure nuclease-free water and molecular grade reagents. Prepare and store all reagents at room temperature (unless indicated otherwise). Diligently follow all waste disposal regulations when disposing waste materials.

2.1   Prepare Plant Material Samples

1. Petioles, leaf veins, or sample prepared by removing the outer bark and scraping off the layer of phloematic tissue with a sterile scalpel.

2. Extraction bags (approximately 12×15 cm) made of heavy duty plastic and with synthetic intermediate layer for filtration of the plant extracts (*see* **Note 1**).

2.2   Grinding Buffer pH 9.6

15 mM Na_2CO_3, 34.9 mM $NaHCO_3$, 2% polyvinyl-pyrrolidone—40 [PVP-40], 1% $Na_2S_2O_5$, 0.05% Tween 20, 0.2% bovine serum albumin [BSA].

1. Dissolve 1.59 g/L Na_2CO_3 with 2.93 g/L $NaHCO_3$ in water. Adjust pH to 9.6 using NaOH solution.

2. Add 20 g/L PVP-40 and 10 g/L $Na_2S_2O_5$ and leave them to dissolve completely in agitation.

3. Add 0.5 mL/L Tween 20 and mix.

   The grinding buffer without BSA can be stored at 4 °C for long time.

4. Before use: Add 2 g/L BSA and stir with magnetic bars until completely dissolved.

2.3   Sample Resuspension Solution

1. Prepare GES buffer: 0.1 M glycine, 0.05 M NaCl, 1 mM EDTA, pH 8.
   Dissolve 7.5 g/L glycine, 2.92 g/L NaCl and 0,372 g/L EDTA in water.
   Adjust pH to 8.0 using NaOH solution.

2. Prepare GES-Polyvinylpolypyrrolidone [PVPP]
   Soak 8 g PVPP for at least 2 h in 250 mL GES buffer
   Leave the suspension after overnight decantation.
   Remove the excess of GES buffer from the top of the precipitated hydrated PVPP.

3. Prepare resuspension solution
   Mix GES buffer with hydrated insoluble PVPP in a ratio 1:1 (v:v)
   Add 0.25% (2.5 mL/L) Triton X-100

   Optional: if extraction from grapevine plant tissues is performed add the following components

   1.0% (10 g/L) Boric Acid
   50% (500 g/L) Betaine.

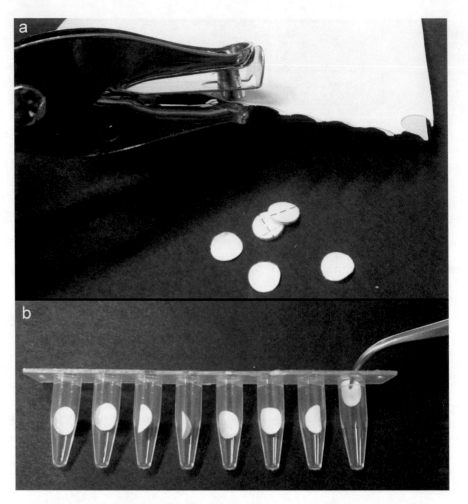

Fig. 1 Nylon membrane discs (6 mm diameter) can be prepared using a standard single-hole puncher (**a**) and transferred in 0.2 mL microtubes using forceps (**b**) avoiding touching with hands

2.4 Nylon Membrane

1. Use positively charged nylon membrane with binding capacity for nucleic acids up to 6 µg/cm^2 recommended for nucleic acid blotting such as northern blot analysis.

2. Cut the necessary 6 mm diameter discs from nylon membrane using a standard single-hole puncher (Fig. 1a). One disc for each sample is needed.

3. Use clean forceps to handle discs avoiding touching with hands.

4. Place each disc in a 0.2 mL microtube (Fig. 1b).

2.5 Primers and Probes

Different qPCR assays can be used, here we report the list of primers and probes tested and validated to work in combination with the crude RNA/DNA extraction protocol (*see* **Note 2**).

Name	Primer/Probe	Specificity	Sequence 5'-3'	Reference
ESFY 16S-F	Forward	*Ca.* P. prunorum	CGA ACG GGT GAG TAA CAC GTA A	[19]
ESFY 16S-R	Reverse		CCA GTC TTA GCA GTC GTT TCC A	
ESFY 16S	Probe		FAM-TAA CCT GCC TCT CAG GCG-MGB	
qAP-16S-F	Forward	*Ca.* P. mali	CGA ACG GGT GAG TAA CAC GTA A	[9]
qAP-16S-R	Reverse		CCA GTC TTA GCA GTC GTT TCC A	
qAP-16S	Probe		FAM-TAA CCT GCC TCT TAG ACG-MGB	
qAP-PD-16S-F	Forward	*Ca.* P. pyri	CGA ACG GGT GAG TAA CAC GTA	Unpublished
qAP-PD 16S-R	Reverse		CCA GTC TTA GCA GTC GTT TCC A	
PD 16S	Probe		FAM-TAA CCT ACC TTT CAG ACG-MGB	
STOL16SF1	Forward	*Ca.* P. solani	GAACGGGTGAGTAACGCGTAA	
STOL16SR	Reverse		CTCCTATCCAGTCTTAGCAGTCG TTT	
STOL 16S	Probe		FAM-CAATCTGCCCCTAAGACG-MGB	
FD16SF	Forward	*Ca.* P. vitis	CGAACGGGTGAGTAACACGTAA	
FD16SR	Reverse		CCAGTCTTAGCAACCGTTTCCG	
FD 16S	Probe		FAM-TAACCTACCTTTAAGACG-MGB	
DiSTA 18S-F	Forward	Plants	TGA CGG AGA ATT AGG GTT CGA	[19]
DiSTA 18S-R	Reverse		CTT GGA TGT GGT AGC CGT TTC	
DiSTA 18S	Probe		NED-CGG AGA GGG AGC CTG-MGB	

2.6 RT-qPCR Reagents

1. Reverse transcriptase enzyme. Moloney Murine Leukemia Virus Reverse Transcriptase (M-MLV RT) is suggested (*see* **Note 3**).

2. Real-time PCR reagent. A 2× qPCR master mix is suggested (*see* **Note 4**).

3. Dithiothreitol (dTT).

4. Recombinant Ribonuclease inhibitor (RNaseOUT).

5. Nuclease-free water.

3   Methods

Carry out all procedures at room temperature unless otherwise specified.

3.1   Preparation of Nylon Membrane Discs

Be sure to have enough nylon membrane discs for the samples to be processed. Eight-tube strips or 96-well plates are recommended for a large number of samples.

3.2   Preparation of the Crude Extract

1. Place 1.0 g petioles, leaf veins, or phloem-rich tissue in a plastic extraction bag.
2. Add 5 mL grinding Buffer to each bag. Final dilution 1:5.
3. Homogenize the plant material in the buffer using a manual or a drill press (Fig. 2a). Keep bags at 0–4 °C.
4. Perform the appropriate dilution of the sap, if needed (*see* **Note 5**). Sap can be stored at −20 °C for at least 4 months.
5. Transfer 5 µL sap onto nylon membrane discs placed inside 0.2 mL microtubes (Fig. 2b).
6. Place nylon membranes under vacuum conditions for 10 min. Ensure that all discs are completely dry.
7. Dried discs can be stored at room temperature or at 4 °C for at least 4 months (*see* **Note 6**).
8. Add 100 µL of resuspension solution to each microtube. Be sure to maintain PVPP in suspension in the resuspension solution by a magnetic stirrer or shaking the bottle by hands.

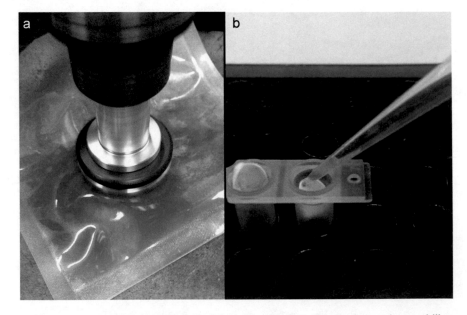

Fig. 2 Sap obtained homogenizing plant material in an extraction plastic bag using a drill press (**a**) is transferred onto the nylon membrane discs placed inside 0.2 mL microtubes (**b**)

9. Vortex the microtubes to favor resuspension.

10. Boil the microtubes at 95 °C for 10 min in a thermocycler and immediately chill on ice.

11. Centrifuge at $16,000 \times g$ for 5 min at 4 °C. The PVPP will pellet at the bottom of each tube and a clear solution containing DNA and RNA will remain on the top, ready to be used for RT-qPCR.

3.3 One Tube One Step Single or Multiplex RT-qPCR

Volumes indicated are referred to a single reaction.

1. Prepare a solution containing the appropriate amount of your favorite qPCR reagents considering a final volume of 25 μL for each reaction (*see* **Note 7**).

2. Add primers and probes according to the phytoplasma assay for which the analysis is required:

Single RT-qPCR	Final concentration (nM)	Stock concentration (μM)	Stock volume (μL)
Forward primer	400	10	1
Reverse primer	400	10	1
Probe	160	4	1

Duplex RT-qPCR Ca. P. prunorum	Final concentration (nM)	Stock concentration (μM)	Stock volume (μL)
ESFY 16S-F	900	10	2.25
ESFY 16S-R	900	10	2.25
ESFY 16S	100	4	0.625
DiSTA 18S-F	150	10	0.375
DiSTA 18S-R	150	10	0.375
DiSTA 18S	75	4	0.47

Duplex RT-qPCR Ca. P. pyri	Final concentration (nM)	Stock concentration (μM)	Stock volume (μL)
qAP-PD-16S-F	900	10	2.25
qAP-PD 16S-R	900	10	2.25
PD 16S	125	4	0.78
DiSTA 18S-F	150	10	0.375
DiSTA 18S-R	150	10	0.375
DiSTA 18S	75	4	0.47

Duplex RT-qPCR Ca. P. mali	Final concentration (nM)	Stock concentration (μM)	Stock volume (μL)
qAP-16S-F	900	10	2.25
qAP-16S-R	900	10	2.25
qAP-16S	125	4	0.78
DiSTA 18S-F	150	10	0.375
DiSTA 18S-R	150	10	0.375
DiSTA 18S	75	4	0.47

Triplex RT-qPCR Ca. P. solani Ca. P. vitis	Final concentration (nM)	Stock concentration (μM)	Stock volume (μL)
STOL16SF1	300	10	0.75
STOL16SR	300	10	0.75
STOL 16S	75	4	0.468
FD16SF	300	10	0.75
FD16SR	300	10	0.75
FD 16S	75	4	0.468
DiSTA 18S-F	150	10	0.375
DiSTA 18S-R	150	10	0.375
DiSTA 18S	50	4	0.3125

3. Add the following reagents at the indicated final concentration or quantity:

Dithiothreitol (dTT)	1 mM
Recombinant Ribonuclease inhibitor (RNaseOUT)	10 U
Reverse transcriptase (M-MLV RT) (*see* **Note 3**)	2 U

4. Add the necessary volume of nuclease-free water to reach the volume of 23 μL for each reaction.

5. Add 2 μL of crude extract (clear supernatant obtained as described in Subheading 3.2) for each sample.

6. Run the RT-qPCR reaction in a real-time PCR thermal cycler under the following conditions (*see* **Note 8**):

48 °C for 30 min.

95 °C for 10 min.

95 °C for 15 s.

60 °C for 1 min.

The last two steps are repeated for 40 cycles.

4   Notes

1. A product with specified characteristics can be purchased from BIOREBA AG, Switzerland.

2. All primers and probes have been tested and validated to work in combination with the crude RNA/DNA extraction protocol. In particular, spot method and CTAB extraction have been compared using RT-qPCR as detection methods to monitor phytoplasma infections in field samples from apricot, plum, peach pear or apple orchards as well as from vineyards. Both methods produced the same results in terms of detection, identifying the same positive and negative samples.

3. A standard, good quality, and routinely used M-MLV RT usually represent a satisfactory compromise between performance and price. We used a product form Promega (Madison, WI, USA); however, different type and units of reverse transcriptase can be tested. As few units of reverse transcriptase are required in the amplification reaction, we suggest preparing a dilution 1:50 or 1:100 in nuclease-free water just before use. If analysis from grapevine plant tissues is performed use 8 units of M-MLV RT in each RT-qPCR reaction.

4. Common master mixes for qPCR are user-friendly in particular for large-scale analysis reducing pipetting and risk of errors or contaminations if compared to single-reagents kits. We used products from bio-rad (Hercules, CA, USA), KAPA biosystems (Wilmington, MA, USA), or Promega (Madison, WI, USA). RT-qPCR master mixes can be used as well but are usually more expensive.

5. Dilution of extracted sap could improve the sensitivity of the protocol. Usually, this is not necessary in the case of samples from stonefruit or pomefruti while a dilution 1:4 or 1:6 (v/v) with grinding buffer is necessary for samples from grapevine plants.

6. Storage property of nylon membrane discs has been evaluated over a period of 4 months for Ca. P. prunorum [19]. A similar experiment has been conducted for samples infected by other phytoplasmas tested by RT-qPCR discs spotted with the same samples immediately or after storing at either 4 °C or room

temperature. No significant variation on Ct values was obtained, confirming that nucleic acids are stable on nylon membrane and suggesting that spotted discs could be used to send samples over long distance by mail.

7. The protocol as well as primers and probes have been tested with several commercial qPCR reagents however, some adjustment on mg^{++} concentration could be necessary for some reagents. Moreover, the reaction volume of 25 μL could be reduced according to the specific reagents, real-time PCR thermal cycler and microtubes used.

8. Amplification conditions may need to be adjusted according to the characteristics of real-time PCR thermal cycler and qPCR reagents used.

References

1. Gundersen D, Lee I, Schaff D, Harrison N, Chang C, Davis R et al (1996) Genomic diversity and differentiation among phytoplasma strains in 16S rRNA groups I (aster yellows and related phytoplasmas) and III (X-disease and related phytoplasmas). Int J Syst Bacteriol 46(1):64–75

2. Heinrich M, Botti S, Caprara L, Arthofer W, Strommer S, Hanzer V et al (2001) Improved detection methods for fruit tree phytoplasmas. Plant Mol Biol Rep 19:169–179

3. Jarausch W, Peccerella T, Schwind N, Jarausch B, Krczal G (2004) Establishment of a quantitative real-time PCR assay for the quantification of apple proliferation phytoplasmas in plants and insects. Acta Horticulturae (657):415–420

4. Torres E, Bertolini E, Cambra M, Montón C, Martín MP (2005) Real-time PCR for simultaneous and quantitative detection of quarantine phytoplasmas from apple proliferation (16SrX) group. Mol Cell Probes 19:334–340

5. Galetto L, Bosco D, Marzachí C (2005) Universal and group-specific real-time PCR diagnosis of Flavescence dorée (16Sr-V), bois noir (16Sr-XII) and apple proliferation (16Sr-X) phytoplasmas from field-collected plant hosts and insect vectors. Ann Appl Biol 147:191–201

6. Bianco PA, Casati P, Marziliano N (2004) Detection of phytoplasmas associated with grapevine Flavescence dorée disease using real-time PCR. J Plant Pathol 86:259–264

7. Christensen NM, Nicolaisen M, Hansen M, Schulz A (2004) Distribution of phytoplasmas in infected plants as revealed by real-time PCR and bioimaging. Mol Plant-Microbe Interact 17:1175–1184

8. Hren M, Boben J, Rotter A, Kralj P, Gruden K, Ravnikar M (2007) Real-time PCR detection systems for Flavescence dorée and bois noir phytoplasmas in grapevine: comparison with conventional PCR detection and application in diagnostics. Plant Pathol 56:785–796

9. Baric S, Dalla-Via J (2004) A new approach to apple proliferation detection: a highly sensitive real-time PCR assay. J Microbiol Methods 57(1):135–145

10. MacKenzie DJ, McLean MA, Mukerji S, Green M (1997) Improved RNA extraction from woody plants for the detection of viral pathogens by reverse transcription-polymerase chain reaction. Plant Dis 81(2):222–226

11. Ahrens U, Seemüller E (1992) Detection of DNA of plant pathogenic mycoplasmalike organisms by a polymerase chain reaction that amplifies a sequence of the 16 S rRNA gene. Phytopathology 2(8):828–832

12. Chang S, Puryear J, Cairney J (1993) A simple and efficient method for isolating RNA from pine trees. Plant Mol Biol Rep 11(2):113–116

13. La Notte P, Minafra A, Saldarelli P (1997) A spot-PCR technique for the detection of phloem-limited grapevine viruses. J Virol Methods 66(1):103–108

14. Dovas CI, Katis NI (2003) A spot nested RT-PCR method for the simultaneous detection of members of the Vitivirus and Foveavirus genera in grapevine. J Virol Methods 107(1):99–106

15. Margaria P, Turina M, Palmano S (2009) Detection of Flavescence dorée and bois noir phytoplasmas, grapevine leafroll associated

virus-1 and-3 and grapevine virus a from the same crude extract by reverse transcription-RealTime Taqman assays. Plant Pathol 58 (5):838–845

16. Bertolini E, Felipe R, Sauer A, Lopes S, Arilla A, Vidal E et al (2014) Tissue-print and squash real-time PCR for direct detection of 'Candidatus Liberibacter' species in citrus plants and psyllid vectors. Plant Pathol 63 (5):1149–1158

17. Osman F, Rowhani A (2006) Application of a spotting sample preparation technique for the detection of pathogens in woody plants by RT-PCR and real-time PCR (TaqMan). J Virol Methods 133(2):130–136

18. Margaria P, Rosa C, Marzachi C, Turina M, Palmano S (2007) Detection of Flavescence doree phytoplasma in grapevine by reverse-transcription PCR. Plant Dis 91 (11):1496–1501

19. Minguzzi S, Terlizzi F, Lanzoni C, Poggi Pollini C, Ratti C (2016) A rapid protocol of crude RNA/DNA extraction for RT-qPCR detection and quantification of "Candidatus phytoplasma prunorum". PLoS One 11(1): e0146515. https://doi.org/10.1371/journal.pone.0146515

Chapter 14

Quantitative Analysis with Droplet Digital PCR

Nataša Mehle and Tanja Dreo

Abstract

Digital PCR-based methods, such as droplet digital PCR, are one of the best tools for determination of absolute nucleic-acid copy numbers. These techniques avoid the need for reference materials with known target concentrations. Compared to real-time PCR, they provide higher accuracy of quantification at low target concentrations, and have higher resilience to inhibitors. In this Chapter, we describe the droplet digital PCR workflow for the detection and quantification of flavescence dorée phytoplasma.

Key words Absolute quantification, Digital PCR, Droplet digital PCR, Flavescence dorée phytoplasma, Phytoplasma reference material

1 Introduction

Quantification of phytoplasma is useful for monitoring the progress of an infection, and the variations in phytoplasma titers through a season and in different plant tissues [1, 2]. These analyses provide both crucial information for epidemiology studies and optimization of sampling for diagnosis. Phytoplasma quantification is also important in screening of plants for resistance against phytoplasma, and for estimation of the number of copies that are carried by an insect vector [3, 4].

For quantification by real-time PCR (qPCR), a standard curve with known concentrations of the target is necessary to transform the output of quantification cycles (Cq) into actual concentrations (i.e., target copies/μL). The lack of standardized or certified reference materials for phytoplasma can lead to significant inter-laboratory and inter-experimental quantification bias. Moreover, many factors, including inhibitors, can influence the efficiency of qPCR. Therefore, the accuracy of qPCR for quantification will continue to vary widely unless the effects of different amplification efficiencies between standard curves and samples are compensated for by a relatively complex analysis [5]. qPCR also shows limitations

Rita Musetti and Laura Pagliari (eds.), *Phytoplasmas: Methods and Protocols*, Methods in Molecular Biology, vol. 1875, https://doi.org/10.1007/978-1-4939-8837-2_14, © Springer Science+Business Media, LLC, part of Springer Nature 2019

when rare variants need to be quantified in a high background of wild-type targets [6].

Digital PCR (dPCR)-based methods, such as droplet dPCR, have been shown to have the potential to improve upon these limitations of qPCR. dPCR provides absolute quantification of target sequences without the need to rely on standard curves. As such, dPCR is a promising method of choice for both detection and calibration of phytoplasma reference materials in laboratories worldwide. The absolute concentration of the target copies in the initial sample is determined from the number of positive and negative partitions after end-point PCR amplification through the application of Poisson statistics [7]. Therefore, the result is independent of variations in PCR amplification efficiency. Droplet dPCR has been shown to be less affected by inhibitory substances and more robust to target sequence variation, and it provides higher quantification accuracy at lower target concentrations, compared to qPCR [8–14].

Here, we describe the dPCR workflow for quantification of DNA of flavescence dorée phytoplasma (FDp) using the QX100 and QX200 droplet dPCR platforms (Bio-Rad). The workflow consists of four main steps: (1) preparation of the reaction mixture; (2) droplet generation; (3) PCR amplification; and (4) droplet reading and analysis of the results. This protocol was developed by Mehle et al. [12]. The primers and probe target the $secY$ gene, and they are the same as for the qPCR assay published by Hren et al. [15]. Droplet dPCR assay has been shown to be comparable to qPCR in terms of sensitivity for the detection of FDp, but to provide higher precision and repeatability for quantification of FDp at low concentrations [12]. In grapevine tissue samples, FDp is usually present at low concentrations, and thus, a higher quantification accuracy at lower concentrations is here a particularly important advantage.

2   Materials

2.1   Samples and Controls

1. Extracted DNA of samples (*see* **Notes 1–3**).

2. Extracted DNA of sample controls and dPCR controls (Table 1; *see* **Notes 1, 4**, and **5**).

2.2   Reagents

1. 2× ddPCR Super Mix for probes (Bio-Rad) (*see* **Note 1**).

2. Droplet generation oil for probes (Bio-Rad).

3. Molecular grade nuclease-free water (*see* **Note 6**).

4. Primers and probe (Table 2; *see* **Notes 1, 7**, and **8**).

Table 1
Common quality controls recommended for dPCR-based quantitative diagnostics/ quantification of phytoplasma (*see* Note 4)

Description		Aim	Recommendation for use
DNA extraction sample controls:			
Negative extraction control	Buffer instead of sample	To reveal contamination of reagents during extraction	Each run
	Healthy material of same host and part	To assess the background signal inherent to the matrix	Validation process
Positive extraction control	Internal positive control: samples spiked with exogenous nucleic acid that has no relation with the target phytoplasma or endogenous nucleic acid (conservative non-pest target nucleic acid that is also present in the sample)	Confirmation that the extractions from different samples have been successful	For each individual sample separately (*see* **Note 5**)
	External positive control: naturally infected host tissue or spiked healthy host tissue	To show that detection/ quantification of the phytoplasma from defined samples is possible	Validation process
Droplet dPCR controls:			
Negative dPCR control	Nuclease-free water instead of sample DNA	To reveal contamination of reaction mix and pipetting	Each run
Positive dPCR control	Sample DNA containing known concentration of the target phytoplasma or synthetic control	To assess dPCR performance	Each run

Table 2
Primers and probe targeting the *secY* gene of FDp [15]

Orientation	Sequence (5′-3′)	Working concentration (μM)
Forward primer	TTA TGC CTT ATG TTA CTG CTT CTA TTG TTA	10
Reverse primer	TCT CCT TGT TCT TGC CAT TCT TT	10
Probe	FAM- ACC TTT TGA CTC AAT TGA- NFQ[a]	2.5

[a]NFQ, non-fluorescent minor groove binder (MGB) quencher

2.3 Equipment

1. QX100 or QX200 droplet digital PCR system (Bio-Rad), including droplet generator (QX100 droplet generator, or automated droplet generator), droplet reader, and QuantaSoft data acquisition and analysis software.

2. Thermal cycler; e.g., T100 or C1000 (Bio-Rad).

3. Droplet generator cartridge holder, DG8 droplet generator cartridges and gaskets for generation of droplets using a QX100 droplet generator, or DG32 cartridges for generation of droplets using an automated droplet generator.

4. Two UV chambers (*see* **Note 9**).

5. Easy to pierce foil plate seals.

6. PCR plates (96-well).

7. PCR plate sealer (*see* **Note 10**).

8. Set of pipettes and tips with aerosol barrier filters. If the droplets are generated using a QX100 droplet generator, the manufacturer recommends Rainin pipettes (L-20, L-50, L8-50, L8-200) and Rainin tips for the steps of droplet generation and droplet handling (*see* **Note 11**).

9. Microcentrifuges and vortex for master mix preparation, and appropriate nuclease-free plastic-ware.

10. Cooling blocks (*see* **Note 12**).

3 Methods

3.1 Preparation of the Reaction Mixture

1. Design the experiment (*see* **Note 13**) and calculate the volume of components needed for the assay master mix according to Table 3 (*see* **Notes 14** and **15**). Include sufficient reagents for the number of samples and controls to be tested. It is good practice to analyze samples in duplicate or in triplicate (*see* **Note 16**).

2. In nuclease-free tubes, add sterile nuclease-free water, 2× ddPCR Super Mix for probe, and primers and probe according to the quantities calculated in **step 1**. After completion, mix all of the reagents well with the vortex, and centrifuge them briefly (for ~5 s) in a microcentrifuge.

3. Distribute the prepared mixture into nuclease-free tubes or strips, or 96-well plates (16 μL/tube or well; *see* **Note 15**). Note that pipetting into 96-well plates is required if the droplets are to be generated using an automated droplet generator, and that there should not be any empty wells left in any of the columns of the plate used (*see* **Note 13**).

4. Add 4 μL of each DNA sample or control (*see* **Note 15**) into each of the tubes containing the ddPCR Master Mix. Mix

Table 3
Reaction setup for DNA amplification (*see* Note 14)

Component	Final concentration	Volume (μL)
Sterile nuclease-free water		0.4
2× ddPCR Super Mix for probe	1×	10.0
Forward primer (10 μM)	900 nM	1.8
Reverse primer (10 μM)	900 nM	1.8
Probe (2.5 μM)	250 nM	2.0
DNA sample[a]		4.0
Final volume (*see* **Note 15**)		20.0

[a]For quantification purposes the target concentration should be within the linear range of the method (*see* **Note 3**)

thoroughly by pipetting up and down, or by vortexing and brief centrifugation. If the droplets are to be generated using an automated droplet generator, the 96-well plates should be heat sealed with pierceable foil. It is recommended to keep the reaction mixture at 4 °C (e.g., using cooling blocks).

3.2 Droplet Generation

3.2.1 Droplet Generation Using a QX100 Droplet Generator

1. Switch on the droplet generator by plugging it into an electrical outlet (as it has no on/off switch).

2. Place a DG8 droplet generation cartridge into the cartridge holder.

3. Transfer 20 μL of each prepared reaction mixture into each of the eight wells indicated as "sample" in the droplet generation cartridge. There should be no empty wells left (*see* **Note 13**). Precautions should be taken to avoid bubble formation at the bottoms of the wells, as bubbles can interfere with droplet generation.

4. Add 70 μL droplet generation oil into each of the wells indicated as "oil."

5. Hook the gasket over the cartridge holder using the holes on each side.

6. Place the holder with the cartridge in the QX100 droplet generator.

7. Initiate droplet generation. The oil and sample are pushed through microfluidic channels and mixed in the cartridge, which forms droplets. During this process, each sample is partitioned into up to 20,000 nanoliter-sized droplets. These droplets are accumulated in the wells indicated as "droplets." This process takes 2–3 min for each cartridge.

8. Remove the holder with the cartridge from the QX100 droplet generator.

9. Remove the gasket from the cartridge holder and transfer 40 μL of the droplet suspension from the cartridge to a 96-well PCR plate on a cooling block. The pipetting of the droplet suspension should be carried out slowly, to protect the integrity of the droplets (*see* **Note 11**).

10. Keep the droplet suspensions at 4 °C using the cooling block.

11. Repeat the process of droplet generation as many times as required according to the number of samples.

12. After completion for all of the samples, the 96-well plate with the suspensions of the droplets is heat sealed with pierceable foil (*see* **Note 17**). Do not vortex or centrifuge the 96-well plate, as to do so would break down the droplets.

3.2.2 Droplet Generation Using an Automated Droplet Generator

1. Configure the sample plate in the automated droplet generator.

2. Place the following consumables into the dedicated holders of the instrument: DG32 cartridges (holder: "DG32 plate"); full pipette tip boxes with box lids removed (holders: "pipette tips"); and the tip waste bin.

3. Place the sealed 96-well PCR plate containing the prepared droplet dPCR reactions into the holder indicated as "sample plate."

4. Place the cooling block into the holder indicated as "droplet plate" (*see* **Note 12**), and then place a clean 96-well PCR plate for droplet collection into the cooling block accessory.

5. Load the bottle of automated droplet generation oil into the tower of the oil delivery system of the instrument. Select the type of oil that is loaded into the instrument (in this particular case, select "probes").

6. Initiate droplet generation. The instrument partitions each sample into approximately 20,000 uniform nanoliter-sized droplets. This process takes approximately 45 min for a full 96-well plate of samples.

7. Remove the 96-well plate containing the dPCR droplets from the instrument. The plate should be heat sealed with pierceable foil (*see* **Note 17**). Do not vortex or centrifuge the plate, as to do so would break down the droplets.

3.3 PCR Amplification

1. Transfer the sealed 96-well plate into the thermocyler and run the PCR under the conditions shown in Table 4 (*see* **Note 14**).

3.4 Droplet Reading and Analysis of Results

1. Switch on the droplet reader in advance to warm it up. Check on the status of the oil in the oil reservoir, and the waste bottle, according to the manufacturer's instructions (*see* **Note 18**).

2. Switch on the connected computer and run the QuantaSoft software. Click "Setup" to enter information about the

Table 4
PCR cycling conditions[a] (*see* Notes 7 and 14)

Cycling step	T (°C)	Time	Number of cycles
Enzyme activation	95	10 min	1
DNA denaturation	94	30 s	45
Annealing and elongation	60	1 min	
Heat deactivation	98	10 min	1
Hold	4	∞	1

[a]Use a heated lid set to 105 °C and set the sample volume to 40 μL. Ramp rates should be adjusted to 2–3 °C/s

samples, assays, and experiments. Double-click anywhere on the plate to open up the view where you can define the information for each well. The necessary information includes type of experiment, type of mastermix ("supermix"), and measuring channel for each well (*see* **Note 19**). After entering this information through the drop-down menus or into the corresponding fields, click "Apply" to assign them to the selected wells. When done, click "Okay." It is good practice to save the template.

3. Place the 96-well PCR plate into the plate holder of the droplet reader, close the cover, and close the lid.

4. Initiate the reading of the droplets. The droplet reader acts as a flow cytometer and reads each droplet to determine their signal and fluorescence level (called "Amplitude" in the software) in the selected detectors.

5. Analyze the data using the QuantaSoft software by clicking the "Analyse" button. Positive droplets that contain amplification products are identified from negative droplets by automatic or manual application of a fluorescence amplitude threshold in the software (*see* **Note 20**). The software offers different ways of viewing the results; e.g., one-dimensional amplitude of one channel (Fig. 1), or copy number in each well or channel. The software also provides a Table with the parameters that result from the analysis, such as the concentration target copies per microliter of reaction (*see* **Notes 21** and **22**), the number of total accepted droplets (*see* **Note 23**), the positive droplets and the negative droplets. The number of positive droplets and the number of accepted droplets can be used to calculate the different parameters (Table 5; *see* **Note 24**).

6. If one of the controls (Table 1) does not perform as expected, or if large differences are seen between the replicates, the test might have to be repeated (*see* **Notes 4** and **16**).

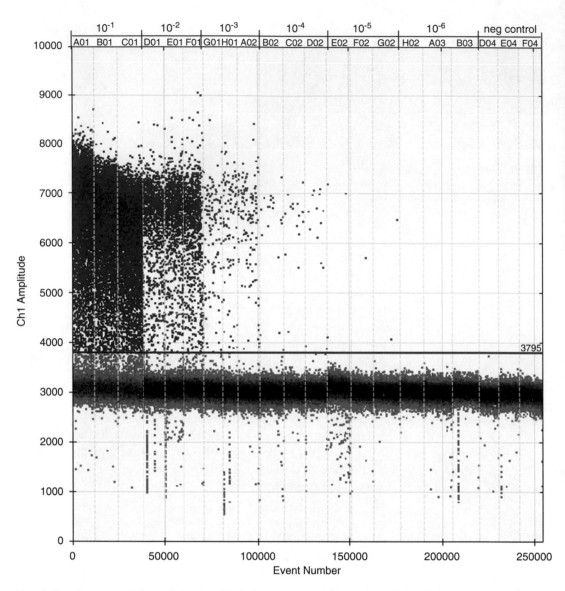

Fig. 1 Visual representation of droplet dPCR data generated by the Bio-Rad QX100 instrument. Positive droplets (upper cluster) and negative droplets (lower cluster) are shown for a range of six serially diluted FDp DNA samples (decimal dilution factors: 10^{-1}–10^{-6}). Each DNA dilution was analyzed as three replicates. The results of three negative dPCR controls (neg control) are also shown. The *Y*-axis shows the fluorescent intensity, with individual droplets shown sequentially as read on the *X*-axis for each sample. The threshold to discriminate between negative and positive droplets was set at a fluorescence of 3795, and was chosen as described in **Note 20**. The concentrations determined and the parameters calculated for the samples shown here are listed in Table 5

Table 5
Selected parameters from droplet dPCR analysis of serial dilutions of FDp DNA

Dilution of FDp-positive DNA sample	Number of positive droplets	Number of accepted droplets	λ[a]	P[b]	FDp DNA copy number in 1 μL dilution of DNA sample[c]
10^{-1}	6590	12,245	0.7726	0.5382	4545
	6766	12,977	0.7369	0.5214	4334
	7359	13,727	0.7681	0.5361	4518
10^{-2}	1004	11,498	0.0914	0.0873	537
	952	9879	0.1013	0.0964	596
	840	10,408	0.0842	0.0807	495
10^{-3}	109	11,873	0.0092	0.0092	54
	99	10,223	0.0097	0.0097	57
	83	8661	0.0096	0.0096	57
10^{-4}	15	13,050	0.0012	0.0011	7
	16	12,173	0.0013	0.0013	8
	12	12,010	0.0010	0.0010	6
10^{-5}	2	12,521	0.0002	0.0002	1
	1	11,958	0.0001	0.0001	0[d]
	1	14,204	0.0001	0.0001	0[d]
10^{-6}	0	15,734	0.0000	0.0000	0
	0	14,471	0.0000	0.0000	0
	0	12,761	0.0000	0.0000	0

[a]Mean number of target copies per droplet $= -\ln$ (1-(number of positive droplets)/(number of accepted droplets))
[b]Fraction of positive droplets (probability that a droplet is full) = (number of positive droplets)/(number of accepted droplets)
[c]Concentration of target copies per microliter of DNA sample = λ/(DNA volume per droplet). In this case, the droplet volume of 0.85 nL is used to calculate the DNA copy numbers in the sample (*see* **Note 22**). DNA volume per droplet = (droplet volume)/(total reaction volume) × (DNA sample volume) = 0.00085 μL/(20 μL × 4 μL)
[d]No droplets with amplification were observed in negative dPCR controls (Fig. 1; *see* **Note 20**), and thus this single positive droplet can be considered as a positive reaction, although it is below the limit of quantification

---

4   Notes

1. Mix all of the DNA samples, controls and reagents well, and centrifuge them briefly (for ~5 s) in a microcentrifuge before use. If they were stored frozen, they should be allowed to thaw and equilibrate to room temperature before mixing.

2. For DNA extraction, the same protocols as those used for other PCR-based methods can be used. For the validation of

FDp-specific droplet dPCR, the DNA extraction was based upon the binding of DNA to magnetic beads [16]. dPCR shows higher resilience than qPCR to amplification inhibitors arising from plant and environmental samples [13], and the possibility of direct droplet dPCR quantification of bacteria without DNA extraction has been shown [11]. Thus, it is expected that highly purified DNA is not necessary for quantification of phytoplasma DNA. Note, however, that less purified DNA preparations might result in more "rain" droplets.

3. As the absolute concentration of target copies in the initial sample is determined from the number of positive and negative partitions after the end-point PCR amplification through the application of Poisson statistics, the theoretical upper limit of the linear range is defined by the number of droplets [17]. With the droplet dPCR instrument suggested in this Chapter, the PCR reaction mixture is theoretically separated into up to 20,000 droplets. Therefore, for very high concentrations of targets, where almost all of the analyzed droplets will contain target copies, Poisson law can no longer be applied. Therefore, under these conditions the exact concentration of targets cannot be determined beyond stating that it is above the highest concentrations that dPCR can quantify (according to the manufacturer's recommendation this is >120,000 copies/20 μL reaction) [17]. Based on experimental data, FDp is usually present in concentration below these limits in symptomatic grapevine tissue samples, and can be readily quantified [12]. However, for higher FDp loads, the DNA samples will need to be diluted before these droplet dPCR measurements. Alternatively, a droplet dPCR instrument that can create larger numbers of droplets can allow direct quantification of samples with higher levels of FDp [18].

4. Each time droplet dPCR is run, the performance is monitored by including positive and negative controls, in line with good laboratory practice. Negative controls are used to monitor for contamination of samples, reagents, or laboratory equipment. Positive controls are mainly used to check whether the test has been performed correctly. The material used as positive controls can consist of "complete" phytoplasma, as well as relevant parts thereof. These can be nucleic-acid extracts, synthetic controls, or cloned PCR products, preferably at a concentration close to the limit of quantification. If the test is performed frequently, the data for the positive dPCR control can be subjected to trend analysis for monitoring validity over time [19]. In addition, for diagnostic purposes, it is crucial to evaluate the performance of any test through validation of several parameters [20]. Running of assays that are not fully validated will require additional controls, to define possible cross-

reactions with sample tissue, and to show that quantification of phytoplasma from defined samples is possible.

5. The quality of the extracted DNA of each individual sample can also be controlled using other methods; e.g., by qPCR specific for the eukaryotic 18S rRNA or the plant cytochrome oxidase gene. In our experience, however, the concentration of such controls is often very high. Consequently, they can be identified as positive in dPCR but are often not quantifiable; i.e., they are beyond the highest quantifiable concentration.

6. Use of molecular grade nuclease-free water is essential. Do not use DEPC-treated water, because the slightly acidic pH can promote primer degradation. It is recommended to aliquot the water for use.

7. The primers and probe used for FDp detection in droplet dPCR format are the same as for the qPCR published by Hren et al. [15]. In addition, the key assay parameters are not modified (i.e., primer and probe concentrations, time and temperature for annealing–elongation step). In general, assays that have been optimized in qPCR format with an amplification protocol similar to that recommended for dPCR are easily transferred to the dPCR format [9, 11, 12, 14]. For assays with considerable differences between the optimal amplification protocol and the dPCR recommended one, optimization might be necessary. Assays that have shown limitations in qPCR (e.g., cross-reactivity, low efficiency of amplification) are more likely to fail in a dPCR format as well. However, at least one test that was problematic in qPCR format has been shown to have significantly improved performance in the dPCR format [11].

8. Purchase lyophilized oligonucleotides from any commercial source that are synthesized at a 25 nM scale. For these, standard desalting is sufficient, and no additional purification is required. Resuspend in molecular grade nuclease-free water to 100 μM. Dilute each stock of primers and probe to appropriate working concentrations (Table 2), and store them at −20 °C. It is recommended that the primers and probe are aliquoted for use. Protect the probe from excessive exposure to light (e.g., put the tubes with probes in dark plastic bags/containers), to prevent photobleaching of the fluorescent dyes and evaporation. In our experience, when stored correctly and subjected to minimal freeze-thaw cycles, the primer and probe aliquots can last up to at least 6 years.

9. The reaction mixtures should be prepared and the addition of the DNA samples should be carried out in two separate rooms or UV chambers, using dedicated laboratory equipment (e.g., pipettes, tips, tubes, microcentrifuge tube opener, racks for

tubes, lab coat, gloves): first, prepare reaction mixtures and load them into the tubes or 96-well plates, or onto the strips (no DNA other than primers and probes should be present here); second, add the DNA samples in another UV chamber. Both steps should be carried out in separate locations from those used for the DNA extraction. Keep a dust-free environment as much as possible. Dust particles often fluoresce and can interfere with droplet formation and readings (e.g., causing streaking).

10. Switch on the PCR plate sealer in advance to reach its optimal temperature. The manufacturer recommends sealing at 180 °C for 5 s. If the pressure is applied for too long, the seal will be broken.

11. In dPCR, the pipetting errors are greater than the errors associated with the method itself. It is therefore highly recommended to use calibrated pipettes that ensure accurate liquid handling. In addition, when handling the droplets, the pipetting should be done slowly (to preserve the integrity of the droplets) and using recommended pipettes and tips.

12. The cooling blocks should be placed in a −20 °C freezer for at least 2 h before being used. The cooling block assembly designed to be used in an automated droplet generator should be a dark purple color, which indicates that it is at the correct working temperature. If the block is pink, it has warmed up and should not be used. The cooling block is used mainly to prevent droplet evaporation.

13. It is recommended that experiments are designed in such a way that droplet generation cartridges are used at their full capacity; i.e., eight samples per cartridge. If there are not enough samples to fill all of the eight wells of a droplet generation cartridge with the reaction mixture, the droplet dPCR buffer control kit (Bio-Rad) should be used to fill the remaining wells. Note that when using an automated droplet generator, one cartridge is required for one column of reaction mixture in the 96-well plate.

14. As primer and probe concentrations and annealing temperatures can have an influence on the fluorescence level and clustering of the droplets, different concentrations of primers and probe and a range of annealing–elongation temperatures between 55 °C and 65 °C can be optimized if the assays do not show the desired performance in the dPCR format. Changing other cycling conditions, such as elongation time or cycle number, might also provide further optimization. Optimal concentrations of primers and probe and optimal cycling conditions are where visual inspection of the droplet readout shows the best resolution between clusters of positive

and negative droplets, and where the lowest number of droplets with intermediate fluorescence are observed (i.e., the rain effect) [21].

15. It is advised to allow for some excess volume for each component (e.g., 10% extra volume of each component, which would result in 22 μL final volume), to ensure the total volume of at least 20 μL of the reaction mixture for transfer into the DG8 or DG32 cartridges.

16. Although the repeatability of this droplet dPCR has been shown to be better than that of qPCR [12], it is recommended to analyze samples in duplicate or in triplicate. In the example shown in Table 5, each dilution of the FDp-positive DNA sample was analyzed in three replicates. Up to a dilution of 10^{-3}, the coefficient of variation (CV) between replicates was between 2.6% and 9.3%. As expected, a slightly higher CV (14.3%) was observed with the lower concentrations of the target DNA (dilution, 10^{-4}). In this study, the replicate measurements within one run were carried out under repeatability conditions, which means the same analyst, the same reaction mixture, chemicals and cartridges from the same batch, and the same instruments.

17. The manufacturer recommends that the thermal cycling is begun within 30 min of sealing a plate, or to store the plate at 4 °C for up to 4 h prior to the thermal cycling.

18. Check the indicator lights on the front of the droplet reader. If the lights are flashing amber, the run cannot be started. Replace the waste bottle with a marked empty oil supply bottle. Replace the droplet reader oil with a new bottle of oil. After changing the oil, click "Prime" in the QuantaSoft software to fill the lines with oil before the system is run.

19. The "supermix" should be defined because it can affect the droplet volume (see **Note 22**). There are two measuring channels: channel 1 for FAM, and channel 2 for HEC/VIC. In this particular case, select FAM. The sample information can also be added after collecting the results.

20. The assay, target, and/or matrix characteristics can affect the separation between the negative and positive droplets, which can result, for example, in the presence of higher fluorescence in negative droplets, or the so-called droplet rain effects [21]. Therefore, automatic analysis can sometimes be misleading, and setting the threshold manually might be necessary. To set up the threshold, it is important to evaluate the expected fluorescence of the positive droplets by testing a large number of different types of negative samples (e.g., negative extraction controls, negative dPCR controls, samples infected by closely related pathogens) and positive samples (e.g., serial dilutions of

target DNA, external positive extraction control). After testing, the threshold can be set at the highest point of the negative droplet cluster (Fig. 1). Even if the threshold is carefully defined, a small number of positive droplets might be observed in the negative samples, and the fluorescence of these droplets can be equal to that of the real positives. In this case, based on the negative samples, a minimum number of droplets should be defined to make a sample positive, and this can exceed the statistically defined minimum of three positive droplets for a positive final result [11].

21. The concentration of the target in the QuantaSoft software is given in copies/μL of reaction (at 20 μL reaction volume). Care should be taken in the conversion of the target DNA copies into concentrations of biological units; e.g., cells. For many assays, there are several copies of targets per each biological unit, and these can differ among strains.

22. Version 1.3.2.0 of the QuantaSoft software uses the preset droplet volume of 0.91 nL for calculations of the concentrations of target copies. The new version of this software, as version 1.7.4, has a new droplet volume of 0.85 nL incorporated into the calculations. In addition, discrepancies have been shown between the droplet volume assigned by the manufacturer and measured by independent laboratories [17, 22–24]. Several factors can affect the droplet volume, such as droplet generators and the supermix. Thus, when high precision of quantification is required, the droplet volumes should be assessed for each system [24]. For the general purposes of determining the phytoplasma concentration, this is not necessary.

23. Low numbers of accepted droplets can negatively affect the precision of the results, and thus several published studies have defined a limit of 10,000 accepted droplets, below which the quantitative results in that well are rejected (*see*, e.g., [9, 11]).

24. The minimum information that is needed for publication of quantitative dPCR experiments is given by Huggett et al. [25], and an R-based script that automates calculations of several parameters has been described [11].

Acknowledgment

This work was supported by the Slovenian Research Agency (grant number P4-0165) and by Euphresco Project 2016-A-215, financed by the Ministry of Agriculture, Forestry and Food through the Administration of the Republic of Slovenia for Food Safety, Veterinary and Plant Protection. The work was performed using droplet dPCR equipment financed by the Metrology Institute of the Republic of Slovenia (MIRS), with financial support from the European Regional Development Fund. The equipment is wholly owned by the Republic of Slovenia.

References

1. Galetto L, Marzachi C (2010) Real-time PCR diagnosis and quantification of phytoplasmas. In: Weintraub PG, Jones P (eds) Phytoplasmas: genomes, plant hosts and vectors. CAB International, Wallingford

2. Prezelj N, Nikolić P, Gruden K et al (2013) Spatiotemporal distribution of flavescence dorée phytoplasma in grapevine. Plant Pathol 62:760–766. https://doi.org/10.1111/j.1365-3059.2012.02693.x

3. Jarausch W, Fuchs A, Jarausch B (2010) Establishment of a quantitative real-time PCR assay for the specific quantification of Ca. Phytoplasma prunorum in plants and insects. In: 21st International Conference on virus and other graft transmissible diseases of fruit crops. Julius-Kühn-Archiv 427:392–394

4. Jarausch W, Peccerella T, Schwind N et al (2004) Establishment of a quantitative real-time PCR assay for the quantification of apple proliferation phytoplasmas in plants and insects. Acta Hortic 657:415–420. https://doi.org/10.17660/ActaHortic.2004.657.66

5. Cankar K, Stebih D, Dreo T et al (2006) Critical points of DNA quantification by real-time PCR effects of DNA extraction method and sample matrix on quantification of genetically modified organisms. BMC Biotechnol 6:37. https://doi.org/10.1186/1472-6750-6-37

6. Sedlak RH, Jerome KR (2013) Viral diagnostics in the era of digital polymerase chain reaction. Diagn Microbiol Infect Dis 75(1):1–4. https://doi.org/10.1016/j.diagmicrobio.2012.10.009

7. Dube S, Qin J, Ramakrishnan R (2008) Mathematical analysis of copy number variation in a DNA sample using digital PCR on a nanofluidic device. PLoS One 3(8):e2876. https://doi.org/10.1371/journal.pone.0002876

8. Hindson CM, Chevillet JR, Briggs HA et al (2013) Absolute quantification by droplet digital PCR versus analog real-time PCR. Nat Methods 10:1003–1005. https://doi.org/10.1038/nmeth.2633

9. Morisset D, Štebih D, Milavec M et al (2013) Quantitative analysis of food and feed samples with droplet digital PCR. PLoS One 8(5):e62583. https://doi.org/10.1371/journal.pone.0062583

10. Strain MC, Lada SM, Luong T et al (2013) Highly precise measurement of HIV DNA by droplet digital PCR. PLoS One 8(4):e55943. https://doi.org/10.1371/journal.pone.0055943

11. Dreo T, Pirc M, Ramšak Z et al (2014) Optimising droplet digital PCR analysis approaches for detection and quantification of bacteria: a case study of fire blight and potato brown rot. Anal Bioanal Chem 406:6513–6528. https://doi.org/10.1007/s00216-014-8084-1

12. Mehle N, Dreo T, Ravnikar M (2014) Quantitative analysis of "flavescence doreé" phytoplasma with droplet digital PCR. Phytopathogenic Mollicutes 4:9–15. https://doi.org/10.5958/2249-4677.2014.00576.3

13. Rački N, Dreo T, Gutierrez-Aguirre I et al (2014) Reverse transcriptase droplet digital PCR shows high resilience to PCR inhibitors from plant, soil and water samples. Plant Methods 10(1):42. https://doi.org/10.1186/s13007-014-0042-6

14. Rački N, Morisset D, Gutierrez-Aguirre I, Ravnikar M (2014) One-step RT-droplet digital PCR: a breakthrough in the quantification of waterborne RNA viruses. Anal Bioanal Chem 406(3):661–667. https://doi.org/10.1007/s00216-013-7476-y

15. Hren M, Boben J, Rotter A et al (2007) Real-time PCR detection systems for Flavescence dorée and Bois noir phytoplasma in grapevine: a comparison with the conventional PCR detection system and their application in diagnostics. Plant Pathol 56:785–796. https://doi.org/10.1111/j.1365-3059.2007.01688

16. Mehle N, Nikolić P, Rupar M et al (2013) Automated DNA extraction for large numbers of plant samples. In: Dickinson M, Hodgetts J (eds) Phytoplasma: methods and protocols, Methods in molecular biology, vol 938. Springer Science and Business Media LLC, New York, pp 139–145

17. Pinheiro LB, Coleman VA, Hindson CM et al (2012) Evaluation of a droplet digital polymerase chain reaction format for DNA copy number quantification. Anal Chem 84:1003–1011. https://doi.org/10.1021/ac202578x

18. Baker M (2012) Digital PCR hits its stride. Nat Methods 9:541–544. https://doi.org/10.1038/nmeth.2027

19. Mehle N, Dreo T, Jeffries C, Ravnikar M (2014) Descriptive assessment of uncertainties of qualitative real-time PCR for detection of plant pathogens and quality performance monitoring. EPPO Bull 44:502–509. https://doi.org/10.1111/epp.12166

20. EPPO (2014) PM 7/98 (2): specific requirements for laboratories preparing accreditation for a plant pest diagnostic activity. EPPO Bull 44:117–147. https://doi.org/10.1111/epp.12118

21. Gutiérrez-Aguirre I, Rački N, Dreo T, Ravnikar M (2015) Droplet digital PCR for absolute quantification of pathogens. In: Lacomme C (ed) Plant pathology: techniques and protocols, Methods in molecular biology, vol 1302. Springer Science+Business Media, New York, pp 331–347

22. Corbisier P, Pinheiro L, Mazoua S et al (2015) DNA copy number concentration measured by digital and droplet digital quantitative PCR using certified reference materials. Anal Bioanal Chem 407:1831–1840. https://doi.org/10.1007/s00216-015-8458-z

23. Dagata JA, Farkas N, Kramer JA (2016) Method for measuring the volume of nominally 100-μm-diameter spherical water-in-oil emulsion droplets. NIST Spec Publ. https://doi.org/10.6028/NIST.SP.260-184 Accessed 14 Dec 2017

24. Bogožalec Košir A, Divieto C, Pavšič J et al (2017) Droplet volume variability as a critical factor for accuracy of absolute quantification using droplet digital PCR. Anal Bioanal Chem 409:6689–6697. https://doi.org/10.1007/s00216-017-0625-y

25. Huggett JF, Foy CA, Benes V et al (2013) The digital MIQE guidelines: minimum information for publication of quantitative digital PCR experiments. Clin Chem 59 (6):892–902. https://doi.org/10.1373/clinchem.2013.206375

Chapter 15

Rapid Sample Preparation and LAMP for Phytoplasma Detection

Jennifer Hodgetts

Abstract

Loop-mediated isothermal AMPlification (LAMP) allows the rapid detection of pathogens by polymerase-mediated amplification of target nucleic acid sequences at a single incubation temperature. LAMP can be combined with very simple sample preparation/crude DNA extraction protocols, allowing the method to be used away from the laboratory for in-field detection. Equally, these benefits can also be leveraged to provide a rapid method suited to high-throughput diagnostic laboratories. In this chapter we described a crude DNA extraction protocol suitable for use in the field and provide a protocol for real-time detection using LAMP.

Key words Loop-mediated isothermal AMPlification (LAMP), Crude DNA extraction, PEG, In-field detection, Rapid testing

1 Introduction

Phytoplasma detection and identification primarily relies upon molecular diagnostic methods such as PCR, RFLP, real-time PCR, and DNA sequencing, all of which can provide sensitive and specific detection. However, these methods are relatively time-consuming and complex to perform, and are therefore primarily deployed within large centralized laboratories. Over recent years, there has been a drive toward simpler and quicker detection methods which can be performed by non-specialists away from a lab, for example in the field or at the point of border inspection. The primary features of Loop-mediated isothermal AMPplification (LAMP) which differentiates its performance for the other diagnostic methods available are the speed of amplification, along with the wide range of detection methods which can be applied to determine the result of the reactions. Furthermore, the robustness of the method to amplification inhibitors enables the use of crude sample preparation methods rather than time-consuming nucleic acid extraction protocols. These benefits mean that phytoplasma

Rita Musetti and Laura Pagliari (eds.), *Phytoplasmas: Methods and Protocols*, Methods in Molecular Biology, vol. 1875, https://doi.org/10.1007/978-1-4939-8837-2_15, © Springer Science+Business Media, LLC, part of Springer Nature 2019

detection can now be undertaken in the field and results obtained within 30 min.

LAMP uses strand displacement polymerases such as *Bst* polymerase in combination with 4 or 6 primers which bind to 6 or 8 locations of DNA. The polymerase used allows amplification to occur at a single temperature, removing the need for expensive and complex thermocyclers needed for PCR-based methods. The nature of the amplification, which results in an array of amplification products containing repeats of the target sequence, means that amplification is highly efficient, with up to 10^9 copies of the amplicon produced [1], providing sensitive detection. Broadly speaking, the sensitivity of a LAMP assay is between conventional PCR and real-time PCR, although this varies depending upon the assay and some tests can be as sensitive as real-time PCR [2, 3].

Fundamental to the successful use of LAMP, as with any molecular approach, is the availability of an assay with the desired specificity. To date, a number of phytoplasma assays with differing specificities have been published in primary literature, as summarized in Table 1. Subject to the availability of DNA sequence with adequate sequence differences between the desired target phytoplasma and other non-target (phytoplasma and/or bacterial) species, new LAMP assays can be developed as and when they are required. The design of new assays requires knowledge of primer design and the inherent nature of LAMP amplification and as such requires a specialist. However, once an assay has been designed and validated it can be readily deployed. Information on the design of new LAMP assays can be found in [14]. The inclusion of the optional loop primers is generally recommended as these can substantially reduce the amplification time of target DNA [15].

In conjunction with a specific phytoplasma assay, it is recommended that a control assay is used to test for the host matrix, for example the plant cytochrome oxidase I (COX) assay [16]. This performs a dual function of demonstrating that the sample extraction was successful and that the resultant extract supports amplification. Without the use of such a test, then samples negative for the phytoplasma assay cannot confidently be interpreted. All of the described LAMP assays (Table 1) have been used to test plant material. However, LAMP can also be applied to the testing of insect vectors. To date this has been demonstrated when combined with laboratory-based DNA extraction methods [8] rather than using rapid in-field extraction methods.

Due to the isothermal nature of the polymerases used, LAMP reaction incubation can be undertaken in a range of devices which can maintain a constant temperature with the required level of accuracy including water baths, heat blocks, and thermocyclers. LAMP reactions can be prepared using either individual component reagents, or alternatively using commercially available master-mixes. Most master-mixes are designed for use with real-time

Table 1
Summary of phytoplasma LAMP assays published in the primary literature (at the time of writing, December 2017)

Described target phytoplasma	Reference	Demonstrated level of specificity[a]	Notes
16SrI (Aster yellows group)	[4]	16Sr group (I)	
'Ca. Phytoplasma asteris' 16Sr I	[5]	16Sr group (I)	Assay specificity only tested against 6 other 16S groups
16Sr II	[6]	16Sr group (II)	
Napier grass stunt phytoplasma from 16Sr III and 16Sr XI	[7]	Assorted 16Sr groups	Assay detects 5 16Sr groups (16SrVI, X, XI, XII and XIV) but does not detect 16Sr III
Flavescence dorée (FD) phytoplasma	[8]	16Sr group (V)	Validated in line with EPPO standard. Assay detects all 16SrV sub-groups (not solely FD)
Waligama Coconut Leaf Wilt Disease (16Sr XI)	[9]	Unknown	No specificity testing against other phytoplasma isolates/16Sr groups described
'Ca. Phytoplasama solani' 16SrXII (bois noir (BN) phytoplasma)	[10]	16Sr group (XII)	Validated in line with EPPO standard. Assay detects all 16SrXII sub-groups (not solely BN)
16Sr XII	[6]	16Sr group (XII)	
16Sr X; 'Ca. P. mali', 'Ca. P. pyri' and 'Ca. P. prunorum' (apple proliferation, pear decline and European stone fruit yellows phytoplasmas)	[11]	Possibly 16Sr group (X)	The three other 16Sr X species ('Ca. P. spartii', 'Ca. P. rhamni' and 'Ca. P. allocasuarinae') were not tested
16SrXXII (cape St Paul wilt group)	[4]	16Sr group (XXII)	
Cassava witches' Broom (CWB) disease	[12]	Unknown	No specificity testing against other phytoplasma isolates/16Sr groups described
Coconut root wilt disease (RWD) and arecanut yellow leaf disease (YLD)	[13]	Unknown	No specificity testing against other phytoplasma isolates/16Sr groups described

[a]Based on the specificity testing data included in the publication

fluorescent monitoring for detection, whereas using component reagents allows the user to select the detection methodology of their choice. As LAMP is a relatively new method, improvements to the polymerases are ongoing, with new iterations being developed by enzyme manufacturers. For example, New England Biolabs currently sells four versions of *Bst* polymerase with iterative improvements to performance and commercially available kits use different proprietary polymerases, for example GspSSD polymerase in OptiGene Ltd. master-mixes. It is important to note that while LAMP assays can be used with different chemistries to that with which they were published, it is essential that the performance characteristics are validated. For example, isothermal master-mixes 001 and 004 from OptiGene Ltd. provide differing speeds of amplification and a prolonged amplification time with the quicker mix may cause late non-target amplification in non-target species to be observed which is not noted with the slower mix.

LAMP detection comes with a choice of detection strategies (reviewed in [17]) that can be selected based on what is most appropriate for the given testing scenario and laboratory. A range of factors may affect the choice, including if the method is open or closed tubed, the required sensitivity, the degree of subjectivity of interpretation of the result [18], and the availability of equipment. When first described, LAMP products were primarily resolved through the use of agarose gel electrophoresis or turbidity due to the accumulation of magnesium pyrophosphate during amplification. However, electrophoresis is not a closed tube system and thus poses risks of contamination, and turbidity can be subjective in its interpretation. Therefore, alternate strategies such as color change or fluorescent monitoring are becoming more common. Most detection strategies have a similar level of sensitivity, so a key variable in choice can be whether the method is open or close tubed. Open-tubed detection methods required the amplification vessel to be opened once amplification is complete (for example for the addition of a reagent), while with closed-tube detection methods all reagents can be added before amplification commences. This simple delineation is very important, as one of the most significant drawbacks of LAMP in any testing scenario is the risk of contamination. For this reason, it is highly recommended that closed-tube detection strategies are deployed wherever possible, and we would not recommend the use of open-tubed detection strategies.

Real-time detection requires the use of a fluorescent dye within the reaction and an instrument that allows fluorescent monitoring. The most common platforms used for this are real-time PCR instruments, or custom-made platforms such as the Genie® devices (OptiGene Ltd.). Detection is achieved via the use of double-stranded DNA intercalating dyes (such as PicoGreen) which monitor the accumulation of product. One advantage of this detection strategy is the additional information which is obtained through the use of melt-curve analysis of the product post-amplification. Each LAMP amplicon has a unique melting temperature based on the nucleotide composition, the confirmation of which confirms the specificity of amplification. While real-time turbidity monitoring can be performed, this is less commonly used and has not been applied to phytoplasma targets to date.

There are a broad range of end-point detection approaches, differing primarily in being open or close tubed. Open tubes methods include agarose gel electrophoresis, color change (for example with SYBR green I dye), and the use of labeled primers and a detection lateral flow device (LFD). While these approaches are widespread within the literature and provide unambiguous results, the risks of contamination of a laboratory outweigh the benefits. The most commonly used closed tube methods are turbidity and hydroxy naphthol blue (HNB) dye, although interpretation of results using either of these approaches is subjective. For example, the HNB color change is from violet to blue, which is less distinctive than with other reagents (although interpretation can be aided by freezing the product). A newer alternative to HNB is the product GeneFinder™ which has an unambiguous color change from orange to green; however at the time of writing this is not widely available. It is worth bearing in mind that with appropriate verification of performance characteristics (see below) published assays can be used with any detection methodology suited to the given testing scenario, not solely the one with which they were developed.

Before the uptake of any molecular assay into a testing regime it is crucial that the validation of the assay is assessed. Unfortunately, validation is not always adequate within publications, and publication alone should not be taken as being indicative of a well validated assay. Guidelines on validation have been published by The European and Mediterranean Plant Protection Organisation (EPPO) [19], but as a minimum the specificity and sensitivity of assays should be confirmed prior to use. Distinct to assay validation, and of particular note when an assay is being transferred to different reagents or a different detection strategy, is the verification of the published assays performance in the testing lab. These checks ensure that the performance characteristics of the assay seen in the lab which developed the test are replicated. When transitioning an assay, the most important variables to consider are the primer concentrations, incubation temperature, and duration. Deviation from these is most likely to impact the assays performance and should be done with caution. Consideration should also be given to the polymerase used in the assay development.

In this chapter we describe a protocol for simple, rapid sample extraction, using an alkaline polyethylene glycol (PEG) lysis method combined with manual homogenization using ball bearings (modified from [20]). We then provide a method for real-time LAMP detection using fluorescent monitoring which can be performed in the field using a portable LAMP device, or in the laboratory using a real-time PCR machine.

2   Materials

2.1 General Molecular Biology Laboratory Plastic-Ware and Equipment

1. Pipettes, a range of sizes.
2. Filtered pipette tips.
3. 1.5 ml microcentrifuge tubes.
4. Vortex.
5. Microcentrifuge.
6. UV PCR cabinet.
7. pH meter (or pH indicator test strips).

2.2 Crude Sample Preparation Using Alkaline PEG Lysis

1. PEG extraction buffer: 60% PEG 200, 20 mM KOH, pH 13.3–13.5. Weigh 60 g PEG 200 (*see* **Note 1**) and add to 0.93 ml 2 M KOH and 39 ml water (*see* **Note 2**) in a glass bottle. Shake to mix and measure the pH (using pH indicator test strips, or a pH meter, *see* **Note 3**) to confirm that it is within the range of pH 13.3 to 13.5. Adjust using KOH if needed (*see* **Note 4**). Store at room temperature (approximately 20 °C).
2. 5–7 ml screw-cap plastic (polypropylene) bottles/tubes (sterile) (*see* **Note 5**).
3. Stainless steel ball bearings (sterile), 3 mm and/or 8 mm diameter.
4. Molecular biology grade water (DNase, RNase, and pyrogen free).

2.3 LAMP Combined with Fluorescence Monitoring in Real Time

1. Isothermal master-mix ISO-001 (OptiGene Ltd.) (*see* **Note 6**).
2. LAMP primers for pathogen and control assays. The internal (FIP and BIP) primers should be HPLC purified while the external (F3 and B3) and loop (F-loop and B-loop) primers can be purified by HPSF (or a higher level of purification).
3. Molecular biology grade water (DNase, RNase, and pyrogen free).
4. 0.2 ml amplification strips/plates compatible with the detection method (e.g., 96-well plate for real-time PCR machines, or OptiGene 8-well strips for the Genie® platform).
5. Real-time PCR machine, or Genie® (II or III) machine (OptiGene Ltd.) (*see* **Note 7**).

3   Methods

It is recommended that a control assay that amplifies the matrix/host DNA is used in parallel to pathogen-specific assays (*see* **Note 8**). This allows pathogen negative tests to be interpreted as true-negative results assuming the control assay has demonstrated that the DNA extract was successful and the extract supports amplification. The most commonly used assay is the plant COX assay [16], which has been demonstrated to amplify a broad range of plant species.

3.1 Crude Sample Preparation Using Alkaline PEG Lysis

1. Label a 5–7-ml screw-cap plastic tube with the sample name and add a sterile 8 mm stainless-steel ball bearing (*see* **Notes 9 and 10**).

2. Add 1 ml PEG buffer (*see* **Note 11**).

3. Add a thumb-nail (approximately 10 mm^2) sized piece of plant tissue (*see* **Notes 12 and 13**).

4. Shake vigorously by hand for 1 min to lyse/homogenise the tissue (*see* **Note 14**).

5. Dilute the sample 1 in 10 in water in a microcentrifuge tube, e.g., add 10 μl extract to 90 μl water, and vortex briefly to mix (*see* **Notes 15 and 16**).

6. Test the sample by LAMP within 2 h of extraction (*see* **Note 17**).

7. Dispose of the used extract following local plant health quarantine and chemical waste disposal procedures (*see* **Note 18**).

3.2 LAMP Combined with Fluorescence Monitoring in Real Time

1. Work in a DNA-free area and in a UV PCR cabinet to prepare the LAMP master-mix (*see* **Note 19**).

2. Remove reagents from the freezer and thaw in the dark at room temperature (*see* **Note 20**).

3. Vortex reagents briefly to mix and centrifuge briefly to pool.

4. Prepare a master-mix in a microcentrifuge tube for the number of reactions to be tested plus controls with a small excess following Table 2 (*see* **Notes 21–23**).

5. Vortex briefly to mix and centrifuge briefly to pool.

6. Aliquot 20 μl of the master-mix into each well of the reaction vessel (*see* **Note 24**).

7. Move to another area/lab (e.g., not the DNA-free area) and add 5 μl of the DNA extract to the reaction avoiding adding bubbles if possible (*see* **Notes 24 and 25**).

8. Vortex briefly to mix and centrifuge briefly to pool.

9. Move to the post-amplification area/lab.

10. Programme the real-time PCR/Genie® machine with the required incubation and anneal curve programme. A typical programme is 62 °C or 65 °C for 30 min (20 min with ISO-004) followed by an anneal curve from 98 °C to 75 °C with the temperature decreasing at 0.05 °C per second with fluorescent monitoring for the duration of the programme (*see* **Notes 26 and 27**).

Table 2
Components of a LAMP reaction (*see* Note 24)

Reagent	Stock concentration	Volume per reaction (µl)	Final concentration
Isothermal master-mix	1.67×	15	1×
F3 primer	10 µM	0.5	200 nM
B3 primer	10 µM	0.5	200 nM
FIP primer	100 µM	0.5	2 µM
BIP primer	100 µM	0.5	2 µM
F-loop primer	100 µM	0.25	1 µM
B-loop primer	100 µM	0.25	1 µM
Molecular grade water		To a final reaction volume of 20 µl	

11. Place the strips/plate into the machine and start the programme.

12. Once the run has completed, examine the results for the phytoplasma and control assays. Examples of a typical positive and negative result are shown in Fig. 1.

A reaction may be considered positive if **ALL** the following criteria are met:
- Amplification profile: an increase in fluorescence is seen with a sigmoidal amplification curve. Record the time to positive (Tp) value (*see* **Notes 28** and **29**).

- Anneal plot: a defined (tall and narrow) peak is present at the assay specific temperature. Record the anneal temperature (Ta) value. Discount small/broad peaks (*see* **Note 30**).

A reaction may be considered negative if **ALL** the following criteria are met:
- Amplification profile: No increase in fluorescence.

- Anneal plot: No anneal peak is present. Discount small/broad peaks.

13. The positive and negative controls for each assay and the extraction blank should conform to the relevant criteria above. Assuming the assay controls have worked as expected, determine the result of each sample using Table 3 (*see* **Notes 31** and **32**).

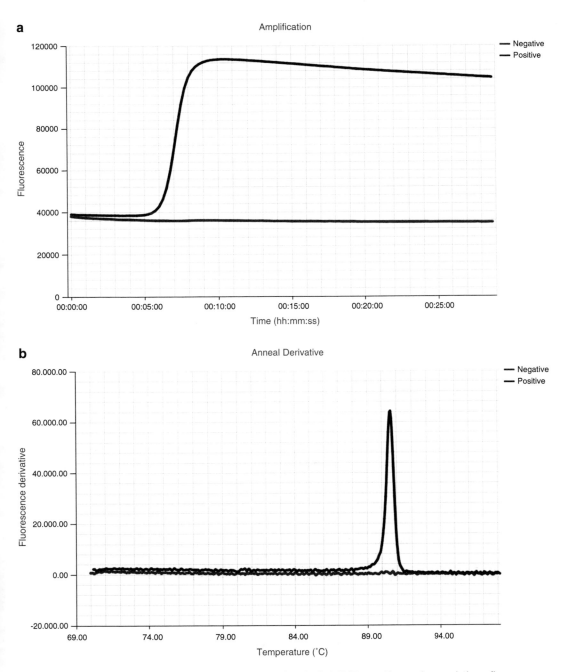

Fig. 1 Typical results of a positive (blue) and negative (red) LAMP reaction using real-time fluorescence monitoring. (**a**) The amplification profile, where the *x*-axis shows time and *y*-axis shows fluorescence, and (**b**) the anneal curve analysis, where the *x*-axis shows temperature and the *y*-axis shows fluorescence

Table 3
Interpretation of LAMP pathogen and control assay results (*see* Note 31)

Control assay	Phytoplasma assay	Overall result
Positive	Positive	Positive for phytoplasma
Positive	Negative	Negative for phytoplasma
Negative	Positive	Positive for phytoplasma but control assay issue
Negative	Negative	Extraction and/or amplification issue; re-test and/or re-extract the sample

4 Notes

1. PEG 200 is measured by mass rather than volume because of the viscosity of the liquid.

2. 2 M NaOH can be substituted for 2 M KOH in the buffer (in the same quantities).

3. The pH can be determined using a pH meter or a pH indicator test strip with the required level of precision.

4. In most cases, the pH should fall within the required range, however due to storage some batches of PEG 200 have an acidic, rather than neutral, pH. In this case add additional KOH as required to reach the target pH range.

5. Ensure that the tubes used for extraction are impact/shatter resistant with the selected ball bearing size(s) prior to use with samples/buffer.

6. OptiGene Ltd. sell two widely used master-mixes, ISO-001 and ISO-004. ISO-004 provides quicker amplification than ISO-001. If ISO-004 is used the incubation duration requires optimization to prevent the observation of nonspecific amplification in non-targets which are the equivalent of very prolonged run times (>1 h) with other master-mixes.

7. The Genie® II and Genie® III allow 16 and 8 reactions to be tested per run respectively. Real-time PCR machines generally allow the testing of 48 or 96 reactions in a run (or 384 reactions when using a reduced reaction volume, typically 10 μl).

8. While it is possible to multiplex LAMP assays to allow the pathogen and control assays (or two pathogen assays) to be combined in a single reaction, the assays must have different annealing temperatures so that they can be differentiated and the results interpreted [10]. Additionally, optimization is required to ensure that the assays amplification efficiencies are not altered (due to the large number of primers in the reaction,

which may interact). Therefore, it is generally easier to run LAMP assays separately.

9. The size of the ball bearing used should be optimized for the plant tissue being tested. We typically find either a single large (8 mm diameter) ball bearing, or 5–7 smaller (3 mm diameter) ball bearings to be effective for most plant tissues. Generally, the tissue should be broken down into smaller pieces after shaking; however, the tissue does not need to be completely ground/homogeneous.

10. For each batch of extracts an extraction blank should be prepared and tested. This should be treated in the same way as the samples but without the addition of plant material. This control confirms that the reagents are not contaminated and that cross-contamination did not occur during processing.

11. The amount of buffer used to process samples should be approximately ten times greater than the sample size, but this can be optimized.

12. The sample size should be optimized for the given plant species being tested. It may benefit from tearing into smaller pieces/roughly cutting up with a scalpel before adding to the bottle, particularly for very fibrous/tough leaves or needles (e.g., pine/fir species).

13. The selection of the sample taken from the plant/leaf should consider the possible distribution of the phytoplasma within the plant/tissue, e.g., phloem-rich tissue such as petioles and mid-ribs should generate the highest probability of detection.

14. The shaking time can be optimized for each given matrix, typically ranging from 10 s to 2 min. Both over and under homogenization can be problematic in terms of inadequate tissue disruption leading to failed release of the pathogen, or excessive homogenization releasing excess inhibitors for the plant tissue. Length of time for homogenization is a key parameter to consider during the development and validation of a test and is a critical point when transferring technology to end users.

15. Dilution is required to maintain the LAMP reaction composition/pH, and PEG extracts should not be added neat to a LAMP reaction.

16. The optimal dilution can range from 1 in 10 to 1 in 500 depending upon the matrix being tested and should be determined for each matrix. During optimization experiments for a given pathogen/host matrix combination we would typically assess the following dilution range; 1 in 10, 1 in 20, 1 in 50, 1 in 100, and 1 in 200. Bear in mind that the optimal dilution may vary for the plant control and phytoplasma assays.

17. Store extracts at 4 °C if not testing immediately, and vortex briefly before use. PEG extracts are not suitable for long-term storage.

18. Ball bearings can be sterilized and reused if required. To allow this, after use first wash to remove any residual tissue, soak in 10% bleach for at least 1 h, wash thoroughly in several changes of water, and then autoclave to sterilize before reuse.

19. To minimize the chance of contamination it is desirable to physically separate the various stages of the process and to use different equipment (pipettes, racks, lab coats, etc.) for each stage. This would preferably be three different labs, or if this is not possible, three distinct areas of a lab. Each lab would be used as follows; (1) a DNA-free pre-amplification lab for reagent preparation/reagent storage and master-mix preparation, (2) a lab for adding DNA to the reaction (and DNA extraction if this cannot done in another different lab), and (3) a post-amplification lab for reaction incubation and detection of results.

20. LAMP master-mix should be stored/thawed in the dark to prevent degradation of the fluorescent dye.

21. The reaction composition provided is a general one that we use for initial testing of new assays. If an assay is being used from the literature it is recommended that the primer concentrations from the publication are used, even if the reagent is being switched (e.g., from *Bst* polymerase to a master-mix).

22. It is good practice to test each sample and control in duplicate, however for in-field testing it may be more practical to test each sample and control once, with confirmation of results in the laboratory if required.

23. A number of controls should be undertaken for each batch of LAMP testing, including (1) a no-template (negative) control to assess for contamination, (2) a positive amplification control to confirm the reagents/amplification were successful, and (3) an extraction blank to confirm the extraction reagents were not contaminated. It is good practice to use a positive control which is diluted to close to the assays limit of detection. It is good practice to test 2 negative controls, where one is sealed after the reagents are added to the reaction vessel (to confirm the reagents are not contaminated) and another which is sealed once the DNA extract has been added to the reaction vessel (to confirm that no contamination occurred during the addition of template DNA to the reactions).

24. When testing crude DNA extracts such as those from the described alkaline PEG lysis method, we generally recommend testing 5 μl of extract per 25 μl reaction. If testing a standard laboratory DNA extract (e.g., from a CTAB or silica spin-

column-based method) then we would recommend testing 1 or 2 μl of extract per 25 μl reaction. Adjust the volume of water in the master-mix based on the volume of template DNA to be tested per reaction.

25. Bubbles should be avoided in the reaction as movement of the bubbles during the reaction may alter the fluorescent profile (resulting in step changes). Gently flicking the tube may aid the removal of bubbles.

26. When using an assay from the literature the incubation temperature should be as described in the paper. The incubation time may be varied depending upon the type of DNA extract tested (pure or crude DNA) and the polymerase used.

27. When LAMP was first described an initial denaturation of template DNA was often included in the incubation profile, however in our experience this step is not necessary.

28. The fluorescent value observed is dependent upon each assay but should be relatively consistent across runs and of a level where the signal to noise can be clearly distinguished. Some assays have an intrinsically low fluorescence level and may benefit from the addition of further intercalating dye to the reaction. This should be optimized and titrated on a per assay basis as an excess of intercalating dye can be inhibitory to the LAMP reaction.

29. The amplification profile should be sigmoidal in shape. Other profiles (e.g., more linear) may be indicative of inhibition of the reaction. In this instance retest the sample with a greater factor dilution of the crude extract.

30. The anneal temperature is unique for each assay and should be determined during assay validation. Anneal curves at a different temperature may indicate nonspecific amplification or degradation of reagents and should be investigated further. Occasionally the presence of primer dimers (particularly in negative reactions) may result in a small, broad anneal curve at a lower temperature to that obtained from positive samples. These can be clearly distinguished from positive samples by the different temperature and lower signal level and are not positive results.

31. Samples that are positive for the pathogen but are negative with the control assay may be considered positive, however an investigation as to the cause of the failure of the control assay should be made.

32. Due to the large amount of amplification product generated in LAMP, measures to prevent laboratory contamination are essential. A primary means to avoid this is physical separation of the stages as described in **Note 19**, and only using closed-tube detection methods so that completed LAMP reactions are

never opened in the lab. Adequate controls (*see* **Note 23**) should always be tested to enable identification of contamination.

Acknowledgments

This work was funded with the support of the Plant Health Division of the UK Department for Environment, Food & Rural Affairs (Defra).

References

1. Notomi T, Okayama H, Masubuchi H, Yonekawa T, Watanabe K, Amino N, Hase T (2000) Loop-mediated isothermal amplification of DNA. Nucleic Acids Res 28(12):e63. https://doi.org/10.1093/nar/28.12.e62

2. Hodgetts J, Hall J, Karamura G, Grant M, Studholme DJ, Boonham N, Karamura E, Smith JJ (2015) Rapid, specific, simple, in-field detection of *Xanthomonas campestris* pathovar *musacearum* by loop-mediated isothermal amplification. J Appl Microbiol 119 (6):1651–1658. https://doi.org/10.1111/jam.12959

3. Bühlmann A, Pothier JF, Tomlinson JA, Frey JE, Boonham N, Smits THM, Duffy B (2013) Genomics-informed design of loop-mediated isothermal amplification for detection of phytopathogenic *Xanthomonas arboricola* pv. *pruni* at the intraspecific level. Plant Pathol 62(2):475–484. https://doi.org/10.1111/j.1365-3059.2012.02654.x

4. Tomlinson JA, Boonham N, Dickinson M (2010) Development and evaluation of a one-hour DNA extraction and loop-mediated isothermal amplification assay for rapid detection of phytoplasmas. Plant Pathol 59 (3):465–471. https://doi.org/10.1111/j.1365-3059.2009.02233.x

5. Sugawara K, Himeno M, Keima T, Kitazawa Y, Maejima K, Oshima K, Namba S (2012) Rapid and reliable detection of phytoplasma by loop-mediated isothermal amplification targeting a housekeeping gene. J General Plant Pathol 78 (6):389–397. https://doi.org/10.1007/s10327-012-0403-9

6. Bekele B, Hodgetts J, Tomlinson J, Boonham N, Nikolić P, Swarbrick P, Dickinson M (2011) Use of a real-time LAMP isothermal assay for detecting 16SrII and XII phytoplasmas in fruit and weeds of the Ethiopian Rift Valley. Plant Pathol 60(2):345–355. https://doi.org/10.1111/j.1365-3059.2010.02384.x

7. Obura E, Masiga D, Wachira F, Gurja B, Khan ZR (2011) Detection of phytoplasma by loop-mediated isothermal amplification of DNA (LAMP). J Microbiol Methods 84 (2):312–316. https://doi.org/10.1016/j.mimet.2010.12.011

8. Kogovšek P, Hodgetts J, Hall J, Prezelj N, Nikolić P, Mehle N, Lenarčič R, Rotter A, Dickinson M, Boonham N, Dermastia M, Ravnikar M (2015) LAMP assay and rapid sample preparation method for on-site detection of flavescence dorée phytoplasma in grapevine. Plant Pathol 64(2):286–296. https://doi.org/10.1111/ppa.12266

9. Siriwardhana PHAP, Gunawardena BWA Millington S (2012) Detection of phytoplasma associated with Waligama coconut leaf wilt disease in Sri Lanka by loop mediated isothermal amplification assay performing alkaline polyethylene glycol based DNA extraction. J Microbiol Biotechnol Res 2(5):712–716

10. Kogovšek P, Mehle N, Pugelj A, Jakomin T, Schroers H-J, Ravnikar M, Dermastia M (2017) Rapid loop-mediated isothermal amplification assays for grapevine yellows phytoplasmas on crude leaf-vein homogenate has the same performance as qPCR. Eur J Plant Pathol 148(1):75–84. https://doi.org/10.1007/s10658-016-1070-z

11. De Jonghe K, De Roo I, Maes M (2017) Fast and sensitive on-site isothermal assay (LAMP) for diagnosis and detection of three fruit tree phytoplasmas. Eur J Plant Pathol 147 (4):749–759. https://doi.org/10.1007/s10658-016-1039-y

12. Vu NT, Pardo JM, Alvarez E, Le HH, Wyckhuys K, Nguyen K-L, Le DT (2016) Establishment of a loop-mediated isothermal amplification (LAMP) assay for the detection of phytoplasma-associated cassava witches' broom disease. Appl Biol Chem 59

(2):151–156.   https://doi.org/10.1007/s13765-015-0134-7

13. Nair S, Manimekalai R, Ganga Raj P, Hegde V (2016) Loop mediated isothermal amplification (LAMP) assay for detection of coconut root wilt disease and arecanut yellow leaf disease phytoplasma. World J Microbiol Biotechnol 32:108. https://doi.org/10.1007/s11274-016-2078-4

14. Tomlinson J (2013) In-field diagnostics using loop-mediated isothermal amplification. In: Dickinson M, Hodgetts J (eds) Phytoplasma methods and protocols, Methods in molecular biology, vol 938, 1st edn. Springer, Humana Press, London, pp 291–300. https://doi.org/10.1007/978-1-62703-089-2

15. Nagamine K, Hase T, Notomi T (2002) Accelerated reaction by loop-mediated isothermal amplification using loop primers. Mol Cell Probes 16(3):223–229. https://doi.org/10.1006/mcpr.2002.0415

16. Tomlinson JA, Dickinson MJ, Boonham N (2010) Rapid detection of *Phytophthora ramorum* and *P. kernoviae* by two-minute DNA extraction followed by isothermal amplification and amplicon detection by generic lateral flow device. Phytopathology 100 (2):143–149. https://doi.org/10.1094/PHYTO-100-2-0143

17. Zhang X, Lowe SB, Gooding JJ (2014) Brief review of monitoring methods for loop-mediated isothermal amplification (LAMP). Biosens Bioelectron 61:491–499. https://doi.org/10.1016/j.bios.2014.05.039

18. Wastling SL, Picozzi K, Kakembo ASL, Welburn SC (2010) LAMP for human African trypanosomiasis: a comparative study of detection formats. PLoS Negl Trop Dis 4(11):e865. https://doi.org/10.1371/journal.pntd.0000865

19. EPPO (European and Mediterranean Plant Protection Organization) (2014) PM 7/98 (2) specific requirements for laboratories preparing accreditation for a plant pest diagnostic activity. EPPO Bull 44(2):117–147. https://doi.org/10.1111/epp.12118

20. Chomczynski P, Rymaszewski M (2006) Alkaline polyethylene glycol-based method for direct PCR from bacteria, eukaryotic tissue samples, and whole blood. BioTechniques 40:454–458. https://doi.org/10.2144/000112149

Chapter 16

Assembly of Phytoplasma Genome Drafts from Illumina Reads Using Phytoassembly

Cesare Polano and Giuseppe Firrao

Abstract

Genome drafts for the phytoplasmas may be rapidly and efficiently assembled from NGS sequence data alone exploiting the proper bioinformatic tools and starting from properly collected samples. Here, we describe the use of the *Phytoassembly* pipeline (https://github.com/cpolano/phytoassembly), a fully automated tool that accepts as input row Illumina data from two samples (a phytoplasma infected sample and a healthy reference sample) to produce a phytoplasma genome draft, using the healthy plant host genome as a filter and profiting from the difference in reads coverage between the genome of the pathogen and that of the host. For phytoplasma infected samples containing >2% of pathogen DNA and an isogenic healthy reference sequence the resulting assemblies span the almost entire genomes.

Key words Illumina, *Candidatus* Phytoplasma, Second generation sequencing, Genome draft

1 Introduction

Genomics of the phytoplasma is made challenging by the fact that they are difficult to cultivate in vitro [1]. To obtain phytoplasma-enriched DNA, protocols typically involve time consuming isolation and purification of nucleic acids from plant or insect infected tissue using CsCl equilibrium buoyant density gradient in the presence of bisbenzimide [2], or physical isolation by pulsed-field gel electrophoresis (PFGE) of entire chromosomes [3]. Currently, only for four phytoplasmas the genomes have been sequenced to completion: "*Ca*. Phytoplasma asteris" Onion Yellows phytoplasma strain M [3], "*Ca*. P. asteris" Aster Yellows phytoplasma strain Witches' Broom [4], "*Ca*. P. mali" strain AT [5], and "*Ca*. P. australiense" strains Paa and SLY [6, 7].

As New Generation Sequencing (NGS) technology has become increasingly common, it has allowed the use of bioinformatics to select the sequences of the pathogen from a library of DNA extracted and random-sequenced from a diseased plant sample. However, selecting the pathogen sequence is not trivial; therefore,

Rita Musetti and Laura Pagliari (eds.), *Phytoplasmas: Methods and Protocols*, Methods in Molecular Biology, vol. 1875, https://doi.org/10.1007/978-1-4939-8837-2_16, © Springer Science+Business Media, LLC, part of Springer Nature 2019

Table 1
Available phytoplasma genome drafts with their accession number for retrieval from the Assembly database of NCBI (https://www.ncbi.nlm.nih.gov/assembly/)

"*Ca*. P. asteris" (Wheat blue dwarf ph.)	GCA_000495255.1
"*Ca*. P. asteris" strain CYP (Chrysanthemum yellows ph.)	GCA_000803325.1
"*Ca*. P. asteris" strain NJAY	GCA_002554195.1
"*Ca*. P. asteris" strain OY-V ("*Chrysanthemum coronarium*" ph.)	GCA_000744065.1
"*Ca*. P. aurantifolia" strain WBDL	GCA_002009625.1
"*Ca*. P. oryzae" strain Mbita1	GCA_001578535.1
"*Ca*. P. phoenicium" strain SA213	GCA_001189415.1
"*Ca*. P. pruni" strain CX	GCA_001277135.1
"*Ca*. P. solani" strain 231/09	GCA_000970395.1
"*Ca*. P. solani" strain 284/09	GCA_000970375.1
"*Echinacea purpurea*" witches'-broom ph. strain NCHU2014	GCA_001307505.1
Italian clover phyllody ph. strain MA1	GCA_000300695.1
Milkweed yellows ph. strain MW1	GCA_000309485.1
Peanut witches'-broom ph. NTU2011	GCA_000364425.1
Phytoplasma sp. strain Vc33	GCA_001623385.2
Poinsettia branch-inducing ph. strain JR1	GCA_000309465.1
Rice orange leaf ph. strain LD1	GCA_001866375.1
Vaccinium witches'-broom ph. strain VAC	GCA_000309405.1

some genome drafts obtained so far using this method are incomplete. At the time of writing, the genome drafts available for the phytoplasmas [8–17] are listed in Table 1 with their accession numbers, which will be a primary source for reference and comparison when producing a new genome draft.

A major obstacle in producing a phytoplasma genome drafts is the need to select the phytoplasma genome sequences by sorting the large amount of sequence information made available by NGS. The *Phytoassembly* pipeline is a dedicated tool developed for this purpose and it is described here, for the use by the non-specialist.

2   Materials

The *Phytoassembly* pipeline is written in the BASH and PERL languages and requires a working installation of BioPerl (http://bioperl.org/), NCBI BLAST+ (https://blast.ncbi.nlm.nih.gov/Blast.cgi) and the A5 pipeline (https://sourceforge.net/projects/

ngopt/) [18]. *Phytoassembly* works on GNU/Linux and macOS. At the moment of this writing, the A5 pipeline does not run on Windows, and therefore *Phytoassembly* could not be used with that OS.

The procedure requires two files as input: a reference genome from an uninfected plant in FASTA format and the sequence reads to be analyzed in FASTQ format. The reference genome can also be provided as FASTQ reads to be assembled; an assembly of the sequence reads to be analyzed, in FASTA format, can also be provided if available.

For best results, the healthy plant should be isogenic to, and grown in the same environment as the diseased specimen, so as to match the plant genome and include the same contaminants; on the other hand, it is also possible to input a collection of reference genomes (simply by joining the relative FASTA files), e.g., to filter out known pathogens.

3   Methods

Phytoassembly exploits on one hand the differential in coverage of the sequences originating from the pathogen and the host, which allows us to discard a significant part of the (under-represented) sequences from the plant, and on the other hand the mapping of the remaining reads on a healthy plant reference, which filters out the rest of the plant sequences.

The pipeline requires both a sequencing of a diseased plant sample and a sequencing of a healthy plant as close as possible to the diseased one. The DNA extracted from the diseased sample should be enriched so that the phytoplasma portion is preferably >2%–10% of the total; if the phytoplasma portion is around or less than 1%, the results will be unreliable. Methods and meterial required to perform this preliminary step are reported in the step by step protocol below as references to other chapters of this book.

The first steps of the pipeline consist in a preassembly, the estimation of pre-contigs coverage and calculation of the cutoff value. Then the Illumina reads belonging to contigs above the cutoff are selected and aligned against the healthy plant genome reference, so that those pertaining to the plant can be discarded and the non-plant reads can be assembled in preliminary phytoplasma assembly. Further polishing is carried out to filter out ambiguous contigs, originating from low-quality reads from the plant. This is based on the percentage of identity of BLAST matches against the healthy plant reference.

Step by step protocol:

1. As mentioned, the data to be analyzed consists in a sample from a plant that presents symptoms of phytoplasma infection. For best results, a sample should also be taken from a healthy plant closest to the diseased one, ideally isogenic.

2. Extract the nucleic acids from the collected samples using the methods described in Chapter 6.
Optionally, the phytoplasma concentration could be evaluated by quantitative PCR using the methods described in Chapter 10 (*see* **Note 1**).

3. Send the nucleic acids to a suitable service for Illumina DNA sequencing: you will need paired-ends reads at least 80 nts long, and 1,000,000,000 nts total sequence (or less if the phytoplasma concentration is high); for best results, use MiSeq 2 × 300 nt runs. A tool like FastQC (http://www. bioinformatics.babraham.ac.uk/projects/fastqc) can provide quality assessment of the sequence data received.

4. Make sure Perl 5.6.1 or later is installed by opening a Terminal window and typing "perl -v" (*see* **Note 2**). Install the BioPerl package (*see* **Note 3**).

5. Download the BLAST+ installer from the NCBI site and install the executables (*see* **Note 4**). Check the setup by typing "blastn -v" in a Terminal window.

6. Download the *A5 pipeline* and install it according to the instructions (*see* **Notes 5** and **6**). Check the setup by typing "a5_pipeline.pl" in a Terminal window: you should get the usage notes.

7. Download *Phytoassembly,* put the files into your preferred bin folder, and add it to your $PATH if required (*see* **Note 6**). Check the setup by typing "phytoassembly.sh" in a Terminal window: you should get the usage notes.

8. Prepare a folder with the sequencing data: reads will be stored as FASTQ files (*.fastq or *.fq), either in two (paired) or one (interleaved) file (*see* **Note 7**). If you have an assembled reference, it should be in nucleotidic FASTA format (*.fasta, *.fna, *. fas or *.fa; not *.faa, which is usually aminoacidic).

9. Launch a Terminal window, navigate to the data folder you created and run *Phytoassembly* by issuing the command (*see* **Note 8**):

phytoassembly.sh [-skipref -ref REFERENCE_GENOME.FASTA | -ref REFERENCE_GENOME.FASTQ] (-ref2 REFERENCE_GENOME. FASTQ_2) [-skipreads -readsfa READS_CONTIGS.FASTA] [-reads READS.FASTQ] (-reads2 READS.FASTQ_2) [-threads THREADS] [-min CUTOFFMIN -max CUTOFFMAX -step CUTOFFSTEP].

10. Examine the results: the most important file in the data folder will be one or more of *Stage3.x.contigs.phyto.fasta.gz,* the assembled phytoplasma genome, where *x* is the cutoff determined by the pipeline or those from the custom interval. Other potentially important files are *Diseased.ec.fastq.gz,* which will contain the error-corrected reads from the diseased plant data, *stats.txt,* which will contain statistics about the files created during the run (*see* **Note 9**), and *infoxxxxxx.txt,* which will contain all the printouts from the pipeline. In the *Other_files* folder, *Stage3.x.contigs.phyto.csv* and *Stage3.x.contigs.plant.csv* will contain the sequences attributed using BLAST to the phytoplasma and to the plant, respectively.

11. Optionally, if the assembly results are not satisfactory, run the pipeline a second time using phytoiterative.sh (the syntax is the same as in **Note 8**, but without the "-min", "-max" and "-step" options).

12. The draft phytoplasma assembly is now ready for gene finding and annotation, e.g., with an annotation server like RAST [19] or MG-RAST [20].

4   Notes

1. Quantitative PCR is necessary if there are several samples and the one containing the highest percent of phytoplasma DNA has to be identified, or if there is no information about the phytoplasma titre in that host: if the phytoplasma DNA is less than 1% total DNA, then it will be difficult to obtain a good draft. Quantitative PCR is not a necessary step, as the cutoff is estimated internally.

2. Perl is normally installed by default in Linux and MacOS. In the unlikely case it is not, or is outdated, you can get the latest version from https://www.perl.org/.

3. Many Linux distributions have a *bioperl* package; you can, e.g., use the Terminal command "sudo apt-get install bioperl" If a package is not available, the simplest method is probably using *perlbrew* and *cpanminus:* first install *perlbrew* (e.g., with "apt-get install perlbrew"), then type "perlbrew install-cpanm," followed by "cpanm Bio::Perl." A complete guide to install BioPerl is available at http://bioperl.org/INSTALL.html.

4. As of this writing, BLAST installers are available at ftp://ftp.ncbi.nlm.nih.gov/blast/executables/blast+/LATEST/. MacOS users should download the *.dmg file, double-click it, then launch the installer. For Linux users, the easiest method is probably using instead the Terminal command "sudo apt-get install ncbi-blast+"

5. To run the *A5 pipeline,* you will need a Java Runtime Environment (https://www.java.com/download/). MacOS users should download the *.dmg file, double-click it, then launch the installer. Linux users with administrative privileges can use the Terminal command "sudo mv /path/to/jre-*xxxx*-linux-x64.tar.gz" then "cd /usr/java/", then "sudo tar zxvf jre-*xxxx*-linux-x64.tar.gz" Without administrative privileges, create the *java* folder within the home folder, follow the above instructions, then add it to $PATH using the Terminal command "cd ; echo 'export PATH=$HOME/java/jre-*xxxx*: $PATH' >> .bashrc"

6. For ease of access to the respective commands, put the *A5 pipeline* and the *Phytoassembly* scripts into, e.g., a "bin" folder inside your home folder ("cd ; mkdir bin"). Then add the *bin* folder to your $PATH, by using the Terminal command "cd ; echo 'export PATH=$HOME/bin:$PATH' >> .bashrc" (replace ".bashrc" with ".profile" on macOS).

7. Make sure to use paired or interleaved reads, as the *A5 pipeline* cannot assemble single-end reads, which are also generally too short to produce good results for the purposes of this method. With interleaved reads, usually file 1 contains the forward and file 2 the reverse reads.

8. If you have an assembly for the healthy plant genome, use "-skipref -ref REFERENCE_GENOME.FASTA" *in place of* "-ref REFERENCE_GENOME.FASTQ". If you have paired FASTQ files, use "-ref REFERENCE_GENOME.FASTQ" with the forward and "-ref2 REFERENCE_GENOME.FASTQ_2" with the reverse file. If you also have an assembly for the diseased plant genome, load it with "-skipreads -readsfa READS_CONTIGS.FASTA" You will need to load the diseased plant reads with "-reads READS.FASTQ" (and "-reads2 READS.FASTQ_2" if the file is paired). You can specify the number of threads with "-threads THREADS", and you can pick a custom range of cutoffs to test with "-min CUTOFFMIN -max CUTOFFMAX -step CUTOFFSTEP"; if none of these last three flags are specified, the pipeline will determine a convenient value.

9. The statistics relative to the phytoplasma assembly are in the *stats.txt* file, in a line that will look like, e.g., "File: Stage3.5. contigs.phyto.fasta sequences: 237 nt|aa: 706848 mean: 2638.7 G+C: 26.1%" Typically, a phytoplasma draft is 400,000–1,000,000 nts long.

Appendix: How *Phytoassembly* Works

1. The procedure uses the *A5 pipeline* to assemble the healthy plant sequence reads (*Healthy.contigs.fasta*), if no assembly was provided. The remaining files are archived. Next, the diseased plant reads are assembled (producing the file *Diseased.contigs. fasta*). A step in the *A5 pipeline* produces error corrected reads (*Diseased.ec.fastq*), which are used in all the subsequent steps.

2. The assembled reference sequence file is then indexed and aligned with the error corrected reads by the *BWA* tool [21] using the *index* and *mem* commands. Using the *samtools* (http://www.htslib.org/doc/samtools.html) commands *view, sort, index* and *idxstats*, a summary of statics is produced (*Diseased.sorted.csv*), consisting of the reference sequence name, sequence length, number of mapped, and unmapped reads.

3. This summary is passed to a *phytocount.pl* to estimate a cutoff value, by running once with cutoff 0, then using a fraction of the ratio between the sum of the lengths of the non-mapping reads at cutoff 0 (*Stage2.0.nonmatch.fastq*, see below) and the sum of the lengths of the error corrected reads (*Diseased.ec. fastq*) of the diseased plant. Alternatively, if the user wants to supply a range of specifies fixed cutoff values, then the pipeline repeats the following steps from the lowest to the highest values provided (represented here as *$cutoffval*).

4. From the summary of statistical data (*Diseased.sorted.csv*), per-contig coverages are calculated and saved in a text file (*Diseased.sorted.cov.csv*).

5. The contigs with a coverage higher than *$cutoffval* are exported (*Diseased.cutoff.$cutoffval.fasta*, where *$cutoffval* is, e.g., "10"). The error-corrected reads from the diseased plant (*Assembly.ec.fastq*) are then aligned to the contigs in that last file using *BWA* (*Stage1.$cutoffval.match.sam*).

6. Using *phytofilter.pl* the reads above the cutoff from the alignment file are extracted and exported (*Stage1.$cutoffval.match. fastq*), using the SAM flag #4 ("the query sequence itself is unmapped") as filter.

7. These reads are now aligned with *BWA* against the healthy plant reference (*Healthy.contigs.fasta*), and the reads that do not align are exported (*Stage2.$cutoffval.nonmatch.fastq*). These non-aligned reads are assembled with the *A5 pipeline* (*Stage3.$cutoffval.contigs.fasta*).

8. A BLAST nucleotide database is created, using *phytoblast.pl*, from the reference healthy plant file (*Healthy.contigs.fasta*, which could also be a combination of different references) and used to query the contigs outputted by the previous stage (*Stage3.*

$cutoffval.contigs.fasta) using *tblastx*. The results are saved in table (*Stage3.$cutoffval.contigs.csv*), which is then filtered according to the identity percentage (IP): entries with an IP greater than 95% are attributed to the plant (*Stage3.$cutoffval. contigs.plant.csv*), while those with a lower IP are attributed to the phytoplasma (*Stage3.$cutoffval.contigs.phyto.csv*). Using this last file the contigs pertaining to the phytoplasma are extracted from the query (*Stage3.$cutoffval.phyto.fasta*).

9. Lastly, the main outputs are archived and moved to a folder (*Results_$timestamp*), statistical data such as contigs size and number are calculated, and intermediate files are moved to a sub-folder (*Other_files*).

References

1. Tran-Nguyen LTT, Gibb KS (2007) Optimizing phytoplasma DNA purification for genome analysis. J Biomol Tech 18:104–112

2. Saeed E, Seemüller E, Schneider B et al (1994) Molecular cloning, detection of chromosomal DNA of the Mycoplasmalike organism (MLO) associated with Faba bean (*Vicia faba* L.) phyllody by southern blot hybridization and the polymerase chain reaction (PCR). J Phytopathol 142:97–106. https://doi.org/10. 1111/j.1439-0434.1994.tb04519.x

3. Oshima K, Kakizawa S, Nishigawa H et al (2004) Reductive evolution suggested from the complete genome sequence of a plant-pathogenic phytoplasma. Nat Genet 36:27–29. https://doi.org/10.1038/ng1277

4. Bai X, Zhang J, Ewing A et al (2006) Living with genome instability: the adaptation of phytoplasmas to diverse environments of their insect and plant hosts. J Bacteriol 188:3682–3696. https://doi.org/10.1128/ JB.188.10.3682-3696.2006

5. Kube M, Schneider B, Kuhl H et al (2008) The linear chromosome of the plant-pathogenic mycoplasma "*Candidatus* Phytoplasma Mali". BMC Genomics 9:306. https://doi.org/10. 1186/1471-2164-9-306

6. Tran-Nguyen LTT, Kube M, Schneider B et al (2008) Comparative genome analysis of "*Candidatus* Phytoplasma australiense" (subgroup *tuf*-Australia I; *rp*-a) and "*Ca*. Phytoplasma asteris" strains OY-M and AY-WB. J Bacteriol 190:3979–3991. https://doi.org/10.1128/ JB.01301-07

7. Andersen MT, Liefting LW, Havukkala I, Beever RE (2013) Comparison of the complete genome sequence of two closely related isolates of "*Candidatus* Phytoplasma australiense" reveals genome plasticity. BMC Genomics

14:529. https://doi.org/10.1186/1471-2164-14-529

8. Mitrovic J, Siewert C, Duduk B et al (2014) Generation and analysis of draft sequences of "Stolbur" phytoplasma from multiple displacement amplification templates. J Mol Microbiol Biotechnol 24:1–11. https://doi.org/10. 1159/000353904

9. Lee I-M, Shao J, Bottner-Parker KD et al (2015) Draft genome sequence of "*Candidatus* Phytoplasma pruni" strain CX, a plant-pathogenic bacterium. Genome Announc 3: e01117–e01115. https://doi.org/10.1128/ genomeA.01117-15

10. Kakizawa S, Makino A, Ishii Y et al (2014) Draft genome sequence of "Candidatus Phytoplasma asteris" strain OY-V, an unculturable plant-pathogenic bacterium. Genome Announc 2:e00944-14. https://doi.org/10. 1128/genomeA.00944-14

11. Fischer A, Santana-Cruz I, Wambua L et al (2016) Draft genome sequence of "*Candidatus* Phytoplasma oryzae" strain Mbita1, the causative agent of Napier grass stunt disease in Kenya. Genome Announc 4:e00297–e00216. https://doi.org/10.1128/genomeA.00297-16

12. Saccardo F, Martini M, Palmano S et al (2012) Genome drafts of four phytoplasma strains of the ribosomal group 16SrIII. Microbiology 158:2805–2814. https://doi.org/10.1099/ mic.0.061432-0

13. Chung W-C, Chen L-L, Lo W-S et al (2013) Comparative analysis of the peanut witches'--broom phytoplasma genome reveals horizontal transfer of potential mobile units and effectors. PLoS One 8:e62770. https://doi.org/10. 1371/journal.pone.0062770

14. Davis RE, Zhao Y, Dally EL et al (2013) "*Candidatus* Phytoplasma pruni", a novel taxon associated with X-disease of stone fruits, *Prunus* spp.: multilocus characterization based on 16S rRNA, *secY*, and ribosomal protein genes. Int J Syst Evol Microbiol 63:766–776. https://doi.org/10.1099/ijs.0.041202-0

15. Quaglino F, Zhao Y, Casati P et al (2013) "Candidatus Phytoplasma solani", a novel taxon associated with stolbur- and bois noir-related diseases of plants. Int J Syst Evol Microbiol 63:2879–2894. https://doi.org/10.1099/ijs.0.044750-0

16. Chen W, Li Y, Wang Q et al (2014) Comparative genome analysis of wheat blue dwarf phytoplasma, an obligate pathogen that causes wheat blue dwarf disease in China. PLoS One 9:e96436. https://doi.org/10.1371/journal.pone.0096436

17. Quaglino F, Kube M, Jawhari M et al (2015) "Candidatus Phytoplasma phoenicium" associated with almond witches'-broom disease: from draft genome to genetic diversity among strain populations. BMC Microbiol 15:148. https://doi.org/10.1186/s12866-015-0487-4

18. Tritt A, Eisen JA, Facciotti MT, Darling AE (2012) An integrated pipeline for *de novo* assembly of microbial genomes. PLoS One 7:e42304. https://doi.org/10.1371/journal.pone.0042304

19. Aziz RK, Bartels D, Best AA et al (2008) The RAST server: rapid annotations using subsystems technology. BMC Genomics 9:75. https://doi.org/10.1186/1471-2164-9-75

20. Glass EM, Meyer F (2011) The Metagenomics RAST server: a public resource for the automatic phylogenetic and functional analysis of metagenomes. In: Metagenomics complement approaches, Handbook of molecular microbial ecology, vol 8, pp 325–331. https://doi.org/10.1002/9781118010518.ch37

21. Li H, Durbin R (2009) Fast and accurate short read alignment with burrows-wheeler transform. Bioinformatics 25:1754–1760. https://doi.org/10.1093/bioinformatics/btp324

Chapter 17

Protocol for the Definition of a Multi-Spectral Sensor for Specific Foliar Disease Detection: Case of "Flavescence Dorée"

H. Al-Saddik, A. Laybros, J. C. Simon, and F. Cointault

Abstract

Flavescence Dorée (FD) is a contagious and incurable grapevine disease that can be perceived on leaves. In order to contain its spread, the regulations obligate winegrowers to control each plant and to remove the suspected ones. Nevertheless, this monitoring is performed during the harvest and mobilizes many people during a strategic period for viticulture. To solve this problem, we aim to develop a Multi-Spectral (MS) imaging device ensuring an automated grapevine disease detection solution. If embedded on a UAV, the tool can provide disease outbreaks locations in a geographical information system allowing localized and direct treatment of infected vines. The high-resolution MS camera aims to allow the identification of potential FD occurrence, but the procedure can, more generally, be used to detect any type of foliar diseases on any type of vegetation.

Our work consists on defining the spectral bands of the multispectral camera, responsible for identifying the desired symptoms of the disease. In fact, the FD diseased samples were selected after establishing a Polymerase Chain Reaction (PCR) confirmation test and then a feature selection technique was applied to identify the best subset of wavelengths capable of detecting FD samples. An example of a preliminary version of the MS sensor was also presented along with the geometric and radiometric required corrections. An image analysis based on texture and neural networks was also detailed for an enhanced disease classification.

Key words Flavescence dorée, Multispectral sensor, Texture analysis, Feature selection, Radiometric/geometric corrections, Classification

---

1 Introduction

Since one of the main causes of loss in the vineyard sector is due to diseases, a continuous protection approach is applied, which means that fungicides/pesticides are sprayed as uniformly as possible in the vineyard according to a regular, frequent calendar. More than ten treatments are executed per season in several of the main wine-producing regions worldwide. Hence, there is an interest in detecting initial symptoms of diseases to selectively target their treatment,

Rita Musetti and Laura Pagliari (eds.), *Phytoplasmas: Methods and Protocols*, Methods in Molecular Biology, vol. 1875, https://doi.org/10.1007/978-1-4939-8837-2_17, © Springer Science+Business Media, LLC, part of Springer Nature 2019

preventing and controlling the establishment of the infection and its epidemic spread to wider patches or to the whole vineyard.

In France, vineyards form a crucial component in territorial and economic development. Nowadays, there are over 2 million acres (800,000 hectares) of vineyards in 70 departments and 13 regions. There are currently 140,000 farms, and the field assures around 100,000 direct employments and 500,000 indirect. Between 7 and 8 billion bottles of wine are produced every year and over 147 million wine cases are exported for a total value of 7 billion euros [1].

There are currently two widely used techniques for vineyard disease detection: symptom monitoring and molecular approaches. Visual inspection is the most popular technique for grapevine monitoring. It is, however, subject to an individual's experience and can be affected by temporal variation; moreover, an expert is needed for permanent monitoring which is, on one hand, expensive and, on the other hand, not practical in wide fields. Under the category of molecular-based methods, the most common method for pathogen detection is PCR analysis. Nevertheless, even if this method, with all its various variants, guarantees reliable and accurate results, a variable amount of time is required for sample preparation (collection and extraction) and analysis.

New technologies, such as Remote Sensing (RS), have progressed tremendously in the last few years, allowing collecting at distance information about an object, area, or phenomenon. Thus, RS technology can be used for monitoring the vegetation status from distance, evaluating the crop spatial features, vigor, and health demands. In the scientific literature, some studies employed remote sensing means for vineyard disease detection. In [2], visible-near infrared spectroscopy was used to detect leafroll-associated virus-3. Reflectance spectra were acquired using a spectro-radiometer from healthy and infected grapevine leaves of two varieties (Cabernet Sauvignon and Merlot). Various vegetation indices and individual wavelength bands were found to be capable of assessing the infection. Under the Multi-Spectral (MS) imaging category, a multispectral sensor was used to calculate seven vegetation indices. These indices, together with the four MS bands (Red, Green, Blue & Near infrared) formed a vector of 11 features. The vector calculation resulted in better discrimination between healthy and grapevine leaf-roll diseased grapevines also in [3]. A MS camera was also used in [4] to take images in three spectral channels (Green, Red, and Near Infrared) with the aim of improving the detection sensitivity of powdery mildew symptoms in grapevine. Furthermore, Hyperspectral airborne imaging was applied to map leafroll infection in cabernet sauvignon grapevines in [5].

Among various grapevine infections, FD is a quarantine disease, particularly contagious and incurable. It can be transmitted from one plant to another by grafting or by a north-American leafhopper *Scaphoideus titanus*, as shown in Fig. 1.

Fig. 1 Development stages of S. titanus, principal vector of the FD disease. From left to right: third lavar stage, fifth lavar stage, adult. (Source [6] modified)

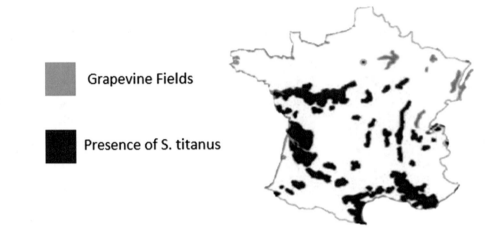

Fig. 2 Presence of FD in France. (Source [7] modified)

The FD is spreading in the South European vineyards causing heavy crop losses and putting in danger the sustainability of the vineyards. The contagious character of FD and its rapid spread make the fight against the disease obligatory and necessary. In France, the contamination by FD is expanding since the 50s and more than half of the French vineyards (450,000 ha) are subject to a mandatory control plan under national and European regulations (Fig. 2).

The French regulations enabled the containment of the disease in affected regions. Nevertheless, the progression of the infection did not stop in France or in Europe. For the moment, there is no definitive cure for the disease. To fight or prevent the FD damage, either insecticide treatments are applied massively or suspected grapevines are spotted after surveillance campaigns and then extracted.

An automated search and detection solution for grapevine diseases with a dedicated processing tool could offer a turnkey for winegrowers, enabling the search for potential FD foci, first, then more generally any type of detectable foliar vine diseases (Esca, Mildew, etc.). Possible advantageous economic consequences of the project include mainly benefit for the winemakers; in fact, they will be able to diagnose early symptoms of FD and, in consequence, implement control methods at a lower cost. The new MS tool could replace the prospection experts, actually, early detection is suggested to limit crop losses and thus increase productivity between 5 and 15%. The tool will diagnose FD outbreaks early to contain them as soon as possible; hence, there will be no need to spray a large number of phytosanitary products uniformly in the field. The localization of the infected zones will allow a local and direct treatment of contaminated vines. Thus, the pollution of soil and water will be reduced and biodiversity in the vine cultivars will be maintained.

Aerial Unmanned Vehicle (UAV) imaging is a recent technology that offers interesting advantages over remote sensing by satellite or manned aircrafts or embedded systems on land vehicles (tractors for example). The acquisition of images from UAVs has become a common practice because it is possible nowadays to install digital cameras on board. A Tetracam ADC-Lite MS camera, delivering a 0.05 m/pixel ground resolution at a flight height of 150 m, was used in [8]. The study resulted in a high correlation between Normalized Differential Vegetation Index (NDVI) acquired by the UAV and Grapevine Leafroll Disease (GLD) symptoms. Other commercial MS cameras proposed by other companies like Cubert are also available. In general, the maximum spatial resolution offered is not sufficient to visualize details on the leaves (<5 mm), essential for the disease detection task.

In this study, we present in the first section the materials needed for acquiring the data; then we introduce the protocols for spectral and image data acquisition and we explain the spectral data processing for choosing the MS sensor bands for FD foliar symptoms detection. Based on the selected subset of wavelengths, a MS test sensor could be installed along with the necessary geometric and radiometric corrections. The texture analysis as well as the Neural Network classification practiced on digital RGB images, can be transferred to the MS images for a better FD disease classification. In the last section, some conclusions and perspectives of the work are stated.

2   Materials

2.1   Spectral Data Acquisition Materials

– Spectro-radiometer: Spectral reflectance measurements from leaf surfaces were acquired using a portable Spectro-radiometer (FieldSpec 3, Analytical Spectral Devices, Boulder, CO, USA), as shown in Fig. 3a. Each sample data was taken every 1 nm from 350 nm to 2500 nm. This was the result of an interpolation performed by the software because the true spectral resolution of the instrument is about 3 nm at 700 nm wavelengths and

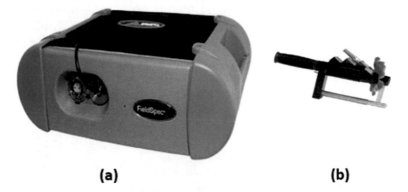

(a)   **(b)**

Fig. 3 Spectro-radiometer FieldSpec 3 (**a**) with the corresponding contact plant probe (**b**)

about 10 nm at 1400 or longer wavelengths. This spectro-radiometer is composed of three different sensors capable of providing spectral information from 350 nm to 2500 nm.

- Contact probe for vegetation: A contact probe has the general advantage of reducing the effect of environmental light scattering to insure better measurement accuracy. Measurements should be made on each leaf using a plant probe, specially designed with a low power source, for sensible vegetation surfaces, leaving no observable damage (Fig. 3b). This is employed instead of the traditional contact probe that could burn leaves due to a much more powerful Halogen bulb.

- Spectro-radiometer batteries for field usage: Two FieldSpec3 batteries should be previously charged and taken to the field to supply necessary power for the spectro-radiometer.

- White reference: The spectral response of a reference white surface is required because the reflectance of a target is computed by dividing its spectral response by that of the reference sample. Using this method, all parameters that are multiplicative in nature and present in both the spectral response of a reference sample of the target are ratioed out such as the spectral irradiance of the illumination source. An inherent assumption is that the characteristics of the illumination are the same for the reference and target materials.

- Black support: a black surface is placed behind each sample leaf prior to spectral measurement acquisition. This approach is used to ensure that the radiation reflected is uniquely due to the leaf since the black support absorbs the transmitted radiation by the leaf and does not reflect any.

- PC fully charged: the PC is used with the spectro-radiometer to acquire the reflectance spectra.

- Software RS3: the software comes with FieldSpec3, it allows the user to adjust the measurement configuration, enables the spectral acquisition with the spectro-radiometer, and saves the corresponding spectral data into memory.
- Connection cables: two from the spectro-radiometer toward the probe, one from the spectro-radiometer toward the batteries, and one ethernet cable from the spectro-radiometer toward the PC.

2.2 Digital Data Acquisition Materials

- Red-Green-Blue (RGB) camera: A digital camera Sony Alpha 5000 with 20 Mpixels resolution.
- Color Checker: The Color Checker Classic target is an array of 24 scientifically prepared natural, chromatic, and primary colored squares in a wide range of colors. Each solid patch is formulated individually to produce a pure, flat, rich color. This is used to perform color calibration of the RGB digital images since the natural illumination can vary during the day.
- A uniform background having a color different than healthy or contaminated leaves is needed (not green, nor yellow or red). It is placed behind each sample leaf prior to image acquisition to facilitate leaf segmentation.

2.3 Extra Materials

- GPS to mark the tested grapevines in the field and enable timely follow-up.
- Colored tapes for marking the grapevines and the leaves to facilitate the timely follow-up.
- A paper and a pen to keep track of some meta-data at each acquisition campaign (Fig. 4).

Error of measurements may have different sources: atmospheric conditions, operator, instrument, and vegetation itself. Since it is difficult to control all parameters, it seems important to save them in a metadata file. If we analyzed the data collected from two acquisition campaigns and had strange results, we could refer to the metadata file and attribute the difference in the result to the modified meta-data parameters between the two acquisition campaigns.

3 Methods

3.1 Location and Weather Requirements

1. Do not limit measurements to one field. Try testing several locations to verify the reliability of the results.
2. Consider many varieties of the vegetation in order to determine if the impact of the disease is the same or if it is function of the

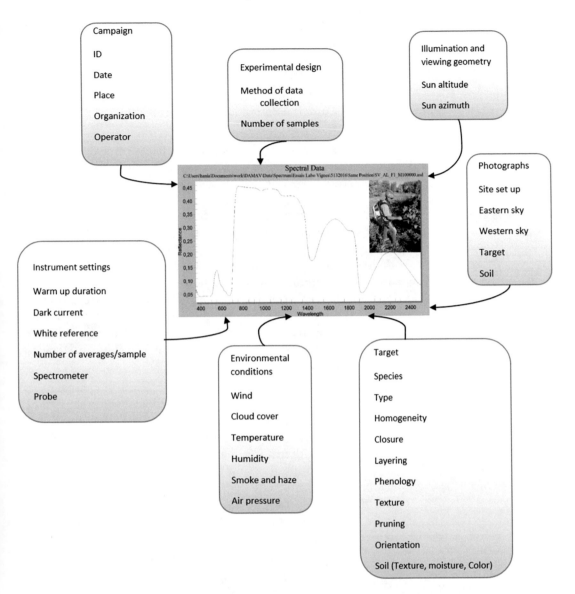

Fig. 4 Meta-data taking into account uncontrollable parameters during spectral data acquisition

variety. In our case four grapevine varieties were tested, two red-berried ones (Marselan, Grenache) and two white-berried ones (Vermentino, chardonnay).

3. Perform measurements several times during the season to see the effect of the severity stage on the results. For the case of FD disease on grapevines, it is important to perform at least two acquisition campaigns. The first one must take place in July when the symptoms affect only some parts of the leaves. Then a second one could be performed in September or in the beginning of October, when FD symptoms appeared clearly on the entire leaves.

4. Do not perform measurements when it is raining or directly after also do not take measurements in case of highly smoky and haze conditions because the water left on leaves can highly impact the spectral reflectance.

5. Do not perform measurements in case of extreme windy conditions because the digital images might be blurred if the leaves are moving fast.

6. Do not perform an acquisition campaign in extreme high temperatures (above 40°) because the DC noise is sensitive to temperature.

3.2 Sample Leaves Selection

1. Choose carefully symptomatic leaves. For FD, three symptoms must be present simultaneously on the same branch to accurately conclude the presence of the FD Yellowing disease: the discoloration of the leaves and their curved appearance, the absence of lignification of the new shoots, and the mortality of the inflorescences and berries. The color modification of leaves changes according to the variety: yellow for white-berried grapevines, red for red-berried ones; this might be the most noticeable symptom and is presented in Fig. 5. Detailed information about FD can be found in [7].

2. Let an expert be present during the acquisition campaign to confirm the occurrence of the disease and its severity stage. The distinction between different crop diseases is a complicated task. In fact, various infestations can cause almost identical symptoms. Furthermore, marks can be the result of a fusion of many infections affecting the plant at the same time. Some factors such as nutrient deficiencies, pollution, and pesticides can produce expressions indistinguishable from those of diseases. That is why it is advised to have a plant pathologist while selecting samples.

3. Select a range of healthy leaves of different ages.

4. Choose a set of infected leaves in order to get a complete and representative range of FD symptoms.

5. Examine at least a total of 30 infested and healthy samples for each variety, this is important to keep the study statistically significant. An example of the different number of samples we considered in our study is presented in Table 1.

6. Test the same number of healthy and diseased samples for each variety to avoid the bias in the analysis.

7. Perform laboratory analysis on tested symptomatic leaves after the end of the acquisition campaigns. The presence of symptoms alone is often not enough to identify the Yellowing in general; additional laboratory tests need to be done in order to confirm the presence of the disease. PCR analysis for example

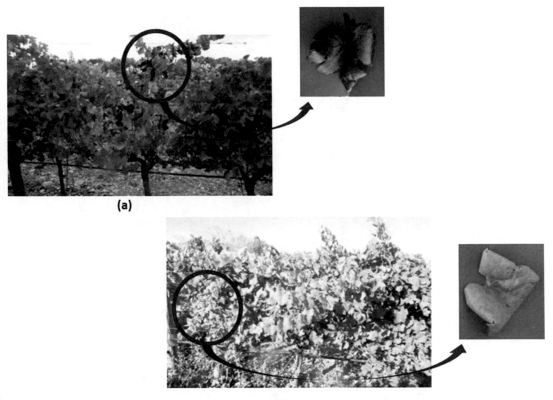

Fig. 5 FD foliar symptoms on Marselan (red) (**a**) and Chardonnay (white) grapevine varieties (**b**)

should be done in the closest laboratory to support the inspector's claim. For more details about plant material collection and DNA extraction, see Chapter 6. For phytoplasma detection, refer to Chapters 8–10. For multiplex-PCR protocol enabling the simultaneous detection of FD and BN (Bois Noir) please refer to [9].

8. Locate the grapevines using a GPS and label all tested leaves in order to ensure timely follow-up.

9. Choose another neighboring leaf when the old one is not available anymore. After testing some leaves in a campaign, they may not be present in the following one. Some of the leaves may either naturally fall or might be cut by the winegrower. Thus, when the leaf is not found, another candidate located on the same branch should be considered.

3.3 Spectral Data Acquisition

Spectral reflectance is the ratio of incident to reflected radiant flux measured from a surface over a defined range of wavelengths. Based on [10, 11], the spectral in-field measurements protocol was made:

1. Let the spectrometer warm up for at least 15 min before starting the measurements.

Table 1
Number of healthy and diseased samples acquired during each acquisition campaign in PACA region

Date		Number of diseased and healthy samples (ND and NH respectively)						
	Marselan	Grenache	Vermentino	Chardonnay	Red (Marselan +Grenache)	White (Vermentino +Chardonnay)	All (Marselan+Grenache +Vermentino+Chardonnay)
27/07/2017 (infestation level 1)	ND = 18 NH = 15	ND = 16 NH = 15	ND = 0 NH = 0	ND = 16 NH = 15	ND = 34 NH = 30	ND = 16 NH = 15	ND = 50 NH = 45
11/09/2017 (infestation level 2)	ND = 15 NH = 15	ND = 15 NH = 15	ND = 15 NH = 15	ND = 15 NH = 15	ND = 30 NH = 30	ND = 30 NH = 30	ND = 60 NH = 60
04/10/2017 (infestation level 3)	ND = 19 NH = 15	ND = 17 NH = 15	ND = 15 NH = 16	ND = 15 NH = 15	ND = 36 NH = 30	ND = 30 NH = 31	ND = 66 NH = 61
Total	ND = 52 NH = 45	ND = 48 NH = 45	ND = 30 NH = 31	ND = 46 NH = 45	ND = 100 NH = 90	ND = 76 NH = 76	ND = 176 NH = 166

2. During the warm up time, use some colored patches to determine the locations of the grapevines and the corresponding leaf samples that will be measured in order to enable a temporal tracking.

3. Connect the PC to the spectrometer and attach the plant probe.

4. Start the RS3 instrument software.

5. Adjust the measurement configuration: Fore optic selection (not used); spectral averaging selection spectrum averaging is the number of samples taken per observation: Spectrum (30), dark current (100), white reference (100). It is advised to use the same instrument settings for each acquisition campaign.

6. Place the probe on the white panel and perform an optimization, the profile will change while the instrument is adjusting, the different regions of the three detectors will be visible.

7. Perform white reference calibration by maintaining the probe on the white panel, a reflectance curve with a near horizontal line at a value of 1 should appear if the setup is correct.

8. Place a black support under each leaf before acquiring the spectral data to avoid interference.

9. Spectral measurements are taken from the four locations on leaves (shown in Fig. 6) and one measurement is taken per location. The locations were chosen according to the disease that has tendency to start growing between the veins first. A range of two-four measurements is considered depending on the leaf surface with respect to the probe diameter. When the leaf is small, only two reflectance spectra from two locations are acquired, and when the leaf is wide enough, four spectral tests

Fig. 6 Locations of the spectral measurements on a sample leaf

are taken from all the four locations. When ready, press the spacebar to save the measurement.

10. Repeat the white reference calibration, whenever there are changes in illumination conditions.

11. At the end of the spectral data acquisition, two, three, or four spectral measurements are acquired from one sample leaf. These are averaged, in order to obtain, in total, one spectra per sample.

3.4 RGB Digital Image Acquisition

To keep track of the spatial appearance of the disease on the leaf, its location, and how it is spreading on the leaf, an RGB image of the sample has to be taken. The protocol for image acquisition is as follows:

1. Prepare a uniform color cardboard and place a color chart on it. This will facilitate the leaf extraction task and the color calibration.

2. Place the cardboard behind each sample leaf.

3. Take one RGB image for each sample leaf (Fig. 7).

3.5 Spectral Data Analysis

After acquiring hyperspectral signatures, a protocol for choosing the optimal wavebands for distinguishing healthy from damaged leaves is presented next.

1. Notice the modifications in spectral data. The spectral reflectance characteristics change when the cellular leaf structure is altered due to the presence of a pathogen [12]. In fact, when comparing spectral signatures of healthy and infected leaves, some differences should be noticed, implicating that the spectral response was affected by the infestation. These depicted changes will help in defining the best spectral bands for disease

Fig. 7 Example of RGB image taken on a healthy Grenache leaf

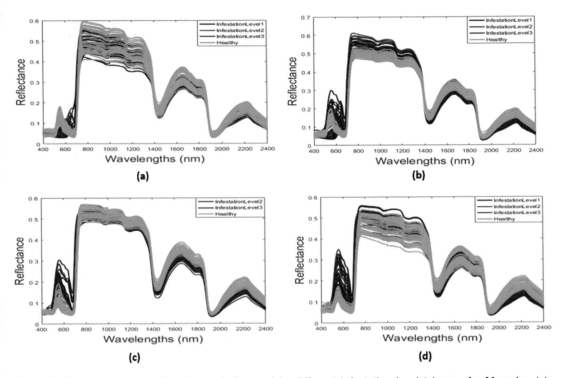

Fig. 8 Reflectance spectra of healthy and diseased (at different infestation levels) leaves for Marselan (**a**), Grenache (**b**), Vermentino (**c**) and Chardonnay (**d**) varieties

characterization. For example, in Fig. 8, the changes between infested and healthy signatures in Marselan (Fig. 8a) and chardonnay (Fig. 8b) prove that the spectral signature depends on the pathogen-host interaction. In other words, we found that the grapevine variety does not express the same pattern when the same infestation occurs.

2. Define the desired number of optimal spectral bands. This will depend on the hardware design limitation imposed by the application. Furthermore, if the sensor will be embedded on a platform, it is important to check the possible payload limit.

3. Apply the Successive Projection Algoritm (SPA) to select optimal wavelengths. Spectral data are described by many wavelengths (reflectance is measured at wavelengths from 350 to 2500 nm giving $W = 2151$ wavelengths); it seems beneficial to reduce the dimensionality of the data and select only some optimal wavelengths features capable of differentiating healthy from infected observations. SPA is a forward feature selection technique that enables locating the best subset of the original features [13]. SPA was used in many studies such as [14] or [15]. It starts with one wavelength and incorporates another one until a specified number is reached. The final selected wavelengths have minimal redundant information content.

4. Implement a Support Vector Machines (SVM) classifier with Radial Basis Function (RBF) kernel for spectral classification. Each subset of spectral features defined by SPA is considered and a model is formed based on the reflectance at these definite wavelengths. Then, each of the models is evaluated and its ability to classify healthy from diseased samples is tested with a cross-validated SVM-model. SVM is widely used for spectral analysis and gave good performance in several applications [16, 17]. The classifier belongs to supervised learning methods, which analyze a given set of labeled observations and then predict the labels of unknown future data. It seeks to find the optimal separating hyper plane between classes. For more details about SVM, refer to [18] who discussed in detail the concept and the application of SVM.

5. Evaluate the classification models by calculating the maximal area under the Receiver Operating Characteristic (ROC). The maximal Area Under the Curve (AUC) was chosen as a criterion of the performance of a given subset, it measures the performance of a classifier and is frequently applied for comparison. A higher AUC value means a better accuracy, accordingly, the subset producing an SVM classification model with the highest AUC is chosen and the wavelengths of that subset are considered as the optimal spectral features.

6. Combine different measurements to simulate the case where different varieties of the same crop were grown together. In our case, four varieties are taken into consideration: Marselan, Grenache, Vermentino, and chardonnay. We tested our approach when two varieties Marselan and Grenache measurements are combined in the red group and when Vermentino and chardonnay varieties measurements are combined in the white group. In the case where different grapevine varieties were grown together, Marselan, Grenache, Vermentino, and chardonnay were combined in the same group. This kind of configuration will define the reliability of the classification in a real field context and will enable possible common FD detection among varieties. Details about the adopted data configuration are given in Table 2.

7. Verify that the accuracy of the classification is higher than 80% in order to be significant. In our case good classification results were achieved and they are presented in Table 3 along with the ROC curves in Fig. 9.

8. Notice if the wavelengths are dependent on the variety considered. In our study, the best wavelengths selected were indeed, different from one case to another. Although the selected subset of wavelengths for each dataset gave good results, there was no single best set in all situation.

Table 2
Data configuration used

	Marselan	Grenache	Vermentino	Chardonnay
Severity of infestation 0 (Healthy)				
Severity of infestation 1 (Slightly infested)				
Severity of infestation 2 (Highly infested)				

———— Marselan   – – – – Grenache   —·— Vermentino   —··— Chardonnay

·········· White   – – – – Red   ▬▬▬ All

Table 3
Results of SVM classification on spectral information and best subset of bands corresponding to each data configuration

	Marselan	Grenache	Vermentino	Chardonnay	Red	White	All
Accuracy (%)	94.8	96.7	96.7	97.8	98.9	96.0	95.3
FM	0.949	0.967	0.965	0.977	0.989	0.958	0.953
AUC	0.997	0.997	0.995	0.984	0.991	0.991	0.990
Best wavelengths	557	450	450	485	450	487	452
	614	509	680	679	543	685	553
	672	620	701	719	597	721	619
	706	679	755	1114	679	1334	680
	790	944	1120	1495	712	1894	715
	1880	1715	1402	1888	1251	1924	816
	1929	1888	1715	1934	1879	2008	1879
	2303	1929	1938	2279	1918	2285	1945

Bold values correspond to wavelengths selected till 1000 nm. When wavelengths don't exceed this value, the sensor material will be made of Silicium and won't be expensive.

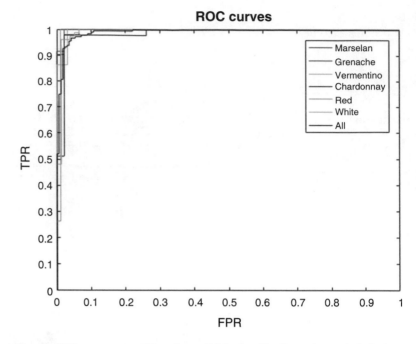

Fig. 9 ROC curves resulting from SVM classification of spectral features corresponding to each data configuration

3.6 MS Test Sensor Design and Corresponding Required Corrections

Once the best wavelengths are chosen, it is possible to implement the MS sensor. Some required geometric and radiometric corrections are needed in order to get good MS images, these are introduced next. The different corrections were implemented in Python programming language.

1. Implement the selected bands from the previous section as filters on the MS sensor. Choose the physical filters that fit the best to the set of optimal wavelengths.

2. Choose identical cameras with a resolution convenient for the application. In fact, the resolution should be chosen carefully with respect to the size of the symptoms of the infestation.

3. Choose corresponding lens, also this choice should depend on the distance between the sensor and the object of interest depending on the application. An example of a primary version of the MS sensor for FD detection is presented in Fig. 10. Here, we limited the number of bands to four.

4. Use a multi-processing technique in order to collect the flux relative to each of the cameras. A program should be implemented to visualize the cameras independently one from another but at a common instant t, MS images can hence be acquired from the sensors and stored in memory. The MS images obtained for our application are displayed in Fig. 11.

Fig. 10 Designed MS test sensor consisting of four identical UI-3370CP-M-GL IDS camera, with a resolution of 4.19 Mpx and a pixel size of 5,5 μm equipped with Fujinon CF16HA-1/F1.4/16 mm lens and four Midopt Narrow band filters {530, 63, 740, 810 nm}

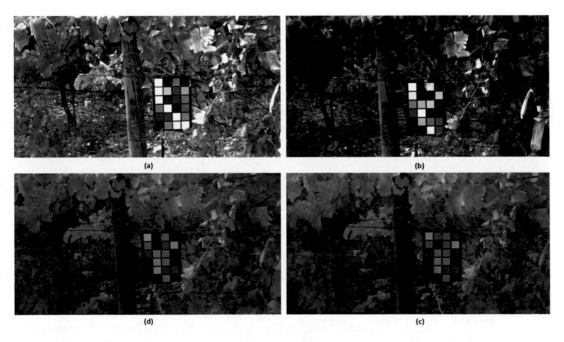

Fig. 11 MS images acquired from the prototype on Chardonnay variety, with images (**a**), (**b**), (**c**), (**d**) corresponding to the four cameras equipped with 530, 63, 740, 810 nm filters respectively

5. Correct the vignetting effect. It is a radial falloff in illumination strength, it appears as a shadowing increasing toward the image periphery. It is due to an occlusion of the detector plane [19]. This was not present in our case so no correction was needed.

6. Perform geometric corrections. Lens distortion [20, 21] is produced by a nonuniform magnification across the lens surface, measurements are radially shifted in consequence. Furthermore, lens distortion can also be generated by a misalignment between the lens and the detector plane, measurements in that case undergo a planar shift. The Brown-Conrady lens distortion model can compensate the effects of both uneven magnification and lens/detector plane misalignment. The model, however, requires the calculation of sensor-specific intrinsic and extrinsic coefficients. A common approach for the calculation of these coefficients is through the use of calibration panels, these are typically planar grids of known geometric properties. For each camera, the lens distortion model is constructed using the extracted coefficients. Once the four models are found, corresponding MS images are corrected.

7. Perform radiometric calibration. Remotely sensed data are affected by environmental conditions, this includes, atmospheric conditions, surface conditions, and illumination changes [22]. In most cases, only temporal change in light, matters, and radiometric calibration is performed to provide consistency in images taken under variable environmental conditions. The empirical line method is the most used approach to radiometric correction. A relationship is assumed to exist between the collected Digital Numbers (DN) measurements from the MS sensor and the reflectance of the surface of the imaged object [23]. For this purpose, a set of targets are used and known as Pseudo Invariant Features (PIFs). The method requires taking at least measurements at two conditions one light and another one dark. A color chart containing 24 regions is used as PIF in our case, it is a spectrally homogeneous surface, invariant with environmental conditions. The at-surface reflectance measurements from each region of the PIF can be measured using the FieldSpec3 portable spectrometer and the contact probe. Then, MS images and radiance of the incident light (acquired using the Photosynthetically active radiation sensor) are taken under different illumination conditions. For each illumination condition, the relationship between the DN measurements recorded within the imagery and the at-surface reflectance acquired from measurements, are calculated and a linear relationship is derived for each band. The slope and the offset of each line obtained are clearly function of

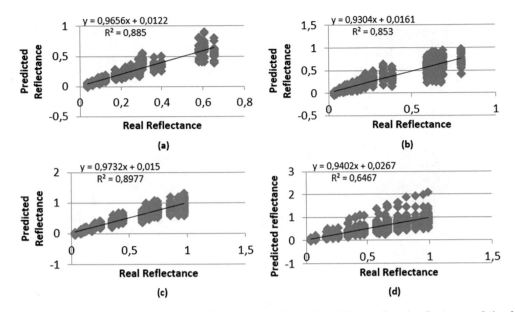

Fig. 12 Graphs presenting the regression line between the real and the predicted reflectance of the four sensors. (**a**), (**b**), (**c**), (**d**), correspond respectively to cameras equipped with 530,630, 740, 810 filters, respectively

the incident light. Two linear relationships were used to calculate both the slope and the offset, as function of the intensity of the incident light. Once the equations are concatenated, a final linear relationship is found between the DN and the at-surface reflectance. An example of what we obtained is displayed in Fig. 12. The final relationships between real reflectance and predicted reflectance by the four sensors, indicating the reliability of each one of them are detailed; R-squared corresponding to each camera are also presented, it is a statistical measure of how close the data are to the fitted regression line. Once these equations are established, MS images can be converted to at-surface reflectance.

8. Register MS images. Image registration is the process of aligning multiple images of the same scene, taken at slightly different times or from different viewpoints or by different sensors into a single integrated image. In our study, the same scene is taken with a little difference in viewpoints. The registration geometrically overlays two images, the sensed image undergoes geometric transformations or local displacements to fit the reference or fixed image. The majority of the registration methods consist of the following four steps [24]: feature detection where characteristic objects are selected; feature matching, here, features found in the sensed image are matched to those from the reference image; transform model estimation consists on estimating the parameters capable of overlaying the sensed

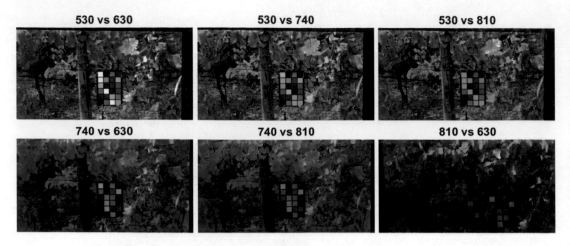

Fig. 13 Image registration on corrected images presented as superposed pairs

and the reference images; image transformation, which means that the sensed image is transformed via the estimated mapping functions. In our work, we calculated the image registration based on a similarity transformation model, it is a nonreflective similarity transformation, consisting of translation, rotation, and scale. After geometrically and radiometrically calibrating the images, images are registered and the superimposed set can be used for further processing. Figure 13 presents the superposition of an example set of four MS images corresponding to the same scene and acquired by the four sensors.

3.7 MS Image Data Analysis

Since spectral information is not sufficient to identify a certain disease because many infections may have similar symptoms, other image processing algorithms need to be implemented on the MS images to make the analysis more robust.

1. Consider the MS images of leaves taken by the sensor.
2. Choose the band displaying the best contrast between the leaves and the background.
3. Apply K-means to the selected band image to segment the corresponding leaves.
4. Segment the rest of the images in the other bands.
5. Construct Symmetrical Grey-Level co-occurrence matrices (GLCMs) for the vegetative area of each segmented channel image and calculate for ($\theta = 0°$, $\theta = 45°$, $\theta = 90°$, $\theta = 135°$) with an offset of 1 pixel. GLCM is one of the most frequently used approaches for texture analysis. It can describe essential characteristics of an image related to second order statistics which were introduced by Haralick [25]. GLCMs can be computed by defining the neighbor relationship between pixels, indeed, the distance step length d and the direction θ should

be determined to create the co-occurrence matrix $G(i, j \mid d, \theta)$. The element (i, j) of the matrix G is the frequency of appearance of gray-tone j near the reference gray-tone i within the image at a defined distance and direction. The GLCM is a square matrix where the number of rows and columns equals the number of gray levels considered. Many scientific studies applied texture parameters from GLCMs for disease classification of various crops [26] employed GLCMs to classify citrus leaves under laboratory conditions. In [27], diseases from ten species were studied (banana, beans, jackfruit, lemon, mango, potato, tomato, and sapota). In [28], downy mildew and powdery mildew are studied.

6. Calculate Haralick texture features from digital images based on GLCMs. Many of these parameters are correlated so in this study, only contrast, correlation, energy, and homogeneity are extracted. As a result, four texture features should be calculated for each channel. These features are then injected to a neural network for classification purposes.

7. Implement a Back Propagation Neural Network (BPNN) classifier for texture classification. There are hundreds of classifiers in the literature and it is often difficult for researchers to choose an appropriate classifier for a certain application. Neural Network (NN) is chosen to classify texture features deduced from MS images. Artificial Neural Networks are the biologically inspired simulations performed on the computer to perform certain specific tasks like clustering, classification, or pattern recognition. NN is self-learning and self-organizing, it is a network of interconnected processing elements operating in parallel to solve specific problems. The neurons are organized into three layers: input, hidden, and output and they operate by two modes: learning or testing modes. The Artificial Neural Network receives input from the external world in the form of pattern. Each input is multiplied by its corresponding weights, the weight represents the strength of the interconnection between neurons inside the neural network. The weighted inputs are all summed up inside computing unit (artificial neuron) and then bias is added to scale up the system response. To limit the response of the NN and let it reach a desired value, the threshold value is set up. For this, the sum is passed through activation function to get desired output. There are linear as well as the nonlinear activation function. NN are used to model complex relationships between inputs and outputs and they were used extensively in the scientific literature. NNs were configured in [29] for automatic detection and classification of plant leaf diseases after texture extraction. The algorithm was tested on five diseases: Early scorch, Cottony mold, ashen mold, late scorch, tiny whiteness. Images of infected

leaves were processed in [30], texture and color features are extracted, and NNs are trained to distinguish healthy from diseased samples appropriately. In this study, a BPNN consisting of one hidden layer with 10 neurons was used.

8. Repeat the classification around 20 times to insure the stability of the results.

9. Use F-Measure (FM) and area under the ROC curve as evaluation measures of classification. FM is the weighted average of Precision and Recall. Therefore, this score takes both false positives and false negatives into account. Precision (P) is the number of true positives divided by the total number of elements considered as belonging to the positive class. On the other hand, Recall (R) is the number of true positives divided by the total number of elements that actually belong to the positive class. The harmonic mean of precision and recall, also called, the traditional F-measure or balanced F-score combines precision and recall. True Positives (TP) are the true positive observations that are correctly classified as belonging to the positive class. False Positives (FP) are the false positive observations that are wrongly classified as belonging to the positive class. False Negatives (FN) are the false negative observations that are not classified as belonging to the positive class while they should have been

$$P = \frac{TP}{TP + FP} \tag{1}$$

$$R = \frac{TP}{TP + FN} \tag{2}$$

$$FM = 2 * \frac{P * R}{P + R} \tag{3}$$

When plotting on a single graph, the False Positive Rate (FPR) values on the abscissa and the True Positive Rate (TPR) values on the ordinate, the resulting curve is called ROC curve, Area Under ROC (AUC) refers to the area under the curve.

$$TPR = \frac{TP}{TP + FN} \tag{4}$$

$$FPR = \frac{FP}{FP + TN} \tag{5}$$

True Negatives (TN) are the true negative observations that are correctly classified as belonging to the negative class.

An example of the classification results on our RGB images is presented in Table 4 along with the ROC curves in Fig. 14. It can be concluded that using texture analysis on digital images for FD detection in grapevine leaves definitely has potential. In our future

Table 4
Results of NN classification of texture features corresponding to each data configuration

	Marselan	Grenache	Vermentino	Chardonnay	Red	White	All
Accuracy (%)	89.3	80.7	92.8	81.2	87.1	88.0	75.3
FM	0.895	0.792	0.791	0.831	0.871	0.905	0.791
AUC	0.956	0.890	0.957	0.900	0.944	0.946	0.844

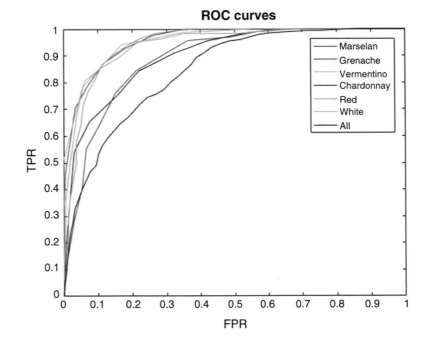

Fig. 14 ROC curves resulting from NN classification of texture features corresponding to each data configuration

works, we plan to extend the analysis and apply it on the MS images acquired from the system.

4 Conclusions

In general, the imaging technology has emerged as a powerful tool in agriculture but it is not much exploited for vines. The effort that has been undergone for the moment to accomplish early detection of grapevine diseases automatically, is minimal, with respect to the major impact of those diseases on grapes yield and wine quality. There is still much space for improvement; we should be able to detect the disease not only in symptomatic hosts in their natural production environment but also in asymptomatic hosts meaning that symptoms are not yet clear. A vine disease must be studied in many cultivars including white and red-berried ones and at different severity stages. This work focused on FD disease; other studies targeted the detection of powdery and downy mildew and grapevine leaf-roll in grapes so it seems essential to explore other diseases and stresses.

The aim of our work is to provide an innovative tool that will enable the diagnosis of the symptoms of FD (and other diseases that have an impact on the foliage) in a systematic way without mobilizing significant human resources and limiting phytosanitary treatments. Our study confirmed the feasibility of choosing some specific spectral bands to characterize infected leaves in vines.

Our study focused on the FD occurrence at the leaf-scale. However, when going from considering leaves to examining a complete branch, or maybe the whole grapevine, some corrections need to be taken into account. Geometric and radiometric improvements capable of solving shadowing problems, branch structures, and interfering reflectance from other surrounding objects are necessary. When adopting a multispectral sensor, we will not only have spectral information but also spatial data. This available spatial data will enable adding advanced image processing algorithms to make the detection of FD more robust. Actually, instead of relying on foliar symptoms, other pattern recognition algorithms can be integrated directly in the sensor to detect the absence of lignification and the berry mortality that are additional indicators of FD presence.

Acknowledgments

We thank all the vine-growers who participated in this study that is part of the DAMAV project. We also would like to thank Alice Dubois and Sylvain Bernard from the Regional Federation of Defense against Pests of Provence Alpes Côtes Azur, Corinne Trarieux from the Interprofessional Office of Burgundy Wine, Jocelyn Dureuil from the Chamber of Agriculture 71, and finally Arnaud Delaherche from Château Pape Clément Bordeaux for their excellent technical assistance.

References

1. Maverick JB (2015) The 4 countries that produce the most wine. https://www.investopedia.com/articles/investing/090915/4-countries-produce-most-wine.asp. Accessed 20 Nov 2017

2. Naidu R, Perry E, Pierce F, Mekuria T (2009) The potential of spectral reflectance technique for the detection of grapevine leafroll-associated virus-3 in two red-berried wine grape cultivars. Comput Electron Agric 66 (1):38–45

3. Hou J, Li L, He J (2016) Detection of grapevine leafroll disease based on 11-index imagery and ant colony clustering algorithm. Precis Agric 17(4):488–505

4. Oberti R, March M, Tirelli P, Calcante A, Iriti M, Borghese AN (2014) Automatic detection of powdery mildew on grapevine leaves by image analysis: optimal view-angle range to increase the sensitivity. Comput Electron Agric 104:1–8

5. MacDonald S, Staid M, Staid M, Cooper M (2016) Remote hyperspectral imaging of grapevine leafroll-associated virus 3 in cabernet sauvignon vineyards. Comput Electron Agric 130(15):109–117

6. Di Gennaro SF, Battiston E, Di Marco S, Facini O, Matese A, Nocentini M, Palliotti A, Mugnai L (2016) Unmanned aerial vehicle (UAV)-based remote sensing to monitor grapevine leaf stripe disease within a vineyard affected by esca complex. Phytopathol Mediterr 55(2):262–275. https://doi.org/10. 14601/Phytopathol_Mediterr-18312

7. Clair D, Larrue J, Aubert G, Gillet J, Cloquemin G, Boudon-Padieu E (2003) A multiplex nested-PCR assay for sensitive and simultaneous detection and direct identification of phytoplasma in the elm yellows group and Stolbur group and its use in survey of grapevine yellows in France. Vitis 42 (3):151–157

8. Chuche J, Thiéry D (2014) Biology and ecology of the Flavescence dorée vector scaphoideus titanus: a review. Agron Sustain Dev 34:381–403

9. Pfitzner K, Bartolo R, Carr G, Esparon A, Bollhöfer A (2011) Standards for reflectance spectral measurements of temporal vegetation plots. https://www.environment.gov.au/system/files/resources/bf8002d0-2582-48a1-820f-8e79d056faed/files/ssr195.pdf. Accessed 11 Sep 2016

10. ASDI (2010) Field Spec 3 user manual http://support.asdi.com/Document/Viewer.aspx?id=108. Accessed 10 Sep 2016

11. Jacquemoud S, Ustin S (2001) Leaf optical properties: a state of the art. In: Proc. 8th International Symposium Physical Measurements & Signatures in Remote Sensing, Aussois, France, CNES, pp. 223–232

12. Araujo M, Kawakami T, Galvao R, Yoneyama T, Chame H, Visani V (2001) The succesive projection algorithm for variable selection in spectroscopic multicomponent analysis. Chemom Intell Lab Syst 57(2):65–73

13. Zhang Y, Tan L, Shi H, He Y (2013) Successive projections algorithm for variable selection on the rapid and non-destructive classification of coolant. JDCTA 7:386–394

14. Yang X, Hong H, You Z, Cheng F (2015) Spectral and image integrated analysis of hyperspectral data for waxy corn seed variety classification. Sensors 15(7):15578–15594

15. Rumpf T, Mahlein A, Dörschlag D; Plümer L (2009) Identification of combined vegetation indices for the early detection of plant diseases. In: Proc. SPIE 7472, Remote Sens for Agric, Ecosystems, and Hydrology XI, 747217, September 2009. doi: https://doi.org/10.1117/12.830525

16. Rumpf T, Mahlein A, Steiner U, Oerke E, Dehne H, Plümer L (2010) Early detection and classification of plant diseases with support vector machines based on hyperspectral reflectance. Comput Electron Agric 74:91–99

17. Ben-Hur A, Weston J (2010) A user's guide to support vector machines. Data mining techniques for the life sciences. Methods Mol Biol 609:223–239

18. Goldman DB (2010) Vignette and exposure calibration and compensation. IEEE Trans Pattern Anal Mach Intell 32(12):2276–2288

19. Hugemann W 2010 Correcting lens distortions in digital photographs. https://www.imagemagick.org/Usage/lens/correcting_lens_distortions.pdf. Accessed 12 Nov 2017

20. Wang A, Qiu T, Shao L (2009) A simple method of radial distortion correction with Centre of distortion estimation. J Math Imaging Vis 35(3):165–172

21. Hadjimitsis DG, Clayton CRI, Hope VS (2004) An assessment of the effectiveness of atmospheric correction algorithms through the remote sensing of some reservoirs. Int J Remote Sens 25(18):3651–3674

22. Smith M, Edward J, Milton G (1999) The use of the empirical line method to calibrate remotely sensed data to reflectance. Int J Remote Sens 20(13):2653–2662

23. Zitova B, Flusser J (2003) Image registration methods: a survey. Image Vis Comput 21:977–1000

24. Haralick RM, Shanmuga K, Dinstein I (1973) Textural features for image classification. IEEE Tans Syst Man Cybern SMC-3(6):610–621

25. Pydipati R, Burks TF, Lee WS (2006) Identification of citrus disease using color texture features and discriminant analysis. Comput Electron Agric 52:49–59

26. Arivazhagan S, Shebiah RN, Ananthi S, Varthini SV (2013) Detection of unhealthy region of plant leaves and classification of plant leaf diseases using texture features. CIGR J 15(1):211–217

27. Sannakki SS, Rajpurohit VS, Nargund VB, Kulkarni P (2013) Diagnosis and classification of grape leaf diseases using neural networks.

Paper presented at the 4th ICCCNT, Tiruchengode, India, July 4–6, 2013

28. Al-Hiary H, Bani-Ahmad S, Reyalat M, Braik M, Alrahamneh Z (2011) Fast and accurate detection and classification of plant diseases. IJCA 17(1):31–38

29. Kulkarni A, Patil A (2012) Applying image processing technique to detect plant diseases. IJMER 2(5):3661–3664

30. Cverkivic T (2004) Recognize vector of phytoplasma Flavescence dorée on grapevine. http://www.chem.bg.ac.rs/~mario/scaphoideus/English/side_8_vector_pub.htm. Accessed 17 Oct 2017

Chapter 18

Transcriptomic Analyses of Phytoplasmas

Davide Pacifico, Simona Abbà, and Sabrina Palmano

Abstract

Transcriptomic analyses addressed to study phytoplasma gene expression may present few difficulties due to the uncultivable nature of these intracellular, obligate pathogens. While RNA extraction from insect vectors does not imply any particular adaptation of the protocols used in most commercial kits, RNA isolation from phytoplasma-infected plants can be a challenging task, given the high levels of polyphenol contents and accumulation of sucrose and starch in the different plant tissues. Here, we describe two different transcriptomic approaches, one focused on RNA phytoplasma sequencing and the other on phytoplasma quantitative gene expression in relation to pathogen load.

Key words RNA sequencing, Quantitative RT-PCR, Gene expression, Phytoplasma load, Expression index

1 Introduction

Phytoplasmas, obligate prokaryotes associated with phloem cells, have a wide geographical distribution and host range, but despite their agricultural importance [1], still remain poorly characterized plant pathogens. The advent of the high-throughput sequencing technology is having a tremendous impact on the study of their biology and by exploiting the potential of Next Generation Sequencing (NGS) important information on phytoplasma genome [2], transcriptome [3, 4], proteome [5], and microRNA profile [6] has been so far obtained. In particular, transcript analysis measuring the expression of an organism gene in different tissues, conditions, and time points gives elucidations on how genes are regulated and reveals basic details of the organism biology and pathogenic behavior. However, transcriptomic has been mainly used to investigate the plant response to phytoplasma infection [7–11] and only few works have been focused on phytoplasma gene expression [3, 4, 12]. Working with phytoplasma is indeed challenging due to their inability of growing on axenic condition, low titer, and uneven spatial distribution in the plant [13, 14];

Rita Musetti and Laura Pagliari (eds.), *Phytoplasmas: Methods and Protocols*, Methods in Molecular Biology, vol. 1875, https://doi.org/10.1007/978-1-4939-8837-2_18, © Springer Science+Business Media, LLC, part of Springer Nature 2019

moreover, a range of different matrices are the objects of the RNA extraction step, each raising specific problems, such as excess of complex carbohydrates, phenolics, pigments, or woody materials [15–17]. Starting with high quality, pure and intact total RNA is therefore critical especially for the experiments focused on the phytoplasma itself, whether they are conducted by microarray [18], quantitative reverse transcription PCR (qRT-PCR) [12], or NGS sequencing [3]. In the present chapter we are considering two main sections of phytoplasma transcript analyses: (1) for RNA sequencing by NGS, and (2) for gene expression analyses by qRT-PCR.

1.1 RNA Sequencing

"Dual RNA-seq" studies that simultaneously capture the transcriptome changes of both the pathogen and the host [19] represent the only way to apply NGS methods to study unculturable bacteria like phytoplasmas. Typically, RNA from the infected host is the vast majority of the extracted RNA. Therefore, the enrichment for phytoplasma RNA is a limiting step, which can be addressed by three affordable and straightforward procedures: (a) by enriching the sample with infected host cells (e.g., sampling leaf veins instead of the whole leaf), (b) by performing host rRNA depletion, and (c) by increasing the sequencing depth, i.e., the number of sequenced reads for a given sample. Before sequencing, it is advisable to first estimate the relative concentration of phytoplasma and host RNA in the sample, for instance by real-time qPCR [3]. This result can influence decisions about the sequencing depth and the choice of biological replicates to be sequenced, as sequencing samples with similar pathogen loads are likely to give more statistically consistent results about differentially expressed genes. Obtaining high-quality RNA is another critical step in RNA-seq library preparations. The RNA extraction protocol used for the RNA-seq analysis of Flavescence dorèe-infected grapevine leaves [3] led to the consistent isolation of very high-quality total RNA suitable for obtaining about 30,000 phytoplasma-mapped paired-end (PE) reads out of 240 million of quality-checked reads per sample of host-phytoplasma mixed RNA. The extraction protocol described below was adapted from Chang et al. [20] and Gambino et al. [21].

1.2 Quantitative RT-PCR

Choosing the correct strategy for normalization of transcript amounts represents an important bottleneck for gene expression experiments. Many different methods are currently available for gene expression studies of cultivable bacteria based on both relative [22] or absolute [23] quantification methods. In case of cultivable bacteria, data normalization can be carried out using the number of cells obtained through cell culture, genomic DNA or RNA content, as well as endogenous unregulated reference genes [24, 25]. In case of phytoplasmas, although the use of internal reference genes has been suggested for transcript normalization [26], selection of such genes requires further analysis to check their stability under different experimental conditions, as some of them may not be suitable for analyses over time or during infection in different hosts.

Here we report a method to measure the phytoplasma gene transcript levels by qRT-PCR, in the absence of a phytoplasma endogenous control mRNA [12]. This protocol correlates the absolute quantity of phytoplasma transcripts of a sample to the bacterial population measured for the same sample by qPCR [27]. The obtained value is referred to as an Expression Index (EI), indicating the transcript copy number per phytoplasma cell. The use of EI is suitable to analyze the gene expression trends over time among different hosts (plants and insects), irrespective of the absolute levels of the individual gene expression.

2 Materials

Prepare all solutions using ultrapure water and analytical grade reagents. Buffers and DEPC (diethyl pyrocarbonate)-treated water should be autoclaved, whereas the sodium acetate solution should be filter-sterilized (0.2 μm). Prepare and store all reagents at room temperature unless indicated differently. Glassware (beakers, graduate cylinders, etc.), spoons, stirrer bars, tubes, and pipette tips must be sterilized by autoclaving (plastic devices) or baking at 180 °C for at least 4 h. Wear gloves to reduce the risk of RNase contamination during sample preparation and even during handling.

2.1 Materials for RNA Sequencing

1. Extraction buffer: 2% (w/v) CTAB (cetyltrimethyl ammonium bromide), 2% (w/v) PVP (polyvinyl pyrrolidone) 40,000, 100 mM Tris–HCl pH 8, 25 mM EDTA disodium salt, 2 M NaCl, 0.5 g/l spermidine (*see* **Note 1**).

2. SSTE (SDS-Sodium Chloride-Tris-EDTA) buffer: 1 M NaCl, 0.5% SDS, 10 mM Tris–HCl pH 8, 1 mM EDTA disodium salt (*see* **Note 2**).

3. Chloroform/Isoamyl alcohol 24:1 v/v.

4. Sodium acetate 3 M pH 5.2.

5. Ethanol 70%.

6. RNase-free DEPC water obtained by addition of 0.1% diethyl-pyrocarbonate (DEPC) to MilliQ water into RNase-free glass bottles (*see* **Note 3**).

7. 2% (v/v) of β-mercaptoethanol added just before use.

8. Water bath (65 °C).

9. Refrigerated microcentrifuge.

2.2 Materials for Quantitative RT-PCR

2.2.1 RNA Extraction from Plant and Insect Samples

For qRT-PCR analyses it is not necessary to follow a challenging and time consuming extraction method as the one described for the RNA sequencing. So we report in this section a method that has proved successful for both plant and insect tissues.

1. TRIzol reagent (Invitrogen, CA, USA).
2. RNase-free DEPC water obtained by the addition of 0.1% diethylpyrocarbonate (DEPC) to MilliQ water into RNase-free glass bottles (*see* **Note 3**).
3. Sterile micro-pestel.
4. Sterilized mortars and pestles.
5. Liquid nitrogen.
6. Refrigerated microcentrifuge.
7. Chloroform.
8. Isopropyl alcohol.
9. Ethanol 75% (in DEPC-treated water).

2.2.2 Dnase Treatment

1. RNAse-free DNAse I with the supplied 10× buffer.
2. Phenol-chloroform solution 1:1 (v/v).
3. Chloroform.
4. Ammonium acetate (NH_4OAc) solution 5 M.
5. Ethanol 100%.
6. RNase-free DEPC water.
7. Ethanol 70% solution in DEPC-treated water.
8. Refrigerated microcentrifuge.
9. Water bath (37 °C).

2.2.3 Reverse Transcriptase Protocol

1. 500 ng of total RNA for each sample.
2. High Capacity cDNA reverse transcription kit (Applied Biosystems, USA).
3. Vortex.
4. Centrifuge.
5. Water bath (37 °C).
6. Fridge or refrigerated chamber (+4 °C).

2.2.4 qPCR Protocol

1. iQ SYBR green Supermix (Bio-Rad/Life Science Research, CA, USA).
2. 100 μM stock solution of each primer, diluted 1:10 to obtain 10 μM working solution.
3. Sterile double-distilled water.

4. Real-time thermocycler and combined data analysis software such as Bio-Rad CFX Connect real-time PCR detection system and CFX Manager software v. 3.0.

5. Vortex.

6. Centrifuge.

7. 96-Well PCR Plates.

8. Optically clear seal for PCR plates such as Microseal® 'B' PCR Plate Sealing Film, MSB1001 (Bio-Rad/Life Science Research, CA, USA).

2.2.5 DNA Extraction from Plant and Insect Samples

CTAB-based extraction procedure is suitable for both plants and insect vectors (*see* Chapter 6 for details).

3 Methods

3.1 Methods for RNA Sequencing

Carry out all procedures keeping the samples chilled, preferably on ice, unless otherwise specified.

3.1.1 RNA Extraction

1. Place 900 µL of Extraction Buffer into a 2 mL tube and add 1.8 µL β-mercaptoethanol (2% v/v final concentration), mix briefly, then heat in a 65 °C water bath.

2. Thoroughly grind leaves in liquid nitrogen.

3. Add 90–100 mg of the grounded tissue to each tube. Cap, shake vigorously, and incubate at 65 °C for 10 min.

4. Centrifuge at $11,000 \times g$ for 10 min at 4 °C and recover the supernatant.

5. Add an equal volume of chloroform/isoamyl alcohol (24:1) to the tube, shake vigorously, and centrifuge at $14,000 \times g$ for 10 min at 4 °C.

6. Transfer the supernatant into a new tube and repeat the previous step twice (three times in total) (*see* **Note 4**).

7. Add one-third volume of 12 M LiCl to the supernatant (3 M final concentration). Briefly mix and place at 4 °C for overnight precipitation or in an ice bucket for 30 min.

8. Heat and stir SSTE buffer at 65 °C on a magnetic hot plate stirrer.

9. Pellet the RNA by centrifugation at $16,300 \times g$ for 30 min at 4 °C.

10. After centrifugation, discard the supernatant and pipette off any remaining solution. Resuspend the pellet in 500 µL of hot (65 °C) SSTE buffer using a pipette.

11. Add an equal volume of chloroform/isoamyl alcohol (24:1). Shake vigorously, centrifuge at 14,200 × g for 5 min at 4 °C, and transfer the aqueous phase to a new tube.

12. Add one tenth volume of sodium acetate 3 M pH 5.2 and 2.5 volumes of absolute ethanol to the aqueous phase. Mix and precipitate at −80 °C for 30 min.

13. Centrifuge at 14,200 × g for 30 min at 4 °C and carefully remove the supernatant. Do not disturb the pellet.

14. Wash the RNA pellet with 1 mL of ice-cold 70% (v/v) ethanol. Centrifuge at 14,200 × g for 5 min and carefully remove the ethanol with a pipet. Uncap the tubes and air-dry pellets on the bench for 3–5 min.

15. Resuspend in DEPC-treated water.

3.1.2 RNA Quantity and Quality Evaluation

1. DNase digestion of the purified RNA with RNase-free DNase is highly recommended (*see* **Note 5**).

2. Use 1.5 μL of the resuspended RNA for spectrophotometer quantitation. Absorbance ratios for 260/280 and 260/230 should be greater than 1.8. RNA isolation yields should be at least 10 μg/100 mg frozen tissue.

3. Analyze the quality of RNA using an Agilent Bioanalyzer according to the manufacturer's instructions. The Bioanalyzer will produce an RNA Integrity Number or RIN, which is an objective measure of RNA quality. The RNA sample should have a RIN value higher than 7 (Fig. 1). RNA should be completely free of DNA.

4. Store RNA at −80 °C.

5. Provide the sequencing facility with RNA of sufficient quality and quantity to produce a library. The removal of host rRNA (either plant or insect) and the sequencing of at least 120 million of PE reads are highly recommended. In case of differential gene expression studies, at least two biological replicates for each condition are strictly required.

6. Whatever pipeline and bioinformatics tools are going to be used (3), an in silico analysis generally requires: (1) quality control checks on raw sequence data; (2) mapping onto reference genomes/transcriptomes (if available) or de novo transcriptome assembly (if not); (3) reads normalization and quantification. The complexity of analyzing dual RNA-seq data derives mainly from working simultaneously with at least two transcriptomes, i.e., the host one and the phytoplasma one. The possible presence of transcripts belonging to other prokaryotes or viruses should also be taken into account. Thus, a correct mapping can be considered the most critical step of the whole pipeline.

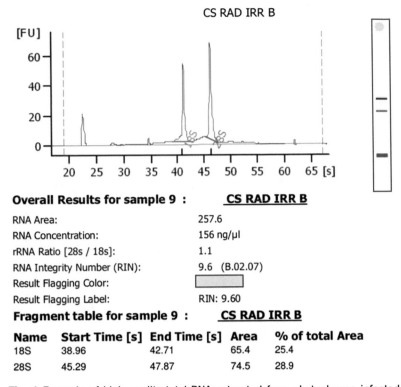

Fig. 1 Example of high-quality total RNA extracted from phytoplasma-infected grapevine leaves

CS RAD IRR B

Overall Results for sample 9 :   **CS RAD IRR B**

RNA Area:	257.6
RNA Concentration:	156 ng/µl
rRNA Ratio [28s / 18s]:	1.1
RNA Integrity Number (RIN):	9.6 (B.02.07)
Result Flagging Color:	
Result Flagging Label:	RIN: 9.60

Fragment table for sample 9 :   **CS RAD IRR B**

Name	Start Time [s]	End Time [s]	Area	% of total Area
18S	38.96	42.71	65.4	25.4
28S	45.29	47.87	74.5	28.9

If the genome sequence of both the phytoplasma and the uninfected host are available, reads can be mapped in parallel against the two genomes, so the critical mapping step can be easily overcome.

If the phytoplasma genome is unavailable or incomplete, there are two possible scenarios, depending on the availability or not of the uninfected host genome/transcriptome. If it is available, reads that map onto the host genome can be identified and separated from the others. Then, a de novo assembly can be performed on the remaining reads, a fraction of which should belong to the phytoplasma. Based on sequence similarity searches with blastx, some of the assembled transcripts will be assigned to the phytoplasma of interest. The worst-case scenario is represented by the unavailability of both the phytoplasma and the host genome. In this case, mapping onto genomes of phylogenetically close organisms could be attempted, but the risk of sequence divergences, especially at the nucleotide level, is quite high. A de novo assembly considering all the reads, both eukaryotic and prokaryotic, is probably the best choice, followed by a blastx analysis and the following assignment of the resulting transcripts to either the phytoplasma or the host.

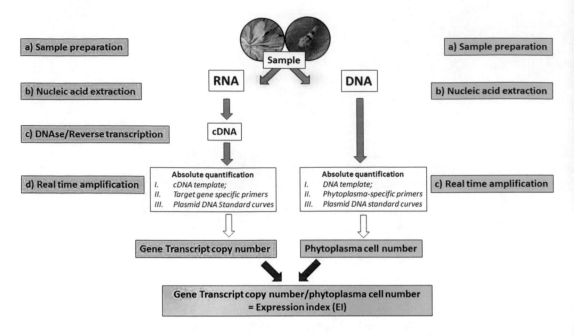

Fig. 2 Summary scheme for quantitative gene expression of phytoplasmas

3.2 Methods for Quantitative RT-PCR

A schematic overview of the steps described below is presented in Fig. 2.

3.2.1 Sample Preparation

1. Plants: pool about 200 mg of leaves and divide into two 100 mg aliquots using sterile blades. Store aliquots at −80 °C before DNA or RNA extraction.

2. Insects: collect single insects and put each of them in sterile 1.5 mL tubes to be stored at 80 °C until DNA or RNA extraction.

3.2.2 RNA Extraction from Plant and Insect Samples

1. Follow the TRIzol reagent protocol to extract RNA from both plant and insect material.

2. Plants: grind 100 mg aliquots of plant material in liquid nitrogen using prechilled mortars and pestles (put in ice before use to reduce uncontrolled tissue thawing), continue grinding until a fine powder is obtained (if necessary add liquid nitrogen until all material is completely processed). Carefully transfer the fine powder to a 1.5 mL sterile tube and add 500 μL TRIzol reagent, then follow the manufacturer's instructions (*see* **Notes 3** and **6**).

3. Insects: add a few liquid nitrogen drops directly into a 1.5 mL tube containing single insects then crush quickly, using a sterile micro-pestel in 200 μL of 1× TE buffer prepared with DEPC water. Transfer the homogenate suspension in two 1.5 mL tubes containing 100 μL aliquots for DNA and RNA extraction,

respectively. For RNA extraction add 400 μL of TRIzol reagent, and then follow the manufacturer's instructions (*see* **Notes 3 and 6**).

3.2.3 Dnase Treatment

1. Total RNA samples extracted from both plants and insects must be treated with RNase-free DNase I, to eliminate residual DNA contaminations. To this aim, add 2 U of DNase I in a 50 μL reaction with the supplied buffer, and incubate at 37 °C for 30 min. Following the digestion, inactivate the DNase by phenol-chloroform extraction according to the following instructions (*see* **Note 7**). Add to the RNA solution an equal volume of phenol-chloroform 1:1 solution, vortex for 30 s then centrifuge for 10 min at 12,300 × *g* to separate the two layers. Rescue the upper aqueous phase and transfer to a new RNAse-free tube. Repeat the purification step using chloroform only, then precipitate the RNA adding ammonium acetate solution to a final concentration of 2.5 M (e.g., add one volume of 5 M ammonium acetate and mix) (*see* **Note 8**). Add 2.5 volumes of cold 100% ethanol to the aqueous phase and incubate for 30 min at −80 °C or overnight at −20 °C. Centrifuge for 10 min at 12,300 × *g* to pellet the RNA and then discard the supernatant. Wash the pellet with 70% ethanol, to remove residual salt (*see* **Note 9**). Air-dry the RNA (or use a speedvac rotor) and suspend it in 30 μL of RNase-free DEPC water.

2. Verify the RNA quality with a spectrophotometer to evaluate the concentration and purity and store at 80 °C.

3.2.4 Reverse Transcription

Transcript accumulation can be measured from a unique cDNA preparation synthesized from 500 ng of total RNA. For reverse transcription step, use the High Capacity cDNA reverse transcription kit (Applied Biosystems, USA) according to the manufacturer's instructions. This protocol allows the conversion of up to 2 μg of total RNA to single-stranded cDNA in 20 μL reactions (*see* **Note 10**).

3.2.5 Real-Time Amplification

1. SYBR green-based quantitative real-time PCR (qPCR) protocols are suitable for absolute quantification of bacterial transcripts using cDNA as template and standard curves obtained from recombinant plasmids (recDNA) at known concentration (copy number) carrying a fragment of target genes.

2. qPCR-specific primers can be selected from the literature or designed using software such as http://primer3.sourceforge.net/. To ensure high amplification efficiency, primers should have a GC content between 40 and 60%, preferably with the 3′ ending in C or G. In case of phytoplasma targets, pick up primers also taking into account their AT-rich genomes and the possible presence of long A or T stretches that may lead to hairpins or primer-dimers formation. The use of specific software may help to reduce this risk.

3. Optimize the PCR conditions for each selected primer pair by testing different primer concentration and annealing temperatures. Keep the final concentrations of forward and reverse primers ranging from 100 nM to 300 nM. To minimize pipetting error, prepare a qPCR master mix containing $1\times$ iQ SYBR green Supermix (Bio-Rad/Life Science Research, Hercules, CA, USA), 100–300 nM (each) primers, and sterile double-distilled water to a final volume of 25 μL. Dispense the qPCR mix in a 96-well qPCR plate (24 μL/well) and then add 1 μL of sample cDNA in each well (*see* **Note 11**). Run each sample in duplicate, together with negative controls obtained by adding total RNA and sterile distilled water instead of cDNA to the complete qPCR mix (*see* **Note 12**).

4. The use of DNA standard curves for absolute quantification allows comparisons of expression data of different genes from the same RNA preparation. To this aim, prepare at least four tenfold serial dilutions of the different plasmids carrying a fragment of the target genes and run the standards and cDNA samples together in the same plate, according to the target. The number of plasmid copies per microliter is derived from the concentration measured at the Nano-Drop spectrophotometer, using the following formula: $M = C\times N/S$, where M is the number of molecules per microliter, C the RNA concentration (ng/μL), S the fragment molecular weight, and N the Avogadro constant.

5. For real-time amplification use the following conditions: 5 min at 95 °C and 45 cycles of 15 s at 95 °C, 30 s at 59 °C, and 30 s at 72 °C. Check the amplification specificity by analysis of melting curves produced at the end of the qPCR reaction and adjust annealing temperatures if necessary. Run qPCR in a thermal cycler supported by a software, which automatically calculates the standard regression curve and the transcript copy number (i.e., the Bio-Rad CFX Connect real-time PCR detection system combined to the CFX Manager software v. 3.0).

3.2.6 DNA Extraction from Plant and Insect Samples

To extract total DNA from plant and insect samples use the CTAB-based procedure, as described in Chapter 6.

3.2.7 Expression Index

Calculate an expression index (EI) to compare the expression analysis of different phytoplasma target genes in one or several samples [12]. Each EI can be obtained by dividing the gene transcript copy number (*see* Subheading 3.2.5, **step 4**) and the phytoplasma cell number of a sample (*see* **Note 13**), obtained from RNA and DNA respectively.

EI = transcript number/phytoplasma cell.

4 Notes

1. Spermidine should not be autoclaved. Prepare the extraction buffer without spermidine, autoclave and add the component only after chilling the solution.

2. SSTE buffer should be autoclaved leaving a magnetic barr inside the bottle. Do not autoclave SDS; during autoclaving SDS irreversibly precipitates and begins to hydrolyze to dodecanol and sulfuric acid. So, add this component only after chilling the solution and pay attention that, at low temperatures, SDS solutions may precipitate: warm at 37 °C and use a magnetic stirrer for complete dissolution, before dispensing the buffer in the tubes.

3. As autoclaving may not be sufficient to eliminate RNase, DEPC treatment of water should be included as precaution for molecular applications based on RNA preparations. Always use DEPC in a fume hood and never add it to aqueous solutions containing ammonia as it would result in production of carcinogen ethyl carbamate. DEPC is unstable in aqueous solutions and it is completely inactivated by autoclaving although its hydrolysis releases ethanol and CO_2. Add 0.1% (w/v) of DEPC to ultrapure water, shake and incubate for 16 h (overnight) at room temperature, then autoclave. To reduce the risk of contamination after autoclaving, prepare several aliquots of DEPC-treated water in RNase-free bottles and eventually divide it in small ready-to-use aliquots.

4. The number of chloroform:isoamyl alcohol extractions can be increased in relation of the RNA quality, i.e., when the final 260/280 and 260/230 ratios are not appropriate.

5. Different DNase commercial treatments are available with good results. In case of sequencing we prefer the use of Kits that may reduce the variabilities introduced by handling. The generic DNAse procedure instead is described in Subheading 3.2.3.

6. Working with TRIZOL Reagent requires protections to avoid any contact with skin or clothes: always wear gloves and work under a chemical hood.

7. Incubation at 75 °C for 10 min can be also applied to inactivate RNase-free DNase I, in this case add EDTA to a final concentration of 5 mM before heating. EDTA is necessary to protect RNA from degradation occurring at high temperature; nevertheless, EDTA may alter the Mg^{2+} concentration during the following reverse transcription reactions.

8. Since ammonium acetate decomposes by loss of ammonia, solutions should be prepared only from the pure salt which

has been kept cool in a closed container. Sterilize only by filtration; do not autoclave Ammonium Acetate Buffer. Store solutions at +4 °C. If precipitates form, warm solution to 37 °C to dissolve it.

9. Decant the ethanol by inverting the tubes on a clean paper towel. Be careful not to lose the small pellet together with the ethanol. Since air-drying takes a long time, use a micropipette and a sterile tip to transfer as much ethanol as possible, then put the tubes in a 65 °C incubator to allow complete ethanol evaporation. Be careful not to disturb the pellet by pipetting out the ethanol.

10. The number of RT reactions can be increased according to the number of target genes in study and the cDNA employed for subsequent PCR steps. Store the cDNA in sterile microtubes until all qPCRs reactions are performed. Fresh cDNA can be stored for hours or few days (1 week) at +4 °C until the experiment is complete. In any case, avoid multiple freeze-thaw cycles that may cause cDNA degradation and consequent under estimation of target gene expression.

11. Following thawing, vortex each reagent for at least 5 s and then speed down the liquid in a centrifuge before PCR mix preparation. Avoid any bubble formation by pipetting the qPCR mix directly to the wall or on the bottom of each well. Mix cDNA by gentle pipetting 5–10 times (up and down) using exclusively retention-free tips, do not vortex the cDNA solution.

12. For a reliable result, add more than one water control, particularly in case of full qPCR plates.

13. The number of phytoplasma cells can be measured according to Marzachì et al. [27], i.e., by comparing the real-time PCR quantification cycles (Cqs) of a phytoplasma-infected sample with those of three dilutions of a plasmid, containing a fragment of the phytoplasma 16S rRNA gene. According to Marzachì et al. [27], 1 fg of the pOP74 plasmid contains 194 molecules of plasmid carrying a single copy of the Chrysanthemum yellows (CYP) 16S rRNA gene. Because this gene is present in two copies in phytoplasma genomes, 1 fg of pOP74 corresponds to $194/2 = 97$ phytoplasma cells.

References

1. Gasparich GE (2010) Spiroplasmas and phytoplasmas: microbes associated with plant hosts. Biologicals 38(2):193–203

2. Kakizawa S, Yoneda Y (2015) The role of genome sequencing in phytoplasma research. Phytopathogen Mollicutes 5(1):19–24

3. Abbà S, Galetto L, Carle P et al (2014) RNA-Seq profile of flavescence dorée phytoplasma in grapevine. BMC Genomics 15:1088

4. Siewert C, Luge T, Duduk B et al (2014) Analysis of expressed genes of the bacterium 'Candidatus phytoplasma Mali' highlights key

features of virulence and metabolism. PLoS One 9:e94391

5. Ji X, Gai Y, Lu B et al (2010) Shotgun proteomic analysis of mulberry dwarf phytoplasma. Proteome Sci 8:20

6. Niu S, Fan G, Deng M et al (2016) Discovery of microRNAs and transcript targets related to witches' broom disease in Paulownia fortunei by high-throughput sequencing and degradome approach. Mol Gen Genomics 291:181–191

7. Hren M, Nikolić P, Rotter A et al (2009) "Bois noir" phytoplasma induces significant reprogramming of the leaf transcriptome in the field grown grapevine. BMC Genomics 10:460

8. Mou H-Q, Lu J, Zhu S-F et al (2013) Transcriptomic analysis of Paulownia infected by Paulownia witches'-broom Phytoplasma. PLoS One 10:e77217

9. Liu LY, Tseng HI, Lin CP et al (2014) High-throughput transcriptome analysis of the leafy flower transition of Catharanthus roseus induced by peanut Witches'-broom phytoplasma infection. Plant Cell Physiol 55:942–957

10. Margaria P, Ferrandino A, Caciagli P et al (2014) Metabolic and transcript analysis of the flavonoid pathway in diseased and recovered Nebbiolo and Barbera grapevines (Vitis vinifera L.) following infection by Flavescence doree phytoplasma. Plant Cell Environ 37:2183–2200

11. Fan G, Xibing Cao X, Niu S et al (2015) Transcriptome, microRNA, and degradome analyses of the gene expression of Paulownia with phytoplasma. BMC Genomics 16:896

12. Pacifico D, Galetto L, Rashidi M et al (2015) Decreasing global transcript levels over time suggest phytoplasma cells enter stationary phase during plant and insect colonization. Appl Environ Microbiol 81:2591–2602

13. Berges R, Rott M, Seemüller E (2000) Range of phytoplasma concentrations in various plant hosts as determined by competitive polymerase chain reaction. Phytopathology 90:1145–1152

14. Constable FE, Gibb KS, Symons RH (2003) Seasonal distribution of phytoplasmas in Australian grapevines. Plant Pathol 52(3):267–276

15. Lepka P, Stitt M, Moll E, Seemüller E (1999) Effect of phytoplasmal infection on concentration and translocation of carbohydrates and amino acids in periwinkle and tobacco. Physiol Mol Plant Pathol 55:59–68

16. Rusjan D, Halbwirth H, Stich K et al (2012) Biochemical response of grapevine variety 'chardonnay' (Vitis vinifera L.) to infection with grapevine yellows (bois noir). European J Plant Pathol 134:231–237

17. Prezelj N, Covington E, Roitsch T et al (2016) Metabolic Consequences of Infection of Grapevine (Vitis vinifera L.) cv. "Modra frankinja" with Flavescence Dorée Phytoplasma. Front. Plant Sci 7:711

18. Oshima K, Ishii Y, Kakizawa S et al (2011) Dramatic transcriptional changes in an intracellular parasite enable host switching between plant and insect. PLoS One 6:e23242

19. Westermann AJ, Gorski SA, Vogel J (2012) Dual RNA-seq of pathogen and host. Nat Rev Microbiol 10:618–630

20. Chang S, Puryear J, Cairney J (1993) A simple and efficient method for isolating RNA from pine trees. Plant Mol Biol Rep 11:113–116

21. Gambino G, Perrone I, Gribaudo I (2008) A rapid and effective method for RNA extraction from different tissues of grapevine and other woody plants. Phytochem Anal 19(6):520–525

22. McMillan M, Pereg L (2014) Evaluation of reference genes for gene expression analysis using quantitative RT-PCR in Azospirillum brasilense. PLoS One 9:e98162

23. Borges V, Ferreira R, Nunes A et al (2010) Normalization strategies for real-time expression data in Chlamydia trachomatis. J Microbiol Methods 82:256–264

24. Vandecasteele SJ, Peetermans WE, Merckx R et al (2002) Use of gDNA as internal standard for gene expression in staphylococci in vitro and in vivo. Biochem Biophys Res Commun 291:528–534

25. Takle GW, Toth IK, Brurberg MB (2007) Evaluation of reference genes for real-time RT-PCR expression studies in the plant pathogen Pectobacterium atrosepticum. BMC Plant Biol 7:50

26. Toruño TY, Music MS, Simi S et al (2010) Phytoplasma PMU1 exists as linear chromosomal and circular extrachromosomal elements and has enhanced expression in insect vectors compared with plant hosts. Mol Microbiol 77:1406–1415

27. Marzachí C, Bosco D (2005) Relative quantification of chrysanthemum yellows (16SrI) phytoplasma in its plant and insect host using real-time polymerase chain reaction. Mol Biotechnol 30:117–127

Part III

Site-Specific Analyses

Chapter 19

Sieve Elements: The Favourite Habitat of Phytoplasmas

Aart J. E. van Bel

Abstract

The sieve elements are the only plant compartments, where phytoplasmas can survive and propagate. Therefore, this chapter is focussed on the specific molecular and cell-biological properties of the sieve element. Sieve element-companion cell complexes arise from (pro)cambial mother cells induced by key genes known to be decisive for sieve-element differentiation. The special anatomy, cell biology, and plasma-membrane outfit of sieve elements allows them to act collectively as a tube system that is able to drive a mass flow against the flow induced by transpiration. Plasmodesmal corridors are vital for the translocation of photoassimilates and systemic signals and for survival of the enucleate sieve elements. Of paramount importance is the Ca^{2+}-dependent gating of plasmodesmata by callose and proteins. Hence, some of the complex, regulatory mechanisms to maintain Ca^{2+} homoeostasis in sieve elements are presented. Finally, the peculiarities of the chemical and physical sieve-element environment offered to phytoplasmas are discussed.

Key words Ca^{2+} regulation, Companion cells, Phytoplasmas, Plasmodesmata, Sieve-element conditions, Sieve elements, Sieve-element cell biology, Sieve-element development, Sieve-element occlusion

1 Introduction: Phytoplasmas and Their Sieve-Tube Habitat

Phytoplasmas are transmitted plant-to-plant by sucking insects such as leafhoppers (Cicadellidae), planthoppers (Cixiidea), and psyllids (Psyllidae) [1, 2] that feed from sieve-tubes. Since phytoplasmas are phloem-limited [2], healthy plants are contaminated by phytoplasmas that have been ingested by previous sieve-tube probing of infected plants. After ingestion of sieve-tube sap, phytoplasmas make their way through the stylet corridor toward the digestive tract, from which phytoplasmas are released into the hemolymph that carries them to the salivary glands [3]. Both stylet interior and digestive tract must provide the chemical and physical conditions, under which phytoplasmas can survive [3]. This must hold to an even greater extent for the sieve-tube environment, where phytoplasmas do not only survive, but multiply and are transported over long distances. Reports on successful efforts to grow phytoplasmas in vitro [4, 5] await further confirmation of feasibility. The

Rita Musetti and Laura Pagliari (eds.), *Phytoplasmas: Methods and Protocols*, Methods in Molecular Biology, vol. 1875, https://doi.org/10.1007/978-1-4939-8837-2_19, © Springer Science+Business Media, LLC, part of Springer Nature 2019

difficulties in cultivation demonstrate how delicate the fine-tuning between the requirements of phytoplasmas and sieve-tube environment must be. The obvious challenges urge to have a closer look at the anatomical, physical, and chemical properties of sieve tubes and sieve-tube content (e.g., [6]). This approach may help to understand the survival and reproductive needs of phytoplasmas.

2   Development of the Sieve Element-Companion Cell Complex

Sieve tubes are composed of longitudinal arrays of sieve-element modules, each of which is flanked by one or a few companion cells [7]. Sieve tubes are extended by sequential development of emerging sieve elements in meristematic zones. In vegetative terminal sinks, protophloem sieve elements are flanked by cells that originate from a different procambial precursor stem cell [8, 9]. The latter cells appear specialized parenchyma cells rather than genuine companion cells, but may accomplish companion cell-like services during the limited life-span of the protophloem. As for the differentiation of sieve elements in procambial metaphloem, the descendance is unclear.

In secondary phloem, sieve elements originate from cambial precursors that divide longitudinally giving rise to two daughter cells that take an entirely different developmental course [10–14]. Transient callose-mediated closure of the cambial plasmodesmata is a likely key event in differentiation of the sieve element-companion cell complex *in statu nascendi* [11]. Given the different developmental course of sieve elements in primary and secondary phloem, it is uncertain, whether the genes involved in specification and differentiation of procambial and cambial sieve elements are identical. This potential dissimilarity is worth to be considered in future investigations.

A few genes have been identified to play a part in early protophloem sieve-element development [15]. They encode for two differently located phosphatidyl inositol 5-phosphate kinases (BRX—BREVIS RADIX [16] and OPS—OCTOPUS [17]), a leucine-rich repeat receptor-like kinase (BAM3—BARELY ANY MERISTEM 3 [18]) and its presumptive peptide ligand (CLE 45—CLAVATA3/EMBRYO-SURROUNDING REGION 45 [19]). Quite recently, the first genes (once again OPS and, as a novelty, OPL2, OCTOPUS-LIKE 2) that are engaged in the early metaphloem sieve-element differentiation were discovered [20]. A couple of genes engaged in the final degradative stages of sieve-element development have also been identified. APL (ALTERED PHLOEM DEVELOPMENT)—a coiled-coil transcription factor [21]—orchestrates nuclear breakdown and reduction of cytoplasmic contents by activation of a family of exonucleases. These NAC45/86-DEPENDENT EXONUCLEASE-DOMAIN PROTEINS (NENs) are expressed in protophloem cells prior to enucleation [22]. Furthermore, genes (GSL7 and CHER) were found that contribute to formation and positioning of sieve pores (*see* the next section on the emergence of plasmodesmal contacts in the sieve element-companion cell complex). Genes involved in the initiation of companion-cell development have not been determined with certainty thus far [9].

3 Ultrastructure and Function-Related Cell Biology of the Sieve Element

After that sieve element and companion cell originate from a common cambial precursor, several essential cell components are degraded by semi-apoptotic events in the sieve-element daughter cell. Again, it should be stressed that the mechanisms of degradation in these sieve elements are not necessarily identical to those in protophloem sieve elements. After semi-apoptosis, the plasma membrane has stayed intact, while the nucleus is absent as well as the ribosomes, the tonoplast, and the Golgi system [7, 14, 23]. What remains is a narrow parietal margin of cytoplasm (called mictoplasm; [7, 24, 25]) that is in direct contact with the sieve-tube sap. The mictoplasm contains small aggregates of smooth ER-stacks that are mostly arranged in parallel with the plasma membrane [26–30]. Rough ER has not been observed, but the existence of free ribosomes may not be excluded (see [23]). The sieve-element ultrastructure could only be elucidated by use of preservative fixation procedures, because the structural integrity is highly sensitive to manipulations (e.g., [30]).

The plasma membrane is indispensable for long-distance transport of materials from sources to sinks. Thanks to batteries of carriers, channels, and pumps in the sieve-element plasma membranes, a turgor gradient is set up that drives mass flow through the sieve tubes. Moreover, selective uptake, release and retrieval of solutes along the sieve tubes allow a precise regulation of the composition of the sieve-tube sap at any location in the phloem system. Prominent representatives of plasma membrane translocators in relation to sieve-tube functioning are sugar carriers build-up of turgor gradients, transport of energy-carrying compounds, and carbon building blocks; (e.g., [25, 31–35]), amino acid carriers long-distance transport of organic nitrogen; (e.g., [36]), aquaporins regulation of turgor gradients; (e.g., [37]), potassium channels osmotic equilibration and regulation of sugar import or release; (e.g., [38–40]), calcium channels Ca^{2+} homeostasis, propagation of electrical potential waves; (e.g., [41–45]), and ATPases and PPases pH homeostasis, energization of carriers; (e.g., [25, 39]).

A reduced number of dilated, sometimes degenerated, small mitochondria occurs in the mictoplasm [23, 30]. The sieve-element mitochondria show a virtual absence of metabolism in contrast to the high metabolic activity of mitochondria in companion cells [46] and are unusually prone to damage to fixation procedures [22]. The observed degree of degeneration may depend on the species and correspond to the fixation procedures [24, 47–49]. As a result of the uncertainties regarding their ultrastructure, genomic outfit, and metabolic impact, the role of sieve-element mitochondria remains obscure for the time being.

Furthermore, unique sieve-element plastids reside along the plasma membrane. Each plant species possesses diverse sieve-element plastids that contain family-specific numbers of protein bodies and/or starch grains of variable size [22, 50]. Clean isolation of intact sieve-element plastids seems virtually impossible so that their function remains puzzling thus far.

Sieve-element plastids, mitochondria, and ER-cisternae are tethered to each other and plasma membrane by numerous minute, possibly proteinaceous anchors [30]. The anchors might prevent the organelles to be displaced by the mass flow in sieve tubes. Despite the original view that a cytoskeleton is absent [8, 12, 51], circumstantial biochemical evidence for cytoskeleton(−associated) components [52–58], theoretical considerations [59], and experimental work [59, 60] render the presence of an actin network in sieve elements likely.

Sieve elements contain structural proteins in a large variety of shapes [61, 62]. Most of the structural phloem-specific proteins have a fibrillar, apparently disorganized appearance [63–65]. A striking exception are the forisomes, structurally ordered giant protein bodies in the sieve elements of Fabaceae [66, 67]. Both fibrillar strands and forisomes are composed of SEO(R) proteins, a ubiquitous plant gene family [63, 64], involved in sieve-tube occlusion. In Fabaceae, SEO1, SEO2, and SEO3 have been identified [68], in *Arabidopsis* only SEO1 and SEO2 [69]. Whether fibrillar proteins can actually occlude sieve pores is still a matter of debate [69–73].

Ubiquitin-related proteolytic activity of sieve-tube sap [57, 58, 74] and more than a hundred proteinaceous components involved in proteasome-associated protein degradation occurring in the sieve tube sap of cucurbits [75] indicate the presence of proteasomes [76], but solid evidence on their occurrence and mode of action is lacking thus far.

4   Ultrastructure and Function-Related Cell Biology of the Companion Cell

The other daughter cell, the companion cell, that is initially just as long as the sieve-element *in statu nascendi*, goes through none, one, or two consecutive transverse divisions during the meristematic stage [77]. Following cell division, scattered fluid vesicles in the cytoplasm fuse to one or a few vacuoles [25]. Apart from the relatively small vacuolar compartment, mature companion cells have a conspicuously dense and extended cytoplasm containing an excessive number of highly active mitochondria [7, 12]. Given the lack of energy produced by the sieve-element mitochondria, it appears that companion cells act as the power suppliers to sustain energy-consuming activities in sieve elements [7, 25]. Moreover, companion cells seem indispensable for the supply of proteins that cannot possibly be manufactured by the enucleate sieve elements [78]. Hence, the interrelationship between companion cell and sieve element has been characterized as one between "a comatose patient and a hyperactive nurse" [12].

That the companion cells have to work for two does not only leave its marks in the (sub)cellular inventory, but also in the unique transcriptome (*see* for a review, [9]). Companion cells must deliver proteins that are needed for the local household in sieve elements. Among others, such proteins maintain the structures and metabolic machineries required for the phloem translocation of thousands of compounds [75] and enable metabolic and protective activities in mictoplasm and sieve-element lumen. Sieve elements contain a broad spectrum of metabolic enzymes, e.g., those involved in carbohydrate processing [79], a complete set of enzymes related to the glutathione and ascorbate cycles to counter oxidative damage [56, 80] and several enzymes related to protein degradation [57, 58, 74, 75]. It remains to be demonstrated, however, that these enzymes are produced by the companion cells and trafficked to the sieve element.

5 Emergence of Diverse Plasmodesmata During Development of the Sieve Element-Companion Cell Complex

The unique intimate relationship between sieve element and companion cells is warranted by several types of plasmodesmata arising during differentiation. During the final developmental stage of the sieve element-companion cell complex, diverse plasmodesmata emerge at the respective interfaces [14]. At the transverse interface between sieve elements, plasmodesmata differentiate to sieve pores with diameters, enlarged up to approx. 10–100 times of the original size [81, 82]. The transition from plasmodesma to sieve pore is preceded by callose deposition under the control of GSL7 in the walls around cambial plasmodesmata, which is necessary for correct positioning of the sieve pores [83, 84]. Subsequent callose dissolution widens the plasmodesmal corridors until the final pore size is reached. Mature sieve pores remain lined by a narrow collar of callose [12]. When the sieve-element Ca^{2+} level is increased as a result of physical or chemical disturbances, the collar is thickened by rapid callose synthesis—probably under the control of GSL12 [85], which leads to constriction of the sieve pores and intermission of mass flow. Widening of plasmodesmata in the walls between the emergent sieve elements is accompanied by a removal of their desmotubules directed by CHER1 (CHOLINE TRANSPORTER-LIKE1) [86], a gene crucial for plasmodesmal maturation [87].

Plasmodesmata at the interface between sieve element and companion cell (pore-plasmodesm units or PPUs) slightly widen at the sieve-element side, while they become branched at the companion-cell side [11, 88]. Mature PPUs are surrounded by a thin callose cell-wall lining that is slightly thicker at the sieve-element side *(Ehlers and van Bel, unpublished results)*. PPUs play a pivotal role in the interaction between sieve element and companion cell, which is crucial for survival of the enucleate sieve elements over many days [7]. Meanwhile, plasmodesmata on the remaining part of the longitudinal sieve-element interfaces disappear. At the interface of the companion cell with phloem parenchyma, plasmodesmata remain predominantly single with diameters that do not deviate much from the original size [89].

6   Functioning of Symplasmic Contacts Between Sieve Elements and Companion Cells

The striking diversity of symplasmic contacts in one cellular complex is likely related to the complexity and multitude of diverse functions to be executed by neighbors, which are interdependent like a Siamese twin. During evolution most of the cellular components of sieve elements were sacrificed to optimize their transport function. The development of a sieve element still reflects the evolutionary developmental course [7]. The plasmodesmata through the transverse walls of cambial sieve elements are widened to sieve pores and large obstacles such as nucleus and vacuole are degraded and removed. Cell components necessary for the transport function such as the plasma membrane are retained [7]. In conclusion, mature sieve elements fully rely on the genetic activities of other cells for protein turnover and many other tasks. The most obvious candidate for such a support is the companion cell, because symplasmic contacts of sieve elements with other neighboring cells are lacking, at least in transport phloem [89].

Lack or paucity of plasmodesmata in sieve-element walls that border non-companion cells [89] forces an exclusive passageway (via PPUs) for macromolecular trafficking between sieve tubes and companion cells. Therefore, the most logical option for macromolecular movement from companion cell to sieve element is trafficking through the PPUs. According to this concept, PPUs are of crucial significance for survival of sieve elements over periods up to years, e.g., in trees.

Apart from the proteins intended for local use in adjacent sieve elements (Subheading 4), sieve tubes contain a wealth of micro- and macromolecular signals, involved in transcriptional and post-transcriptional events in remote tissues. Among the macromolecules in sieve-tube sap are numerous proteins [75] engaged in systemic signaling and several classes of RNA [90]. Micromolecular signals include jasmonic acid [91–93] and salicylic acid [94–97] and their derivatives and, among others, oxylipins, azealic acid, dehydrobietinal, and pipercolic acid (reviewed in [98, 99].

In sieve-tube sap, more than a thousand different proteins have been identified [75]. They encompass (almost) complete sets of proteins necessary for proteolysis via the ubiquitin-26S proteasome pathway [57, 58, 74, 75] and for detoxification of reactive oxygen species [56]. While they are involved in the maintenance of sieve-element function, other proteins probably participate in systemic signaling cascades.

Diverse RNA species in sieve-tube sap constitute another class of macromolecules having a systemic impact [100]. First, several thousands of phloem-mobile mRNAs have been identified [101–103]. In addition, numerous small RNA (sRNA) sequences such as siRNAs (small interfering RNAs) and miRNAs (microRNAs) were found in sieve-tube sap [103–106]. Surprisingly, tRNAs occur in sieve-tube sap [107] despite the long-hold paradigm that sieve elements are incapable of protein synthesis.

Expectedly, macromolecular transport through PPUs is a highly selective and controlled process given the vast range of diverse macromolecules in the sieve-tube sap. PPU-passage was assumed to discriminate between cell-autonomous macromolecules that are retained within the companion cells and non-cell autonomous proteins that are released into the sieve elements. Such a discriminative mode of PPU-transfer was confirmed by selective binding of non-cell autonomous mRNAs to CmPP16. The presence of a specific 36-amino acid RNA-binding motif in CmPP16 further corroborated a concept of selective transfer [108]. CmPP16 increased the plasmodesmal molecular exclusion limit and trafficked through the widened cytoplasmic sleeve as a ribonucleoprotein complex [109]. In addition, the passage only took place after that the ribonucleoprotein complex has interacted with a membrane protein located at the PPU orifice [110]. Likewise, the flowering proteins FTL1 and FTL2 [111] were proposed to be delivered into sieve elements after binding to the ER-membrane-bound protein FTIP1 [112].

Despite its apparent logic, selective symplasmic transfer of macromolecules through PPUs has become a matter of debate [113]. A strictly regulated macromolecular trafficking was questioned by RNA-sequencing studies showing that many graft-transferable mRNA-species were associated in stock and scion with house-keeping activities [100, 101]. Therefore, alternative models were proposed, in which nonselective release of cell-autonomous proteins concurred with selective release of non-cell autonomous proteins [100]. These models again encountered fierce opposition by claims that all mRNAs are nonselectively delivered into sieve elements [114, 115].

7 Occlusion and Constriction of Sieve Pores and Plasmodesmata

It has been demonstrated since a long time that sieve pores of damaged sieve tubes are subject to "plugging by callose clots." However, at closer observation, callose does not occlude the sieve pores, but is deposited extracellularly around the pores and constricts the sieve-pore corridor [22]. Later, forisomes appeared to be able to plug sieve pores by Ca^{2+}-dependent dispersion in response to wounding [116]. Since Ca^{2+} also played a prominent part in triggering callose deposition around the sieve pores in severed sieve tubes [117], the in situ, real-time effects of Ca^{2+} on sieve-pore plugging and constriction were investigated in intact Vicia faba plants [118]. The Ca^{2+} concentration in sieve elements under microscopic examination was enhanced by distant wounding that triggered electrical potential waves associated with Ca^{2+}-influx along the sieve tubes [44, 118, 119]. Forisomes almost instantly dispersed in response to the passage of the Ca^{2+} wave [44, 119, 120] and occluded the sieve plate in a reversible manner. By the time, that the forisomes had re-condensed and sieve-pore plugging was lifted after some 20 min, callose deposition, and thus sieve-pore constriction, reached its maximum [118]. In intact sieve tubes, callose was degraded at a speed in the order of hours, dependent on the plant species [62].

Gating of plasmodesmata (and their ontogenic descendants, the sieve pores) depends on the extracellular deposition of β-1,3 glucans (callose) around the necks of plasmodesmal corridors [121, 122] and sieve pores [117]. Increased callose synthesis constricts the symplasmic passageway as shown by the inverse relationship between plasmodesmal permeability and callose deposition [85, 123–125]. Recent reviews [126–128] advance that callose deposition results from an equilibrium imposed by the activities of plasma transmembrane β-1,3 glucan synthases (GSLs) and plasma-membrane anchored extracellular β-1,3 glucanases (BGs). GSLs constitute a multigene family of large (200–220 kDa) plasma-membrane spanning proteins with both the N- and C-terminus residing in the cytoplasm [129, 130]. The BG family comprises representatives of β-1,3 glucan degrading enzymes or glucanases [131]. GSLs are incorporated in huge protein complexes (CalS holoenzyme complexes; [132, 133]. Callose synthesis is stimulated by elevated Ca^{2+} concentrations [117, 121], but compelling proof of molecular interaction between Ca^{2+} and the CalS holoenzyme complex is lacking thus far.

A similarly quick sieve-pore plugging, perhaps by another type of proteins (PP proteins), and an overlapping slower constriction of sieve pores by callose was found after distant wounding in cucurbits [119]. These dual blockage mechanisms were proposed to represent a safety design that may guarantee hermetic blockage of sieve pores for some time [118, 119]. It is uncertain [69] whether the dual blocking occurs in non-fabacean and non-cucurbit species that contain filamentous SEO proteins [63–65], but it is likely for various reasons [70, 71, 73].

Since sieve pores, PPUs, and single plasmodesmata have the same ontogenic and developmental roots, both occlusion mechanisms might operate at all intercellular symplasmic bridges. Protein-mediated occlusion of PPUs and plasmodesmata has not been investigated thus far, but deposition of callose was observed at any disturbed symplasmic corridor. Occlusion and constriction seem to be important for the putative, transient symplasmic reorganization after passage of a sufficiently vigorous electric potential wave [45].

8 Regulation of the Ca^{2+} Level Engaged in Sieve-Pore Occlusion and Constriction

In view of the correlation between phytoplasma infection and Ca^{2+} [134], control of the Ca^{2+} level deserves special attention. The Ca^{2+} level is of paramount importance for putting up and releasing sieve-tube barriers and its action as a second messenger in a plethora of signaling cascades. The network underlying Ca^{2+} influx and Ca^{2+} homeostasis is dazzlingly complex. Batteries of Ca^{2+} permeable channels are engaged in the regulation of the cytosolic Ca^{2+} level [135]. They reside on probably each membrane system and their collective action creates cytosolic Ca^{2+} signatures, specific time- and space-correlated intracellular Ca^{2+} patterns [136, 137]).

Ca^{2+} permeable channels are sensitive to a vast range of stimuli dependent on their location. For instance, plasma membrane-bound Ca^{2+} permeable channels are either voltage-dependent, mechanosensitive, or ligand-activated, whereas Ca^{2+} permeable channels on ER-membranes are ligand- or Ca^{2+} activated. Despite the reduced number of membrane systems (plasma membrane and ER membranes), regulation of the Ca^{2+} level remains critical and complex in sieve elements. Ca^{2+} permeable channels were identified, but not specified, on both sieve-element membrane systems [44].

Most likely, HACCs (hyperpolarization-activated Ca^{2+} permeable channels), DACCs (depolarization-activated Ca^{2+} permeable channels), and VICCs (voltage-independent Ca^{2+} permeable channels) occur on the plasma membrane of sieve elements. As a speculation, these channels may be involved in the propagation of electric potential waves along the sieve tubes by mediation of a short transient Ca^{2+} influx in response to an external stimulus (DACCs, [41]), by prolonged Ca^{2+} influx during sustained depolarization plateaus (HACCs, [42, 138]) and for maintenance of Ca^{2+} homeostasis in general (VICCs, [43]). Drastic turgor changes due to damage or to abrupt variations in photosynthate supply may be perceived by mechanosensitive Ca^{2+} permeable channels in the plasma membrane. MCAs (mid 1- complementing activity proteins) mediate Ca^{2+} influx in cells experiencing mechanical stress or touch [139, 140]. Since rapid changes in osmolarity provoke mechanical stress or relaxation of the plasma membrane, MCAs may be functional homologs of the osmo-stimulated Ca^{2+} permeable CSCs (calcium permeable stress-gated cation channels, [141]) and OSCAs (reduced hyperosmolarity-induced Ca^{2+} increase channels, [142]). Serious injuries may not only be accompanied by loss of turgor, but also by release of various substances which function as ligands for ligand-activated Ca^{2+} permeable channels such as CNCGs cyclic nucleotide-gated channels, [143, 144–146] and GLRs glutamate receptor homologues, [147]. GLRs are required for systemic electric wound responses including jasmonate signaling [148]. One GLR was localized to the phloem without further specification of the cellular location [149]. Although most of these channels have not been localized to sieve elements as yet, it is likely that several of them are involved in the dynamics of Ca^{2+} signatures in sieve elements and companion cells.

In particular, callose synthesis may require Ca^{2+} levels far above the 50–100 nM Ca^{2+} found in the sieve-element lumen [44]. Therefore, the existence of Ca^{2+} hotspots along sieve-element was postulated, based on the spatial distribution of Ca^{2+} channels [120]. Hotspots would occur in regions, in particular around the sieve pores and PPUs [44], where Ca^{2+} permeable channels are aggregated in the plasma membrane. Ca^{2+} ions accumulate and, in turn, activate ER-located CICR (Ca^{2+} induced Ca^{2+} release) channels in the membranes of the ER stacks [120, 150] that are anchored to the plasma membrane [30]. Concerted actions of various Ca^{2+} permeable channels may lead to local Ca^{2+} concentrations in the range of several hundreds of micromoles in the close vicinity of sieve pores and PPUs [120]. Other ligand-activated ER-membrane-bound Ca^{2+} permeable channels such as animal $InsP_3$ (inositol 1,4,5-trisphosphate) receptors would also be a useful tool for mictoplasmic Ca^{2+} regulation. However, the presence of $InsP_3$ receptors has not been convincingly demonstrated [151] and $InsP_3$ homologues were reported to be missing in the sequenced land plant genomes [135]. Nevertheless, remarkable effects were obtained with the $InsP_3$ inhibitor mastoparan in *Setcreasea purpurea* staminal hairs; mastoparan injection prevented plasmodesmal closure assumed to be due to an inhibited release of Ca^{2+} from ER stacks [121].

Restoration of the original situation—reopening of the sieve pores due to removal of protein plugs and breakdown of callose—was ascribed due to removal of excess Ca^{2+} from the mictoplasm [45] by several Ca^{2+} pumps [135]. These pumps relocate Ca^{2+} ions from the cytosol to the storage compartments. Some P-type Ca^{2+} ATPases (ECAs, $P_{2a-type}$ ER Ca^{2+} ATPases) may mediate Ca^{2+} accumulation in ER [152], while others (ACAs, $P_{2b-type}$ Ca^{2+} ATPases) traffic Ca^{2+} ions into the cell wall compartment [153].

Due to their intimate connection, changing Ca^{2+} levels in the companion cell cytosol may affect events in sieve elements. Companion cells contain active vacuoles acting as abundant Ca^{2+} reservoirs. Vacuolar release and storage are regulated by several Ca^{2+} permeable channels and pumps. A major function in Ca^{2+} release into the cytosol has been attributed to the tonoplast two-pore channel TPC1 [154–156]. While the Ca^{2+}-pumps ACA8, ACA9, and ACA10, which reside in the plasma membrane, are responsible for Ca^{2+} extrusion into the cell wall, tonoplast-bound ACA4 and ACA11 are involved in vacuolar Ca^{2+} storage [157]. Apart from the Ca^{2+} pumps, members of the CAX H^+/cation antiporter family are involved in vacuolar Ca^{2+} sequestration [158, 159] and, by consequence, in the Ca^{2+} homeostasis of cytosol and cell wall compartments [160, 161].

9 Ca^{2+} Homeostasis and ROS (Reactive Oxygen Species) Production

Ca^{2+} homeostasis also depends on ROS-mediated regulatory processes. Release of Ca^{2+} into the cytoplasmic compartment is often linked with an outburst of ROS. Release of Ca^{2+} via Ca^{2+}-permeable channels generally precedes ROS production [94]. These events interact in a complicated fashion: Ca^{2+} release is affected by ROS production (RICR, ROS-induced Ca^{2+} release) and vice versa (CIRP; Ca^{2+}-induced ROS production) [162].

CIRP may be mediated by Ca^{2+} activation of ROS producing RBOHD (RESPIRATORY BURST OXIDASE HOMOLOG D) proteins in various ways. Ca^{2+} ions can bind to the EF hand motifs of RBOHDs [163, 164], induce kinase-mediated phosphorylation of RBOHDs [164, 165], or stimulate the production of phosphatidic acid that binds to N-terminal motifs of RBOHs. In turn, ROS may interact with Ca^{2+}-permeable channels as demonstrated in root cells [166–168] and in guard cells [138, 169].

As a last note, mechanisms of Ca^{2+} homeostasis in sieve elements remain subject of pure speculation as long as Ca^{2+} channels and proteins involved in ROS production in plasma membranes and ER-membranes of sieve-elements have not been located and identified.

10 The Special Nature of Mass Flow in Sieve Tubes, Putative Molecular Hopping and the Consequences for Systemic Signaling

In contrast to mass flow in tubules having impermeable walls, "mass flow" in sieve tubes is a much more dynamic process. Fractions of solvent (water) and carbohydrates in sieve tubes are permanently released sideways to the apoplasm along the phloem pathway [37, 170–174]. The release may be accomplished by facilitated diffusion via SUT1 sucrose carriers, when the "substrate-motive force" of the sucrose solution in the sieve tubes surpasses the proton-motive force [175]. Remarkably, carbohydrate release from transport phloem is decreased in calcium-deprived plants [176]. Sucrose leakage is nearly compensated by concomitant active retrieval by sieve elements [170, 173, 174] so that the driving force for mass flow is maintained. As a consequence, part of the solutes and solvent proceed by hopping along the outskirts of sieve tubes, while they are retrieved from the apoplasmic space by either sieve element-companion cell complexes or phloem parenchyma cells [173]. The intensity of hopping was proposed to depend on the frequency of release/retrieval events [177].

Another form of molecular hopping may occur at the interface between sieve element and companion cell, where macromolecules released into the sieve tubes via PPUs are retrieved by downstream companion cells [178]. "Polar transfer" of phloem-specific proteins via cucurbit graft junctions renders credit to withdrawal of proteins from the sieve-tube stream, most likely by companion cells. Interspecific grafts partners frequently failed to exchange species-specific proteins, which was ascribed to a ready protein degradation in companion cells of the graft partner [179, 180]. In addition, selective distribution of systemically distributed RNAs in sinks [181] is hardly conceivable without selective transfer from sieve tubes into companion cells.

Perhaps, intervention by companion cells also plays a role in the remarkably asymmetric distribution of photoassimilates and SAR-associated signals. Sucrose distribution was incongruent with the expression of pathogenesis-related genes [182]. The discrepancy may be due to phloem-to-xylem transfer of JA-related [183] and SA-related signals [184]. Phloem-to-xylem transfer may also explain the destination-selective transport of phloem-specific pumpkin proteins introduced into the phloem of rice plants via cut stylets [185]. One set of proteins was translocated to the root, the other to the shoot. It is possible that the latter fraction is retrieved from the sieve-tube stream by companion cells and then transported radially via the vascular symplasm to the xylem vessels. Presence of proteins in vessel fluid was documented [186–188], but an exocytotic step at some point has to be invoked to explain transfer from the symplasmic compartment to the apoplasmic space.

11   Physical Conditions Inside Sieve Elements: Osmotic Potential, Pressure Gradients, Sieve-Pore Sizes, Hypoxic Conditions

To maintain the required driving forces for mass flow, sieve-tube sap must be highly concentrated, but offer a solute source-to-sink gradient along the sieve tubes. Even in transport phloem, a concentration slope was found in the direction of the sinks, since not all solutes and solvent escaping from the sieve tubes are retrieved [37, 170–174]. Pioneer experiments using exudates collected from cut stylets of aphids sitting on diverse locations along 1 m long *Salix viminalis* twigs disclosed solute gradients from 890–650 mmol to 700–650 mmol [189]. A similar sucrose basipetal gradient (285–215 mmol) showed up in the sieve-tube exudate of 85 cm tall *Ricinus communis* plants [190]. Conversely, the apoplasmic sucrose concentration near sieve tubes rose from 25 to 60 mmol in the basipetal direction in bean plants [191]. The sucrose concentration in sieve-tube sap is positively correlated with the turgor pressure. A sudden drop in sieve-tube sucrose induced by nitrogen gas supply was accompanied by a drop in turgor pressure [192].

Another point of physiological interest for survival of phytoplasmas is the low partial oxygen pressure in sieve tubes. High alcohol dehydrogenase and ethanol levels in the vascular cambium of tree stems provided circumstantial evidence for hypoxic conditions near the phloem [193]. Later, the partial oxygen pressure in sieve tubes of *Ricinus communis* was indeed determined to be one-third (approx. 7%) of the atmospheric partial oxygen pressure having profound consequences for the metabolism [194].

Sieve-tube diameters increase with a decreased vein order [195], but sieve-pore sizes in veins have been poorly documented. In contrast to extensive documentation of the sieve-pore diameters in transport phloem [81, 82, 196], explicit measurements on sieve-pore diameters in the finest veins were absent in numerous consulted publications on vein anatomy. Pioneer studies on the vein endings in *Beta vulgaris* suggested that the sieve-plate diameter was adapted to the size of the narrow sieve elements, but that the few pores had the same diameter as those in the transport phloem of the major veins [197].

12   Chemical Conditions Inside Sieve Elements: pH, Composition of the Sieve-Tube Sap

Sieve-tube sap was collected from a few spontaneously exuding species [198–200], as sieve-tube drops pressed through stylets cut by stylectomy [201–204] or from cut stems or petioles exuding into EDTA solutions (e.g., [205–207]). EDTA is meant to bind

Ca^{2+} that is set free by stem cutting and would confer sieve-tube blockage [208]. Sieve-tube sap is rich and variable in chemical composition and varies strongly between [198–201, 203, 204] and within plant species [190, 209, 210], and moreover depends on diurnal rhythms, seasonal impact [211], and environmental conditions.

Exudates of *Ricinus communis* [190, 199] and *Nicotiana glauca* [200] collected via stem incisions contain high concentrations of sucrose in the range of several hundreds of mMolar and aggregate amino acid concentrations in the order of 100 mM. Potassium and chloride are also abundant, the other minerals and organic substances occur at much lower concentrations. Stylet-mediated exudates were mainly successfully collected from mono-cotyledons such as rice [201], wheat [203], and maize [204]. As in the bleeding saps of dicotyledons [190, 200, 208], sucrose (250–900 mM) and amino acids (260–600 mM) are the main constituents along with potassium (300–500 mM) and chloride (25–270 mM), while the other minerals and metabolites occur at much lower concentrations [201, 203, 204]. EDTA-facilitated and stylet-mediated exudation yielded similar values for amino acid concentrations [205–207], but in EDTA-probes, part of the sucrose was replaced by hexoses, which was interpreted to be an EDTA-induced artifact [206, 207].

As vacuoles are lacking, sieve elements may be considered protoplasmic units composed of a solid (mictoplasm) and a fluidic state (sieve-element lumen) that form an elongate enucleate syncytium. The pH of sieve-tube exudates [173, 190, 198, 200] infers that the lumen possesses an alkaline pH (7 to 8) similar to that in the cytoplasm of nucleate, vacuolated cells [212–215]. Sieve-tube pH values in the order of 7.5 were confirmed using microprobes of sieve-tube sap exuding from cut aphid stylets [173]. The pH can gradually vary with the sieve-tube location as demonstrated by the differences between basal (7.2) and apical (7.8) *Ricinus* internodes [190].

13   Consequences of Sieve Tubes as a Habitat for the Lifestyle of Phytoplasmas. Implications for Future Studies on Phytoplasmas and for the Composition of Artificial Media for Phytoplasma Cultivation

1. Phytoplasma diameters vary between 200 and 800 nm [216], while the diameters of the sieve pores range between 300 and 2500 nm [81, 82]. These are average values; it should be noted that sieve pores in the center of the sieve plates of *Cucurbita* and *Phaseolus* are appreciably larger than those at the margins [81]. All in all, it seems obvious that phytoplasmas are carried by mass flow through sieve elements and that sieve pores would

hardly present a hindrance to phytoplasma transport. Nevertheless, longitudinal progress of phytoplasmas lags behind solute mass flow [134, 217], which may be ascribed to occasional and transient connection of phytoplasmas to structural sieve-element components. Images in which phytoplasmas are appressed to the sieve-element membrane [218, 219] suggest some mode of docking for unknown purposes.

2. Since the sieve elements are enucleate, responses to phytoplasmas essentially depend on actions of the companion cells. Hence, future molecular studies on plant responses to phytoplasma infection should focus on gene expression in companion cells.

3. Another focus issue may pertain to the macromolecular content of the sieve elements. It is conceivable that the assortment of RNA-molecules in sieve elements plays an auxiliary role in the fabrication of proteins in phytoplasmas. In such a concept, proteins translated from plant RNA may provide a welcome supplement to the proteins produced in organisms such as phytoplasmas having a restricted number of genes [220]. Phytoplasmas may absorb plant RNAs by endocytosis facilitated by the absence of a cell wall.

4. Given the absence of a cell wall, phytoplasmas must maintain internal osmotic values identical to the osmotic potential of sieve elements. Solely under iso-osmotic conditions, phytoplasmas will be able to retain or modulate their shape. In view of the location-dependent differences in sieve-tube sap composition and the occurrence of phytoplasmas throughout the phloem network, phytoplasmas must possess a high potential to osmotic adaptability. For instance, phytoplasmas must permanently adapt to the dropping environmental osmotic potential on their way from source to sink. Therefore, cultivation media should provide osmotic and nutritional conditions in the medium range found in sieve tubes of the phytoplasma host. Phytoplasmas will burst otherwise under hypo-osmotic and shrink under hyper-osmotic conditions.

References

1. Weintraub PG, Beanland L (2006) Insect vectors of phytoplasmas. Annu Rev Entomol 51:91–111

2. Hogenhout SA, Oshima K, Ammar ED et al (2008) Phytoplasmas: bacteria that manipulate plants and insects. Mol Plant Pathol 9:403–423

3. Bosco D, D'Amelio R (2010) Transmission specificity and competition of multiple phytoplasmas in the insect vector. In: Weintraub P, Jones P (eds) Phytoplasmas: genomes, plant hosts and vectors. CABI, Wallingford, pp 293–308

4. Contaldo N, Bertaccini A, Paltrinieri S et al (2012) Axenic culture of plant pathogenic phytoplasmas. Phytopathol Mediterr 51:607–617

5. Contaldo N, Satta E, Zambon Y et al (2016) Development and evaluation of different complex media for phytoplasma isolation and growth. J Microbiol Meth 127:105–110

6. Van Bel AJE, Hafke JB (2005) Physiochemical determinants of phloem transport. In: Holbrook NM, Zwieniecki M (eds) Vascular transport in plants. Elsevier, Amsterdam, pp 19–44

7. Van Bel AJE (2003) The phloem, a miracle of ingenuity. Plant Cell Environ 26:125–150

8. Mähönen AP, Bonke M, Kauppinen L, Riikonen M, Benfey PN, Helariutta Y (2000) A novel two-component hybrid molecule regulates vascular morphogenesis of *Arabidopsis* root. Genes Dev 14:2938–2943

9. Otero S, Helariutta Y (2017) Companion cells: a diamond in the rough. J Exp Bot 68:71–78

10. Esau K (1977) Anatomy of seed plants, 2nd edn. Wiley New York, Santa Barbara, London, Sydney, Toronto

11. Esau K, Thorsch J (1985) Sieve plate pores and plasmodesmata, the communication channels of the symplast: ultrastructural aspects and developmental relations. Am J Bot 72:1641–1653

12. Evert R (1990) Dicotyledons. In: Behnke H-D, Sjolund RD (eds) Sieve elements: comparative structure, induction and development. Springer, Berlin, pp 103–137

13. Van Bel AJE, Knoblauch M (2000) Sieve element and companion cell: the story of the comatose patient and the hyperactive nurse. Austral J Plant Physiol 27:477–487

14. Van Bel AJE, Hess P (2003) Phloemtransport. Kollektiver Kraftakt zweier Exzentriker. Biol Unserer Zeit 33:220–230

15. Rodriguez-Villalon A (2016) Wiring a plant: genetic networks for phloem formation in *Arabidopsis thaliana* roots. New Phytol 210:45–50

16. Mouchel CF, Briggs GC, Hardtke CS (2004) Natural genetic variation in *Arabidopsis* identifies BREVIS RADIX, a novel regulator of cell proliferation and elongation in the shoot. Genes Dev 18:700–714

17. Truernit E, Bauby H, Belcram K et al (2012) OCTOPUS, a polarly localised membrane-associated protein, regulates phloem differentiation entry in *Arabidopsis thaliana*. Development 139:1306–1315

18. Rodriquez-Villalon A, Gujas B, Kang Y et al (2014) Molecular genetic framework for protophloem formation. Proc Natl Acad Sci U S A 111:11551–11556

19. Depuydt S, Rodriquez-Villalon A, Santuari L et al (2013) Suppression of *Arabidopsis* protophloem differentiation and root meristem growth by CLEA45 requires the receptor-like kinase BAM3. Proc Natl Acad Sci U S A 110:7074–7079

20. Ruiz-Sola MA, Coiro M, Crivelli S et al (2017) OCTOPUS-LIKE 2, a novel player in *Arabidopsis* root and vascular development, reveals a key role for OCTOPUS family genes in root metaphloem sieve tube differentiation. New Phytol 216:1191–1204. https://doi.org/10.1111/nph.14751

21. Bonke M, Thitamadee S, Mähönen AP et al (2003) APL regulates vascular tissue identity in *Arabidopsis*. Nature 426:181–186

22. Furuta KM, Yadav SR, Lehesranta S et al (2014) *Arabidopsis* NAC45/86 direct sieve element morphogenesis culminating in enucleation. Science 345:933–937

23. Esau K (1969) The phloem. Borntraeger, Berlin (Encyclopedia of Plant Anatomy, vol 5.2)

24. Engleman E (1965) Sieve elements of *Impatiens sultani*. II. Developmental aspects. Ann Bot 29:103–104

25. Patrick JW, Tyerman SD, van Bel AJE (2015) Long distance transport. In: Buchanan BB, Gruissem W, Jones RL (eds) Biochemistry and molecular biology of plants. Wiley, Cichester, pp 658–710

26. Esau K, Cronshaw J (1968) Plastids and mitochondria in the phloem of *Cucurbita*. Can J Bot 46:877–880

27. Thorsch J, Esau K (1981) Changes in the endoplasmic reticulum during differentiation of a sieve element in *Gossypium hirsutum*. J Ultrastruct Res 74:183–194

28. Thorsch J, Esau K (1981) Nuclear degeneration and the association of endoplasmic reticulum with the nuclear envelope and microtubules in maturing sieve elements of *Gossypium hirsutum*. J Ultrastruct Res 74:195–204

29. Sjolund RD, Shih CY (1983) Freeze-fracture analysis of phloem structure in plant tissue cultures. I. The sieve element reticulum. J Ultrastruct Res 82:111–121

30. Ehlers K, Knoblauch M, van Bel AJE (2000) Ultrastructural features of well-preserved and injured sieve elements: minute clamps keep the phloem conduits free for mass flow. Protoplasma 214:80–92

31. Patrick JW (1997) Phloem unloading. Sieve element unloading and post-sieve element transport. Annu Rev Plant Physiol Plant Mol Biol 28:165–190

32. Sauer N (2007) Molecular physiology of higher plant sucrose transporters. FEBS Lett 581:2309–2317

33. Ayre BG (2011) Membrane-transport systems for sucrose in relation to whole-plant carbon partitioning. Mol Plant 4:377–394

34. Milne RJ, Perroux MJ, Rae AL et al (2016) Sucrose transporter localization and function in phloem unloading in developing stems. Plant Physiol 173:1330–1341

35. Julius BT, Leach KA, Tran TM et al (2017) Sugar transporters in plants: new insights and discoveries. Plant Cell Physiol 58:1442–1460

36. Tegeder M (2014) Transporters involved in source to sink partitioning of amino acids and ureides. J Exp Bot 65:1865–1878

37. Stanfield R, Hacke U, Laur J (2017) Are phloem sieve tubes leaky conduits supported by numerous aquaporins? Am J Bot 104:719–732

38. Thompson M, Zwieniecki M (2005) The role of potassium in long-distance transport in plants. In: Holbrook NM, Zwieniecki M (eds) Vascular transport in plants. Elsevier, Amsterdam, pp 221–240

39. Dreyer I, Gomes-Porras JL, Riedelsberger J (2017) The potassium battery: a mobile energy source for transport processes in plant vascular tissues. New Phytol 216 (4):1049–1053. https://doi.org/10.1111/nph.14667

40. Rogiers SY, Coetzee ZA, Walker RR et al (2017) Potassium in the grape (*Vitis vinifera* L.) berry: transport and function. Front Plant Sci 8:1629

41. Thion L, Mazars C, Nacry P et al (1998) Plasma membrane depolarization activated calcium channels, stimulated by microtubule-depolymerizing drugs in wild-type *Arabidopsis thaliana* protoplasts, display constitutively large activities and a longer half-life in *ton* 2 mutant cells affected in the organization of cortical microtubules. Plant J 13:603–610

42. Hamilton DW, Hills A, Kohler B et al (2000) Ca^{2+} channels at the plasma membrane of stomatal guard cells are activated by hyperpolarization and abscisic acid. Proc Natl Acad Sci U S A 97:4967–4972

43. White PJ, Davenport RJ (2002) The voltage-independent cation channel in the plasma membrane of wheat roots is permeable to divalent cations and may be involved in cytosolic homeostasis. Plant Physiol 130:1386–1395

44. Furch ACU, van Bel AJE, Fricker MD et al (2009) Sieve element Ca^{2+} channels as relay stations between remote stimuli and sieve tube occlusion. Plant Cell 21:2118–2132

45. Van Bel AJE, Furch ACU, Will T et al (2014) Spread the news: systemic dissemination and local impact of Ca^{2+} signals along the phloem pathway. J Exp Bot 65:1761–1787

46. Van Bel AJE, Kempers R (1991) Symplastic isolation of the sieve element-companion cell complex in the phloem of *Ricinus communis* and *Salix alba* stems. Planta 183:69–76

47. Buvat R (1960) Observations sur l'infrastructure du cytoplasma au cours de la différenciation des cellules criblées de *Cucurbita pepo* L. C R Acad Sci 250:1528–1530

48. Behnke H-D (1965) Über das Phloem der Dioscoreaceen unter besonderer Berücksichtigung ihrer Phloembecken II Elektronenoptische Untersuchungen zur Feinstruktur des Phloembeckens. Z Pflanzenphysiol 53:214–244

49. Esau K, Cronshaw J (1968) Endoplasmic reticulum in the sieve element of *Cucurbita*. J Ultrastr Res 23:1–14

50. Behnke HD (1991) Distribution and evolution of forms and types of sieve-element plastids in the dicotyledons. Aliso 3:167–182

51. Parthasarathy MV, Pesacreta TC (1980) Microfilaments in plant vascular cells. Can J Bot 58:807–815

52. Schobert C, Baker L, Szederkenyi J et al (1998) Identification of immunologically related proteins in sieve-tube exudate collected from monocotyledonous and dicotyledonous plants. Planta 206:245–252

53. Schobert C, Gottschalk M, Kovar DR et al (2000) Characterization of *Ricinus communis* phloem profilin, RcPRO1. Plant Mol Biol 42:719–730

54. Kulikova AL, Puryaseva AP (2002) Actin in pumpkin phloem exudate. Russ J Plant Physiol 49:54–60

55. Barnes A, Bale J, Constantinidou C et al (2004) Determining protein identity from sieve element sap in *Ricinus communis* L. by quadruple time of flight (Q-TOF) mass spectrometry. J Exp Bot 55:1473–1481

56. Walz C, Giavalisco P, Schad M et al (2004) Proteomics of cucurbit phloem exudate reveals a network of defence proteins. Phytochemistry 65:1795–1804

57. Giavalisco P, Kapitza K, Kolasa A et al (2006) Towards the proteome of *Brassica napus* phloem sap. Proteomics 6:896–909

58. Aki T, Shigyo M, Nakano R et al (2008) Nano scale proteomics revealed the presence of regulatory proteins including three FT-like proteins in phloem and xylem saps from rice. Plant Cell Physiol 49:767–790

59. Hafke JB, Ehlers K, Föller J et al (2013) Involvement of the sieve element cytoskeleton in electrical responses to cold shocks. Plant Physiol 162:707–719

60. Furch ACU, Buxa SV, van Bel AJE (2015) Similar intracellular location and stimulus reactivity, but differential mobility of tailless (*Vicia faba*) and tailed forisomes (*Phaseolus vulgaris*) in intact sieve tubes. PLoS One 10: e0143920

61. Cronshaw J, Sabnis DD (1990) Phloem proteins. In: Behnke H-D, Sjolund RD (eds) Sieve elements: comparative structure, induction and development. Springer, Berlin, pp 255–283

62. Furch ACU, Hafke JB, van Bel AJE (2008) Plant- and stimulus-specific variations in remote-controlled sieve-tube occlusion. Plant Sign Behav 3:858–861

63. Rüping B, Ernst AM, Jekat SB et al (2010) Molecular and phylogenetic characterization of the sieve element occlusion family in Fabaceae and non-Fabaceae plants. BMC Plant Biol 10:219

64. Anstead JA, Froelich DR, Knoblauch M et al (2012) *Arabidopsis* P-protein filament formation requires both AtSEO1 and AtSEO2. Plant Cell Physiol 53:1089–1094

65. Batailler B, Lemaitre T, Vilaine F et al (2012) Soluble and filamentous proteins in *Arabidopsis* sieve elements. Plant Cell Environ 35:1258–1273

66. Knoblauch M, Ehlers K, Peters WS et al (2001) Reversible calcium-regulated stopcocks in legume sieve tubes. Plant Cell 13:1221–1230

67. Peters WS, Haffer D, Hanakam CB et al (2010) Legume phylogeny and the evolution of a unique contractile apparatus that regulates phloem transport. Am J Bot 97:797–808

68. Pélissier H, Peters WS, Collier R et al (2008) GFP tagging of sieve element occlusion (SEO) proteins results in green fluorescent forisomes. Plant Cell Physiol 49:1699–1710

69. Froelich DR, Mullendore DL, Jensen KH et al (2011) Phloem ultrastructure and pressure flow: sieve-element-occlusion-related agglomerations do not affect translocation. Plant Cell 23:4428–4445

70. Ernst AM, Jekat SB, Zielonka S et al (2012) Sieve element occlusion (SEO) genes encode structural proteins involved in wound sealing of phloem. Proc Natl Acad Sci U S A 109: E1980–E1989

71. Jekat SB, Ernst AM, von Bohl A et al (2013) P-proteins are heteromeric structures involved in rapid sieve tube sealing. Front Plant Sci 4:225

72. Van Bel AJE, Will T (2016) Functional evaluation of proteins in watery and gel saliva of aphids. Front Plant Sci 7:1840

73. Pagliari L, Buoso S, Santi S et al (2017) Filamentous sieve element proteins are able to limit phloem mass flow, but not phytoplasma spread. J Exp Bot 68:3673–3688

74. Schobert C, Großmann P, Gottschalk M et al (1995) Sieve-tube exudate from. *Ricinus communis* L. seedlings contains ubiquitin and chaperones. Planta 196:205–210

75. Lin MK, Lee YJ, Lough TJ et al (2009) Analysis of the pumpkin phloem proteome provides insights into angiosperm sieve tube function. Mol Cell Proteomics 8:343–356

76. Ingvardsen C, Veierskov B (2001) Ubiquitin- and proteasome-dependent proteolysis in plants. Physiol Plant 112:451–459

77. Chavan RR, Braggins J, Harris PJ (2000) Companion cells in the secondary phloem of Indian dicotyledonous species: a quantitative study. New Phytol 146:107–118

78. Kühn C, Franceschi VR, Schulz A et al (1997) Macromolecular trafficking indicated by localization and turnover of sucrose transporters in enucleate sieve elements. Science 275:1298–1300

79. Eschrich W, Heyser W (1975) Biochemistry of phloem constituents. In: Zimmermann MH, Milburn JA (eds) Encyclopedia of plant physiology. Transport in plants I Phloem transport. Springer, Heidelberg, pp 101–136

80. Szederkenyi J, Komor E, Schobert C (1997) Cloning of the cDNA glutaredoxin, an abundant sieve-tube exudate protein from *Ricinus communis* L and characterization of the glutathione-dependent thiol-reduction system in sieve tubes. Planta 202:349–356

81. Mullendore DL, Windt CW, van As H et al (2010) Sieve tube geometry in relation to phloem flow. Plant Cell 22:579–593

82. Bussières P (2014) Estimating the number and size of phloem sieve plate pores using longitudinal views and geometric reconstruction. Sci Rep 4:4929

83. Barratt DH, Kolling K, Graf A et al (2011) Callose synthase GSL7 is necessary for normal phloem transport and inflorescence growth in *Arabidopsis*. Plant Physiol 155:328–341

84. Xie B, Wang X, Zhu M et al (2011) CalS7 encodes a callose synthase responsible for callose deposition in the phloem. Plant J 65:1–14

85. Vaten A, Dettmer J, Wu S et al (2011) Callose biosynthesis regulates symplastic trafficking

during root development. Dev Cell 21:1144–1155

86. Dettmer J, Ursache R, Campilho A et al (2014) CHOLINE TRANSPORTER-LIKE1 is required for sieve plate development to mediate long-distance cell-to-cell communication. Nat Commun 5:4276

87. Kraner ME, Link K, Melzer M et al (2017) Choline-transporter-like (CHER 1) is crucial for plasmodesmata maturation in *Arabidopsis thaliana*. Plant J 89:394–406

88. Deshpande BP (1975) Differentiation of the sieve plate of *Cucurbita*: a further view. Ann Bot 39:1015–1022

89. Kempers R, Ammerlaan A, van Bel AJE (1998) Symplasmic constriction and ultrastructural features of the sieve element/companion cell complex in the transport phloem of apoplasmically and symplasmically phloem loading species. Plant Physiol 116:271–278

90. Lucas WJ, Groover A, Lichtenberger R et al (2013) The plant vascular system: evolution, development and functions. J Integr Plant Biol 55:294–388

91. Wasternack C, Hause B (2013) Jasmonate : biosynthesis, perception, signal transduction and. action in plant stress response, growth and development. An update to the 2007 review in *Annals of Botany*. Ann Bot 111:1021–1058

92. Campos ML, Kang J-H, Howe GA (2014) Jasmonate-triggered plant immunity. J Chem Ecol 40:657–675

93. Das TA, Uddin M, Khan MMA et al (2015) Jasmonates counter plant stress: a review. Env Exp Bot 115:49–57

94. Gilroy S, Bialasek M, Suzuki N et al (2016) ROS, calcium, and electric signals: key mediators of rapid systemic signaling in plants. Plant Physiol 171:1606–1615

95. Klessig DF, Tian M, Choi HW (2016) Multiple targets of salicylic acid and derivatives in plants and animals. Front Immunol 7:206

96. Gaupels F, Durner J, Kogel K-H (2017) Production, amplification and systemic propagation of redox messengers in plants. The phloem can do it all! New Phytol 214:554–560

97. Wang N, Pierson EA, Setubal JC et al (2017) The *candidatus* Liberibacter-host interface: insights into pathogenesis mechanisms and disease control. Annu Rev Phytopathol 35:451–482

98. Dempsey DA, Klessig DF (2012) SOS–too many signals for systemic acquired resistance? Trends Plant Sci 17:538–545

99. Shah J, Chaturvedi R, Chowdhury Z et al (2014) Signaling by small molecules in systemic acquired resistance. Plant Cell 79:645–658

100. Ham B-K, Lucas WJ (2017) Phloem-mobile RNAs as systemic signaling agents. Annu Rev Plant Biol 68:173–195

101. Thieme CJ, Rojas-Triana M, Stecyk E et al (2015) Endogenous *Arabidopsis* messenger RNAs transported to distant tissues. Nat Plants 1:15025

102. Yang Y, Mao L, Jittayasothorn Y et al (2015) Messenger RNA exchange between scions and rootstocks in grafted grapevines. BMC Plant Biol 15:251

103. Zhang Z, Zheng Y, Ham B-K et al (2016) Vascular-mediated signalling involved in early phosphate stress response in plants. Nature Plants 2:16033

104. Yoo B-C, Kragler F, Varkonyi-Gasic E et al (2004) A systemic small RNA signaling system in plants. Plant Cell 16:1979–2000

105. Buhtz A, Springer F, Chappell L et al (2008) Identification and characterization of small RNAs from the phloem of *Brassica napus*. Plant J 53:739–749

106. Ham B-K, Li G, Jia W et al (2014) Systemic delivery of siRNA in pumpkin by a plant PHLOEM SMALL RNA-BINDING PROTEIN1–ribonucleoprotein complex. Plant J 80:683–694

107. Zhang W, Thieme CJ, Kollwig G et al (2016) t-RNA related sequences trigger mRNA transport in plants. Plant Cell 28:1237–1249

108. Taoka K, Ham B-K, Xoconostle-Cazares B et al (2007) Reciprocal phosphorylation and glycolysation recognition motifs control NCAPP1 interaction with pumpkin phloem proteins and their cell-to-cell movement. Plant Cell 19:1866–1884

109. Xoconostle-Cazares B, Xiang Y, Ruiz-Medrano R et al (1999) Plant paralog to viral movement protein that potentiates transport of mRNA into the phloem. Science 283:94–98

110. Lee J-Y, Yoo B-C, Rojas MR et al (2003) Selective trafficking of non-cell-autonomous proteins mediated by NtNCAPP1. Science 299:392–396

111. Lin MK, Belanger H, Lee YJ, Varkonyi-Gasic E et al (2007) FLOWERING LOCUS T protein may act as the long-distance florigenic signal in the cucurbits. Plant Cell 19:1488–1506

112. Liu L, Liu C, Hou X et al (2012) FTIP1 is an essential regulator required for florigen transport. PLoS Biol 10:e1001313

113. Schulz A (2017) Long-distance trafficking: lost in transit or stopped at the gate? Plant Cell 29:426–430

114. Paultre DSG, Gustin M-P, Molnar A et al (2016) Lost in transit: long-distance trafficking and phloem unloading of protein signals in *Arabidopsis* homografts. Plant Cell 28:2016–2025

115. Calderwood A, Kopriva S, Morris TJ (2016) Transcript abundance explains mRNA mobility data in *Arabidopsis thaliana*. Plant Cell 28:610–615

116. Knoblauch M, van Bel AJE (1998) Sieve tubes in action. Plant Cell 10:35–50

117. Kauss H (1987) Some aspects of calcium-dependent regulation in plant metabolism. Annu Rev Plant Physiol 38:47–71

118. Furch ACU, Hafke JB, Schulz A et al (2007) Calcium-mediated remote control of reversible sieve-tube occlusion in *Vicia faba*. J Exp Bot 28:2827–2838

119. Furch ACU, Will T, Zimmermann MR et al (2010) Remote-controlled stop of mass flow in *Cucurbita maxima*. J Exp Bot 61:3697–3708

120. Hafke JB, Furch ACU, Fricker MD et al (2009) Forisome dispersion in *Vicia faba* is triggered by Ca^{2+} hotspots created by concerted action of diverse Ca^{2+} channels in sieve elements. Plant Sign Behav 4:968–972

121. Tucker EB, Boss WF (1996) Mastoparan-induced intracellular Ca^{2+} fluxes may regulate cell-to-cell communication in plants. Plant Physiol 111:459–467

122. Holdaway-Clarke TL, Walker NA, Hepler PK et al (2000) Physiological elevations in cytoplasmic free calcium or ion injection result in transient closure of higher plant plasmodesmata. Planta 210:329–335

123. Lee J-Y, Lu H (2011) Plasmodesmata: the battleground against intruders. Trends Plant Sci 16:201–210

124. Rinne PL, Welling A, Vahala J et al (2011) Chilling of dormant buds hyperinduces FLOWERING LOCUS T and recruits GA-inducible 1,3 beta-glucanases to reopen signal conduits and release dormany in *Populus*. Plant Cell 23:130–146

125. Zavaliev R, Ueki S, Epel BL et al (2011) Biology of callose (β-1,3 glucan) turnover at plasmodesmata. Protoplasma 249:117–130

126. De Storme N, Geelen D (2014) Callose homeostasis at plasmodesmata: molecular regulators and developmental relevance. Front Plant Sci 5:138

127. Kumar R, Kumar D, Hyun TK et al (2015) Players at plasmodesmal nano-channels. J Plant Biol 58:75–86

128. Tilsner J, Nicolas W, Rosada A et al (2016) Staying tight: plasmodesmal membrane contact sites and the control of cell-to-cell connectivity. Annu Rev Plant Biol 67:337–364

129. Farrokhi N, Burton RA, Brownfield L et al (2006) Plant cell wall biosynthesis. Genetic, biochemical and functional genomics approach to the identification of key genes. Plant Biotech J 4:145–167

130. Brownfield L, Ford K, Doblin MS et al (2007) Proteomic and chemical evidence links the callose synthase in *Nicotiana alata* pollen tubes to the product of the NASGL 1 gene. Plant J 52:147–156

131. Doxey AC, Yaish MWF, Moffatt BA et al (2007) Functional divergence in the *Arabidopsis* beta-1,3 glucanase family inferred by phylogenetic reconstruction of expression states. Mol Biol Evol 24:1045–1055

132. Amor Y, Haigler CH, Johnson S et al (1995) A membrane-associated form of sucrose synthase and its potential role in synthesis of cellulose and callose in plants. Proc Natl Acad Sci U S A 92:9353–9357

133. Hong Z-L, Zhang Z-M, Olson JM et al (2001) A novel UDP-glucose transferase is part of the callose synthese complex and interacts with phragmoplastin at the forming cell plate. Plant Cell 13:769–779

134. Musetti R, Buxa SV, De Marco F et al (2013) Phytoplasma-triggered Ca^{2+} influx is involved in sieve-tube blockage. Mol Plant-Micr Interact 26:379–386

135. Tang R-J, Luan S (2017) Regulation of calcium and magnesium homeostasis in plants: from transporters to signaling network. Curr Opin Cell Biol 39:97–105

136. Demidchik V, Maathuis FJM (2007) Physiological roles of non-selective cation channels in plants: from salt stress to signalling and development. New Phytol 175:387–404

137. McAinsh MR, Pittman JK (2009) Shaping the calcium signature. New Phytol 181:275–294

138. Pei Z-M, Murata Y, Benning G et al (2000) Calcium channels activated by hydrogen peroxide mediate abscisic acid signalling in guard cells. Nature 406:731–734

139. Nakagawa Y, Katagiri T, Shinozaki K et al (2007) *Arabidopsis* plasma membrane protein crucial for Ca^{2+} influx and touch sensing in roots. Proc Natl Acad Sci U S A 104:3639–3644

140. Yamanaka T, Nakagawa Y, Mori K et al (2010) MCA1 and MCA2 that mediate Ca^{2+} uptake have distinct and overlapping roles in *Arabidopsis*. Plant Physiol 152:1284–1296

141. Hou C, Tian W, Kleist T et al (2014) DUF221 proteins are a family of osmosensitive calcium-permeable cation channels conserved across eukaryotes. Cell Res 24:632–635

142. Yuan F, Yang H, Xue Y et al (2014) OSCA1 mediates osmotic-stress-evoked Ca^{2+} increases vital for osmosensing in *Arabidopsis*. Nature 514:367–371

143. Clough SJ, Fengler KA, Yu I-C et al (2000) The *Arabidopsis dnd1* 'defense, no death' gene encodes a mutated cyclic nucleotide-gated ion channel. Proc Natl Acad Sci U S A 97:9323–9328

144. Gobert A, Park G, Amtmann A et al (2006) *Arabidopsis thaliana* cyclic nucleotide gated channel3 forms a non-selective ion transporter involved in germination and cation transport. J Exp Bot 57:791–800

145. Ali R, Ma W, Lemtiri-Chlieh F et al (2007) Death don't have mercy and neither does calcium: *Arabidopsis* CYCLIC NUCLEOTIDE GATED CHANNEL2 and innate immunity. Plant Cell 19:1081–1095

146. Gao F, Han X, Wu J et al (2012) A heat-activated calcium-permeable channel–*Arabidopsis* cyclic nucleotide ion channel 6–is involved in heat shock responses. Plant J 70:1056–1069

147. Mousavi SA, Chauvin A, Pascaud F et al (2013) GLUTAMATE RECEPTOR-LIKE genes mediate leaf-to-leaf wound signalling. Nature 500:422–426

148. Farmer EE, Gasperini D, Acosta IF (2014) The squeeze cell hypothesis for the activation of jasmonate synthesis in response to wounding. New Phytol 204:282–288

149. Vincill ED, Bieck AM, Spalding EP (2012) Ca^{2+} conduction by an amino acid-gated ion channel related to glutamate receptors. Plant Physiol 159:40–46

150. Evans MJ, Choi W-G, Gilroy S et al (2016) A ROS-assisted calcium wave dependent on the AtRBOHD NADPH oxidase and TPC1 cation channel propagates the systemic response to salt stress. Plant Physiol 171:1771–1784

151. Kudla J, Batistic O, Hashimoto K (2010) Calcium signals: the lead currency of plant information processing. Plant Cell 22:541–563

152. Sze H, Liang F, Hwang I et al (2000) Diversity and regulation of plant Ca^{2+} pumps: insights from expression in yeast. Annu Rev Plant Physiol Plant Mol Biol 51:433–462

153. Frei dit Frey N, Mbuengue M, Kwaaitaal M et al (2012) Plasma membrane calcium ATPases are important components of receptor-mediated signaling in plant immune responses and development. Plant Physiol 159:798–809

154. Peiter E, Maathuis FJ, Mills L et al (2005) The vacuolar Ca^{2+}-activated channel TPC1 regulates germination and stomatal movement. Nature 434:404–408

155. Hedrich R, Marten I (2011) PPC1-SV channels gain shape. Mol Plant 4:426–441

156. Choi W-G, Miller G, Wallace I et al (2017) Orhestrating rapid long-distance signaling in plants with Ca2+, ROS, and electrical signals. Plant J 90:698–707

157. Dodd AN, Kudla J, Sanders D (2010) The language of calcium signaling. Annu Rev Plant Biol 61:593–620

158. Schumaker KS, Sze H (1985) A Ca^{2+}/H^+ antiport system driven by the proton electrochemical gradient of a tonoplast H^+-ATPase from oat roots. Plant Physiol 79:1111–1117

159. Hirschi KD, Zhen RG, Cunningham KW et al (1996) CAX1, an H^+/Ca^{2+} antiporter family from *Arabidopsis*. Proc Natl Acad Sci U S A 93:8782–8786

160. Cheng N-H, Pittman JK, Shigaki T et al (2005) Functional association of *Arabidopsis* CAX1 and CAX3 is required for normal growth and ion homoiostasis. Plant Physiol 138:2948–2060

161. Wang Y, Kang Y, Ma CX et al (2017) CNGC2 is a Ca^{2+} influx channel that prevents accumulation of apoplastic Ca^{2+} in the leaf. Plant Physiol 173:1342–1354

162. Gilroy S, Suzuki N, Miller G et al (2014) A tidal wave of signals: calcium and ROS at the forefront of rapid systemic signaling. Trends Plant Sci 19:623–630

163. Takeda S, Gapper C, Kaya H et al (2008) Local positive feedback regulation determines cell shape in root hair cells. Science 319:1241–1244

164. Kimura S, Kaya H, Kawarazaki T et al (2012) Protein phosphorylation is a prerequisite for the Ca^{2+}-dependent activation of NADPH oxidases and may function as a trigger for the positive feedback regulation of Ca^{2+} and reactive oxygen species. Biochim Biophys Acta 1823:398–405

165. Dubiella U, Seybold H, Durian G et al (2013) Calcium-dependent protein kinase/NADPH oxidase activation circuit is required for rapid

defense signal propagation. Proc Natl Acad Sci U S A 110:8744–8749

166. Demidchik V, Shabala SN, Davies JN (2007) Spatial variation in H_2O_2 response of Arabidopsis thaliana root epidermal Ca^{2+} flux and plasma membrane Ca^{2+} channels. Plant J 47:377–386

167. Garcia-Mata C, Wang J, Gajdanowicz P et al (2010) A minimal cysteine motif required to activate the SKOR K^+ channel of Arabidopsis by the reactive oxygen species H_2O_2. J Biol Chem 285:29286–29294

168. Richards SI, Laohavisit A, Mortimer JC et al (2014) Annexin 1 regulates the H_2O_2-induced calcium signature in *Arabidopsis thaliana* roots. Plant J 77:136–145

169. Allen GJ, Chu S-P, Schumacher K, Shimazaki CT et al (2000) Alteration of stimulus-specific guard cell calcium oscillations and stomatal closing in *Arabidopsis* det2 mutant. Science 289:2338–2342

170. Minchin PEH, Thorpe ME (1987) Measurement of unloading and reloading of photoassimilate within the stem of bean. J Exp Bot 38:211–220

171. Van Bel AJE (1993) The transport phloem: specifics of its functioning. Prog Bot 54:134–150

172. Ayre BG, Keller F, Turgeon R (2003) Symplastic continuity between companion cells and the translocation stream: long-distance transport is controlled by retention and retrieval mechanisms in the phloem. Plant Physiol l 131:1519–1528

173. Hafke JB, van Amerongen JK, Kelling F et al (2005) Thermodynamic battle for photosynthate acquisition between sieve tubes and adjoining parenchyma in transport phloem. Plant Physiol 138:1527–1537

174. Minchin PEH, Lacointe A (2017) Consequences of phloem pathway unloading/reloading on equilibrium flows between source and sink: a modelling approach. Funct Plant Biol 44:507–514

175. Carpaneto A, Geiger D, Bamberg E et al (2005) Phloem-localized, proton-coupled sucrose carrier ZmSUT1 mediates sucrose efflux under control of sucrose gradient and pmf. J Biol Chem 280:21437–21443

176. Schulte-Baukloh C, Fromm J (1993) The effect of calcium starvation on assimilate partitioning and mineral distribution of the phloem. J Exp Bot 44:1703–1707

177. Van Bel AJE, Furch ACU, Hafke JB et al (2011) (Questions)[n] on phloem biology. 2. Mass flow, molecular hopping, distribution patterns and molecular signalling. Plant Sci 181:325–330

178. Fisher DB, Wu K, Ku MSB (1992) Turnover of soluble proteins in the wheat sieve tube. Plant Physiol 100:587–592

179. Golecki B, Schulz A, Castens-Behrens U et al (1998) Evidence for graft transmission of structural proteins in heterografts of Cucurbitaceae. Planta 206:630–640

180. Golecki B, Schulz A, Thompson GA (1999) Translocation of structural P-proteins in the phloem. Plant Cell 11:127–140

181. Foster TM, Lough TJ, Emerson SJ et al (2002) A surveillance system regulates selective entry of RNA into the shoot apex. Plant Cell 14:1497–1508

182. Kiefer IW, Slusarenko AJ (2003) The pattern of systemic acquired resistance induction within the *Arabidopsis* rosette in relation to the pattern of translocation. Plant Physiol 132:840–847

183. Thorpe MR, Ferrieri AP, Herth MM et al (2007) ^{11}C-imaging: methyl jasmonate moves in both phloem and xylem, promotes transport transport of jasmonate, and of photoassimilate even after proton transport is decoupled. Planta 226:541–551

184. Rocher F, Chollet J-F, Jousse C et al (2006) Salicylic acid, an ambimobile molecule exhibiting a high ability to accumulate in the phloem. Plant Physiol 141:1684–1693

185. Aoki K, Suzui N, Fujimaki S et al (2005) Destination-selective long-distance movement of phloem proteins. Plant Cell 17:1801–1814

186. Biles CL, Abeles FB (1991) Xylem sap proteins. Plant Physiol 96:597–601

187. Buhtz A, Kolasa A, Arlt K et al (2004) Xylem protein composition is conserved among different plant species. Planta 219:610–618

188. Kehr J, Buhtz A, Giavalisco P (2005) Analysis of xylem sap proteins from *Brassica napus*. BMC Plant Biol 5:11

189. Rogers S, Peel AJ (1975) Some evidence for the existence of turgor pressure gradients in the sieve tube of willow. Planta 126:259–267

190. Vreugdenhil D, Koot-Gronsveld EAM (1989) Measurements of pH, sucrose and potassium ions in the phloem sap of castor bean (*Ricinus communis*) plants. Physiol Plant 77:385–388

191. Minchin PEH, Ryan KG, Thorpe MR (1984) Further evidence of apoplastic unloading into the stem of bean: identification of the phloem buffering pool. J Exp Bot 35:1744–1753

192. Gould N, Thorpe MR, Koroleva O et al (2005) Phloem hydrostatic pressure relates to solute loading rate: a direct test of the Münch hypothesis. Funct Plant Biol 32:1019–1026

193. Kimmerer TW, Stringer MA (1988) Alcohol dehydrogenase and ethanol in the stems of trees. Evidence for anaerobic metabolism in the vascular cambium. Plant Physiol 87:693–697

194. Van Dongen JT, Schurr U, Pfister M et al (2003) Phloem metabolism and function have to cope with low internal oxygen. Plant Physiol 131:1529–1543

195. Carvalho MR, Turgeon R, Owens T et al (2017) The scaling of the hydraulic architecture in poplar leaves. New Phytol 214:145–157

196. Esau K, Cheadle VI (1959) Size of pores and their contents in sieve elements of dicotyledons. Proc Natl Acad Sci U S A 45:156–162

197. Trip P, Colvin JR (1970) Sieve elements of minor veins in the leaves of Beta vulgaris. Ann Bot 34:1101–1106

198. Milburn JA (1970) Phloem exudate from castor bean: induction by massage. Planta 95:272–276

199. Smith JAC, Milburn JA (1980) Osmoregulation and the control of phloem-sap composition in Ricinus communis L. Planta 128:28–34

200. Hocking PJ (1980) The composition of phloem exudate and xylem sap from tree tobacco (Nicotiana glauca Grah.). Ann Bot 45:633–643

201. Fukumorita T, Chino M (1982) Sugar, amino acid and inorganic contents in rice phloem sap. Plant Cell Physiol 23:273–283

202. Fisher DB, Frame JF (1984) A guide to the use of the exuding-stylet technique in phloem physiology. Planta 161:385–393

203. Hayashi H, Chino M (1986) Collection of pure phloem sap from wheat and its chemical composition. Plant Cell Physiol 27:1387–1393

204. Ohshima T, Hayashi H, Chino M (1990) Collection and chemical composition of pure phloem sap from Zea mays L. Plant Cell Physiol 31:735–737

205. Weibull J, Ronquist F, Brishammer S (1990) Free amino acid composition of leaf exudates and phloem sap. A comparative study in oats and barley. Plant Physiol 92:222–226

206. Girousse C, Bonnemain J-L, Delrot S et al (1991) Sugar and amino acid composition of phloem sap of Medicago sativa: a comparative study of two collecting methods. Plant Physiol Biochem 29:41–48

207. Van Helden M, Tjallingii WF, van Beek TA (1994) Phloem sap collection from lettuce (Lactuca sativa L.): chemical composition among collection methods. J Chem Ecol 20:3191–3206

208. King RW, Zeevaart JAD (1974) Enhancement of phloem exudation from cut petioles by chelating agents. Plant Physiol 53:96–103

209. Richardson PT, Baker DA (1982) The chemical composition of cucurbit vascular exudates. J Exp Bot 33:1239–1247

210. Hayashi H, Chino M (1990) Chemical composition of phloem sap from the uppermost internode of the rice plant. Plant Cell Physiol 31:247–251

211. Leckstein PM, Llewellyn M (1975) Quantitative analysis of seasonal variation in the amino acids in phloem sap of Salix alba L. Planta 124:89–91

212. Felle H, Bertl A (1986) The fabrication of H^+-selective liquid membrane microelectrodes for use in plant cells. J Exp Bot 37:1416–1428

213. Felle H (2001) pH: signal and messenger in plants cells. Plant Biol 3:577–591

214. Frohmeyer H, Grabov A, Blatt MR (1998) A role for the vacuole in auxin-mediated control of cytosolic pH in Vicia mesophyll and guard cells. Plant J 13:109–116

215. Hafke JB, Neff R, Hütt M-T et al (2001) Day-to-night variations of cytoplasmic pH in a crassulacean acid metabolism plant. Protoplasma 216:164–170

216. Bertaccini A, Duduk B, Paltrinieri S et al (2014) Phytoplasmas and phytoplasma diseases: a severe threat to agriculture. Am J Plant Sci 5:1763–1788

217. Wei W, Kakizawa S, Suzuki S et al (2004) In planta dynamic analysis of onion yellows phytoplasma using localized inoculation by insect transmission. Phytopathology 94:244–250

218. Buxa SV, Degola F, Polizzotto R et al (2015) Phytoplasma infection in tomato is associated with re-organization of plasma membrane, ER stacks and actin filaments in sieve elements. Front Plant Sci 6:650

219. Musetti R, Pagliari L, Buxa SV et al (2016) OHMS: phytoplasmas dictate changes in sieve-element ultrastructure to accommodate their requirements for nutrition, multiplication and translocation. Plant Sign Behav 11:e1138191

220. Kube M, Morovic J, Duduk B et al (2012) Current view on phytoplasma genomes and encoded metabolism. Sci World J 2012:185942

Chapter 20

Laser Microdissection of Phytoplasma-Infected Grapevine Leaf Phloem Tissue for Gene Expression Study

Simonetta Santi

Abstract

Phytoplasmas have been found confined mainly in leaf phloem sieve elements. In spite of this, few researches have been focused on the infected phloem tissue, whereas the plant response at the infection site could be quite different compared to distal parts and almost completely masked when whole organs are considered. Herein, we provide a protocol for the isolation of leaf phloem from paraffin-embedded samples by Laser Microdissection, followed by RNA purification and RNA amplification to generate cDNA libraries. Our protocol, which has been set up for phytoplasma-infected field-grown grapevine and successfully used for gene expression profiling, can be modified according to different plant species.

Key words Laser microdissection pressure catapulting, Phloem, RNA analysis, Gene expression, Bois Noir

1 Introduction

Laser Microdissection (LM; also called Laser Capture Microdissection, LCM) is a technology for selection and collection of morphologically defined cell populations from a tissue section under the guide of the microscope [1, 2]. LM provides a tool for cell-specific resolution of gene expression, which is critical for the analysis of responses that are restricted to a particular tissue or cell. The single-cell approach, in fact, allows the detection of low-abundance or highly cell-specific transcripts and reduces complexity associated with the expression of surrounding tissues. As regards phytoplasmas, it has been shown that they are confined mainly in leaf phloem sieve elements [3], but few researches have been focused on the plant response at the infected phloem tissue. In a previous preliminary study, we used LM to isolate and collect leaf phloem tissue (including sieve elements, companion cells, and phloem parenchyma cells) from leaves of *Vitis vinifera* L. (Chardonnay) that was affected by Bois Noir (BN) [4]. Gene expression analysis at phloem tissue showed the inhibition of a sucrose loading

Rita Musetti and Laura Pagliari (eds.), *Phytoplasmas: Methods and Protocols*, Methods in Molecular Biology, vol. 1875, https://doi.org/10.1007/978-1-4939-8837-2_20, © Springer Science+Business Media, LLC, part of Springer Nature 2019

transporter gene and the upregulation of a sucrose cleavage (sucrose synthase) gene, suggesting the establishment of a phytoplasma-induced switch from carbohydrate source to sink that was not as clearly highlighted when analyzing the whole leaf [5].

Herein, we provide a protocol for the LM-assisted isolation of phloem tissue from leaves of phytoplasma-infected grapevine (BN has been associated with '*Candidatus* Phytoplasma solani', belonging to stolbur group, subgroup XII-A, according to [6]). Leaf samples have been collected from field-grown plants and paraffin embedded. As a brief overview, leaf samples are first fixed in acetic acid-ethanol (Farmer's solution), then paraffin embedded after a few dehydration steps. Paraffin-embedded sections instead of cryo-sections maintain their morphological integrity and are generally used for tissues with large vacuoles and air spaces such as leaves [1]. Once fixed and paraffin embedded, samples can be stored and sectioned repeatedly over a period of days or a few weeks. A PALM MicroBeam system (P.A.L.M. Zeiss) for laser microdissection pressure catapulting (LMPC) is used to cut and catapult phloem tissue specimens under the microscope. The LMPC system is based on an ultraviolet (UV) laser to photo volatilize cells surrounding a selected area of interest [7]. The UV cutting system is particularly useful to microdissect tissue sections up to 200 μm thick, such as plant tissue sections [8]. After LM, RNA can be extracted, amplified, and used to synthesize cDNA for downstream expression analysis experiments.

2 Materials

As aimed to isolate RNA and analyze gene expression, all materials (containers, solutions, scalpels, tweezers, and other tools) and methods (handling from tissue fixation to RNA isolation and downstream analysis) should be RNase-free. Prepare all solutions with ultrapure DEPC-treated water, molecular biology grade reagents, when possible, or analytical grade reagents (*see* **Note 1**).

2.1 Tissue Fixation and Processing

1. Farmer's fixative: 75% ethanol—25% glacial acetic acid. Prepare fresh and keep it on ice.

2. Dehydration solution 1: 80% (v/v) ethanol.

3. Dehydration solution 2: 90% (v/v) ethanol.

4. Dehydration solution 3: 100% ethanol.

5. Dehydration solution 4: pure xylene.

6. Paraffin: Paraplast plus (Mc Cormick Scientific, St. Louis, MO, USA): melt paraffin at 59 °C.

7. Glass vials.

8. Histological cassettes.

9. Disposable base molds.

10. Vacuum pump.

11. Orbital low speed shaker.

2.2 Tissue Sectioning, Slide Preparation, and LMPC Cutting

1. RNase Decontamination Solution.

2. Polyethylene naphthalate (PEN)-covered glass slides (Carl Zeiss GmbH, Germany).

3. Tubes with adhesive cap (Carl Zeiss GmbH, Germany).

4. Rotary microtome.

5. Hot plate.

6. P.A.L.M. Laser-Microbeam System for LMPC (P.A.L.M. Microlaser Technologies, Carl Zeiss MicroImaging GmbH, Germany).

2.3 RNA Extraction and Amplification

1. Polyvinylpyrrolidone-40 (PVP-40).

2. β-mercaptoethanol (β-ME).

3. *N*-lauroylsarcosine detergent. Prepare a 20% (v/v) solution with RNase-free water, divide into small aliquots, and store at room temperature.

4. Sulfolane. Prepare 80% (v/v) sulfolane by diluting 100% sulfolane with RNase-free water (*see* **Note 2**).

5. RNA Extraction Kit: Absolutely RNA Nanoprep Kit (Agilent Technologies, La Jolla, CA, USA). Add 2.5% (w/v) PVP-40 to lysis buffer. We store this solution at room temperature for at least 1 month. Add 0.7% (v/v) β-ME just before RNA extraction (*see* **Notes 3** and **26**). Chloroform-isoamyl alcohol (24:1).

6. One-Step RT-PCR Kit.

7. Real-Time PCR Detection System.

8. Agilent RNA 6000 Pico Kit (Agilent Technologies, Santa Clara, CA, USA).

9. DNA/RNA low-binding tubes.

10. Agilent Bioanalyzer 2100 (Agilent Technologies, Santa Clara, CA, USA).

11. NanoDrop ND-1000 UV-Vis Spectrophotometer (Thermo Fisher Scientific, Inc. MA, USA).

3 Methods

3.1 Tissue Fixation and Processing

Samples are fixed, dehydrated, and then paraffin-embedded. To minimize RNA degradation it is critical to shorten times as much as possible and maintain reduced temperature during the whole procedure, whenever possible. Moderately degraded samples with RNA Integrity Number (RIN, Agilent Bioanalyzer) higher than 5.0 can still yield RNA compatible with gene expression profiling [9]. Fixation is the most critical step to ensure high-quality yield, which depends on length of time for fixative penetration, temperature, and sample size [2].

1. Collect 4–5 leaves per plant, possibly not detached from the cane, keep them at 4 °C. Once reached the lab, wash leaves with distilled water, dry them gently by wiping with soft clean paper, then proceed with tissue fixation (*see* **Note 4**). Cut leaves in 4–5 mm wide and 2–3 mm long pieces including vein and part of the leaf blade directly in the ice-cold Farmer's fixative (*see* **Note 5**). Transfer specimens to vials containing fresh ice-cold fixative, apply vacuum for 1–2 min and release. Repeat three times or until air is completely removed and specimens lie on the bottom of the vial (*see* **Note 6**). Change old fixative with a fresh ice-cold one and put at 4 °C overnight.

2. Replace the fixative with ice-cold 80% ethanol and incubate for not more than 60 min with gentle shaking. Keep on ice (*see* **Note 7**).

3. Replace the solution with ice-cold 90% ethanol and incubate for 50–60 min with gentle shaking. Keep on ice.

4. Replace with 100% ethanol at room temperature for 60 min twice. Then replace with xylene for 50–60 min twice (*see* **Note 8**).

5. Transfer specimens into histological cassettes and incubate the cassettes in 1:1 xylene: paraffin (Paraplast) melted at 59 °C for 50–60 min (*see* **Notes 9** and **10**).

6. Transfer cassettes from the xylene: Paraplast mixture into pure Paraplast melted at 59 °C. Replace the paraffin three times at intervals of 50–60 min (*see* **Note 11**).

3.2 Tissue Embedding

1. Transfer specimens into disposable base molds for embedding with Paraplast and forming blocks (*see* **Note 12**).

2. Blocks are first cooled to RT and then placed at 4 °C for easy unmoulding. The blocks can be kept in plastic bags at 4 °C (*see* **Note 13**).

3.3 Tissue Sectioning

Microtome, hot plate, and LMPC microscope surfaces should be RNase-free (*see* **Note 14**). Tissue slices are transferred onto Polyethylene naphthalate (PEN)-covered glass slides (*see* **Note 15**). Paraffin blocks should be cut and samples de-paraffinized just before LMPC to maintain samples dry.

1. Cut 12–13 μm-thick slices on a rotary microtome (*see* **Note 16**).

2. Stretch slices for a few seconds on a drop of DEPC-water directly delivered on PEN-slides warmed at 42 °C on the hot

plate. Water should be rapidly removed by slightly tapping slides on a clean soft paper towel (*see* **Note 17**).

3. Dry slides on hot plate at 42 °C for 15–20 min (*see* **Note 18**).

4. De-paraffinize sections by pipetting the xylene (0.5 ml aliquotes) until paraffin is completely removed (*see* **Note 19**). Air-dry and immediately use for LMPC (*see* **Notes 20** and **21**).

3.4 Laser Microdissection Pressure Catapulting

Tissue is isolated using a P.A.L.M. Microbeam LMPC system [7]. Clean stage and cap holder with RNase-decontamination solution. Harvesting the greatest number of specimen areas in the shortest possible time is critical for both RNA quality and yield.

1. Scan the slides with a 5× or a 10× objective lens to quickly identify cross sections and phloem tissue (Fig. 1a). Sieve elements, companion cells together with a few phloem parenchyma cells are intended for phloem tissue (*see* **Note 22**). Then select the best objective lens for the laser dissection. We usually use a 20× lens.

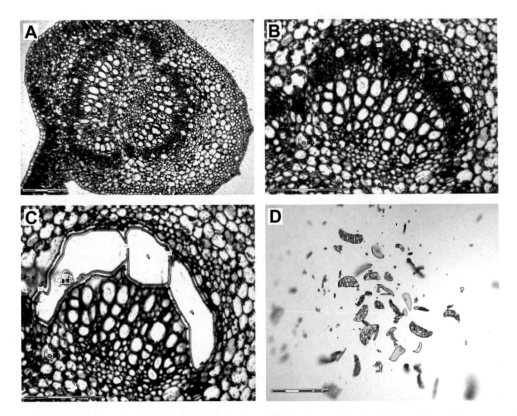

Fig. 1 Laser microdissection and pressure catapulting (LMPC) of grapevine leaf phloem tissue. Representative cross section (12 μm) of leaf before (**a**, **b**) and after (**c**) cutting and catapulting into the adhesive bar. (**b**, **c**) Magnification of the targeted site before and after cutting. Bars = 150 μm. (**d**) Captured areas adhering on the cap. Bars = 300 μm

2. Select the type of laser function to be used to dissect and catapult the sample. We usually use the "CloseCut+AutoLPC" function, setting the dot distance in such a way as to have only a couple of dots per area. Laser energy level and LPC energy level depend on slice thickness and sample. For a 12 μm-thick slice on PEN-slide we indicatively use a laser energy level from 50 to 55 and a laser focus of 69; an LPC energy of 79–84 and a LPC delta focus of 3, with a laser speed of 6–10 (*see* **Note 23**).

3. Select areas on screen by P.A.L.M tool (Fig. 1b, c). Phloem areas from leaves are cut and catapulted into 0.2 ml tubes with adhesive cap (P.A.L.M. Carl Zeiss). Cut and catapult areas from several different sections of each sample. Record the total dissected area by the element list tool. Usually we collect areas from 4 to 5×10^5 μm^2 in total on each adhesive cap (Fig. 1d) (*see* **Note 24**).

4. Deliver on the cap 30 μl of lysis buffer of the RNA extraction Kit. Invert the tube a few times and vortex it. Collect the material at the bottom of the tube by brief centrifugation ($14,000 \times g$) and store at −80 °C (*see* **Note 25**).

5. Proceed catapulting further areas into a new tube. For each plant sample, phloem areas of at least 3×10^6 μm^2 in total are collected, corresponding on average to 480–500 areas. Samples can be stored for several months at −80 °C (six, according to RNA extraction Kit's instructions).

3.5 RNA Isolation

Total RNA can be purified with Absolutely RNA Nanoprep Kit (*see* **Note 26**). We have introduced minor changes to the manufacturer's instructions according to [10] to break bonds between RNA and phenolic substances [4]. The minimum total RNA yield from phloem areas of 3×10^6 μm^2 in total is 7–8 ng and can consistently vary among replicates.

1. Pool areas of the same sample and adjust lysis buffer volume to 200 μl (see **Note 27**). The lysis buffer contains 2.5% (w/v) PVP-40 and 0.7% (v/v) β-ME. Add 1/10 vol of 20% (v/v) N-lauroylsarcosine and incubate at 70 °C for 5 min. After vortexing and briefly spinning, add an equal volume of chloroform–isoamyl alcohol (24:1), mix gently, and centrifuge at $12,000 \times g$ for 10 min. Add an equal volume of 80% sulfolane to the supernatant. Transfer this mixture to the RNA-binding column, and then perform filter washing and on-column DNase treatment according to Kit's instructions. Elute on filter DNase-treated RNA with 14 μl of RNase-free water heated to 60 °C (*see* **Note 28**).

2. Determine RNA quantity and quality by Agilent RNA 6000 Pico Kit at the Agilent 2100 Bioanalyzer. RIN should be higher than 5.0 [9] and 28S:18S ratio higher than 2.0 (Fig. 2) (*see* **Note 29**).

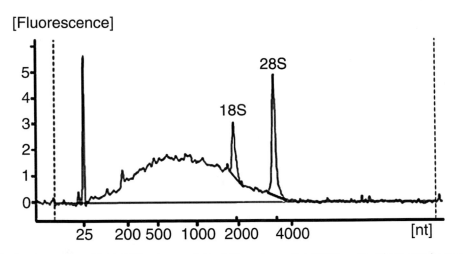

Fig. 2 Electropherogram of total RNA using an Agilent Bioanalyzer. Total RNA purified from specimen areas of about 3×10^6 μm^2 in total from leaf phloem tissue and eluted in 14 μl of water, 1 μl of which analyzed at the Bioanalyzer. The total concentration was estimated at 0.56 ng/μl and the RIN value at 5.1. The 28S and 18S peaks are indicated, with an estimated ratio of 2.1

3.6 Phytoplasma Detection

Phytoplasmas can be detected analyzing its gene expression by one-step real-time RT-PCR as already described for '*Candidatus* Phytoplasma solani' phytoplasma [4]. Detection can be carried out starting from 1 μl of RNA purified from scraped whole tissue sections as well as from LMPC-isolated cells. Gene-specific primers are required for one-step real-time RT-PCR.

1. According to manufacturer's instructions dilute RNA to a final volume reaction of 10 μl containing the buffer provided by the Kit, forward and reverse primer at a final concentration of 400 nM each, and the enzyme mix (*see* **Note 30**). Set up reactions on ice. Keep the plate on ice until ready to run (*see* **Note 31**).

2. Program your real-time instrument according to the manufacturer's instructions and primer characteristics (*see* **Note 32**).

3. Collect data and analyze results.

3.7 RNA Amplification

Gene expression profiling, by both quantitative RT-PCR and Next Generation Sequencing (NGS) RNA sequencing (RNA-seq), requires RNA linear amplification. The choice of the protocol and the Kit to be used depends on downstream experiment. For qPCR analysis of a restricted number of genes (about ten), we performed an in vitro transcription from poly(dA) RNA using a poly(dT) oligo primer containing the T7 phage promoter [4]. We carried out an one-round amplification, but it is possible to perform a two-round amplification to obtain aRNA in a μg scale amount. The obtained aRNA conserves the initial proportions of mRNAs.

For transcriptome profiling by RNA-seq, alternative amplification methods can be used to detect noncoding RNAs, including small RNAs. These methods usually use a mix of examers and poly (dT) oligo primer to initiate amplification at the 3′ end as well as randomly throughout the transcriptome. In this way reads will be distributed across the transcript.

4   Notes

1. Glass (for example: Petri dishes for tissue cutting and glass vials for specimen incubation, if reused, or glass pots for infiltration with paraffin) should be baked at 180 °C for at least 8 h. Plastic ware (for example: histological cassettes) can be soaked in a 3% (v/v) hydrogen peroxide solution for at least 10 min, then rinsed thoroughly with RNase-free water. Clean the equipment with reagents for nucleases removal like RNaseZap (Thermo Fisher) or analogous.

2. Sulfolane (Sigma-Aldrich Co., St. Louis, MO, USA) is required by the Agilent RNA extraction Kit (see below). According to kit's instructions, melt the 100% sulfolane in a 37 °C water bath before diluting. Aliquot the 80% solution and store at room temperature.

3. Add β-ME under a fume hood.

4. Leaf samples should be collected from plants grown under the same condition. At least three biological replicates are required for expression gene profiling, but the value of the experiment considerably increases with more replicates. Time necessary to collect leaf samples should be as shortest as possible.

5. Fixative can depend on plant and tissue [1]. In the case of grapevine leaves, ice-cold 100% acetone was also used [4].

6. Fixation is the most crucial step to ensure high-quality RNA yield. Vacuum is necessary to favor the penetration of the fixative in the tissue, and so it should be applied as soon as possible (a few minutes after cut of leaf pieces). When using Farmer's fixative, Falcon tubes can be also used instead of glass vials. The volume of the fixative should be at least five-fold the volume of the sample.

7. The use of acetone requires a lower number of dehydration steps before infiltrating and embedding, as it can be directly mixed with xylene (1:1 and then 1:3), then replaced with pure xylene. From xylene the two protocols proceed in the same way.

8. If necessary, we stop the procedure and incubate specimens overnight at 4 °C in 100% ethanol.

9. Following infiltration steps require high temperatures, but incubation at high temperatures probably damages RNA, thus work as quickly as possible. RNase-free conditions are mandatory. For this reason we process specimens by transferring the cassettes into RNase-free glass pots containing the melted xylene: Paraplast mixture or pure Paraplast.

10. Paraffin inclusion provides good histological quality of sections, but, on the other hand, high temperature and long processing steps required by this preparation can degrade RNA. To shorten the embedding process, some authors used a microwave method [11–13]. In our protocol, we do not use a microwave oven but, however, we reduce processing time at 1 day.

11. Xylene has to be completely removed; if not, add one more Paraplast step.

12. To obtain the appropriate orientation we usually carry out a double embedding. First, specimens are laid on the bottom of the base mold with midribs possibly orientated along the same direction. Second, once hardened (at 4 °C overnight), blocks are embedded perpendicularly in a new clean base mold. In this way it will be possible to cut a large number of different sections and, at the same time, obtain cross sections from almost all specimens. We usually prepare two blocks per each biological replicate.

13. Because of the mild fixation, RNA could degrade if blocks are stored for too prolonged period. We usually keep blocks at 4 °C for not more than 6 months.

14. Damp a soft towel paper with an RNase decontamination solution and treat all surfaces. We usually use RNaseZAP (Thermo Fisher). Remove RNaseZAP-traces with RNase-free water and then ethanol. The same treatment should be applied to PEN-covered glass slides.

15. LMPC can be used on both normal glass and membrane-covered slides. When normal glass slides are used, multiple pulses are necessary, each catapulting one cell or a small fragment of the tissue. When membrane-mounted slides are used, one single shot is sufficient to catapult the target tissue, avoiding a potential damage of the sample. Moreover, collecting time is shorter and this favorably plays for high-quality yield.

16. Adjust block orientation to obtain cross sections. Choice of thickness is a compromise among the necessity to isolate specific cells, the minimum amount of captured samples needed to extract enough RNA and the energy level of LMPC-laser that is required for tissue cutting, as it increases with thickness (*see* below). To easily cut slices, keep the microtome blade clean using ethanol.

17. We usually transfer 5–6 slices per PEN-slide and prepare several slides per biological replicate, cutting from both blocks.

18. Drying period should be as short as possible, compatibly with time necessary for the slices to adhere to the slide membrane. Usually 15 min is enough.

19. Often, the PEN membrane lifts up forming air bubbles when PEN-coated slides have been immerged in the xylene. We prefer to carefully pipette xylene directly on the membrane, avoiding borders. This operation should be carried out under fume hood.

20. De-paraffine only a couple of slides, then immediately use them for tissue cutting.

21. In order to verify the amount and integrity of RNA prior to LMPC, and to check for the presence of phytoplasmas (by one-step RT-PCR, as described below), LMPC step can be skipped and whole tissue sections scraped off the microscope slide into a tube for RNA extraction.

22. The correct identification of phloem can be facilitated using UV fluorescence, because of autofluorescence of xylem wall. Once identified, areas containing phloem tissue can be cut under light. At low magnification the laser is not powerful enough and even not precise. On the other hand, a high magnification (e.g., 40×) slows down the collection time considerably.

23. To cut 13 μm-thick slices, the laser energy should be higher. If the required energy is too high, areas should be smaller.

24. When too many, pieces could fall on the microscope slide. Control that pieces are adhering at the cap with the dedicated microscope-tool.

25. We have modified the manufacturer's instructions of the RNA extraction Kit adding 2.5% (w/v) PVP-40 to the lysis buffer.

26. Several RNA extraction methods for laser-microdissected cells are available. However, the method of RNA extraction can be crucial, thus the recovery rate of RNA samples of known concentration should be tested before using a new kit [7].

27. For the RNA extraction from sections that have been scraped off microscope slides, 200 μl of lysis buffer is used delivering it directly on slide. Subsequent steps are carried out as described for LMPC-cells.

28. Depending on planned downstream experiments and required amount of RNA, it is possible to concentrate the sample in a SpeedVac system.

29. The minimum RNA/DNA concentration that can be detected by a NanoDrop Specrophotometer is 0.5 ng/μl, thus it can be

used for RNA purified from scraped sections but not from LMPC-cells. The LMPC-cells-RNA can be quantified and its quality controlled by the Agilent Bioanalyzer 2100 (sensitivity: 50 pg/μl in water). Unfortunately sometimes, probably because of both low RNA concentration and sensitivity of the kit to salts and other contaminants, RNA quantity and quality cannot be checked. In this case, one step RT-PCR (*see* Subheading 3.6) can be carried out to check for the expression of known genes. By carefully choosing the annealing position of primers it is possible to obtain some information about both RNA amount and integrity. Anyway, in our experience a low amount of RNA and the impossibility to assess the RNA quality does not preclude a successful amplification, since many amplification protocols can start from RNA quantities less than 1 ng.

30. We usually use the EXPRESS One-Step SYBR GreenER^tm Kit (Invitrogen, Life Technologies, UK). According to the manufacturer, starting material can be as low as 1 pg. We usually use 1 μl of RNA sample. Control reactions should be prepared to test both genomic DNA contamination (no reverse transcriptase control—NRT) and mix contamination (no template control—NTC). To avoid false negatives, we usually set up a reaction including a plant reference gene.

31. To prevent the synthesis of nonspecific cDNA it is crucial to keep on ice reaction mix, samples, and plate until the thermal cycler starts the run protocol. Start the run and pause it when the temperature of cDNA synthesis is reached; only at this time place the plate and resume the run.

32. The temperature of initial incubation step can be set among 50 and 60 °C.

Acknowledgments

We would like to thank Michele Bellucci (IBBR-CNR, Perugia, Italy) for the use of the LMPC system. Research funded by AGER, project No. 2010-2106 "Grapevine yellows: innovative technologies for the diagnosis and the study of plant/pathogen interactions."

References

1. Nelson T, Tausta SL, Gandotra N et al (2006) Laser Microdissection of plant tissue: what you *see* is what you get. Annu Rev Plant Biol 57:181–201

2. Espina V, Wulfkuhle JD, Calvert VS et al (2006) Laser-capture microdissection. Nat Protoc 1:586–603

3. Christensen NM, Axelsen KB, Nicolaisen M, Schulz A (2005) Phytoplasmas and their interactions with hosts. Trends Plant Sci 10:526–535

4. Santi S, Grisan S, Pierasco A et al (2013) Laser microdissection of grapevine leaf phloem infected by stolbur reveals site-specific gene

responses associated to sucrose transport and metabolism. Plant Cell Environ 36 (2):343–355

5. Santi S, De Marco F, Polizzotto R et al (2013) Recovery from stolbur disease in grapevine involves changes in sugar transport and metabolism. Front Plant Sci 4:171

6. Firrao G, Gibb K, Stereten C (2005) Short taxonomic guide to the genus '*Candidatus* Phytoplasma'. J Plant Pathol 87:249–263

7. Micke P, Östman A, Lundeberg J et al (2005) Laser-assisted cell microdissection using the PALM system. In: Murray GI, Curran S (eds) Laser capture microdissection: methods and protocols, Methods in molecular biology, vol 293. Humana Press Inc., Totowa, NJ, pp 151–166

8. Nakazono M, Qiu F, Borsuk LA et al (2003) Laser-capture microdissection, a tool for the global analysis of gene expression in specific plant cell types: identification of genes expressed differentially in epidermal cells or vascular tissues of maize. Plant Cell 15:583–596

9. Gallego Romero I, Pai AA, Tung J et al (2014) RNA-seq: impact of RNA degradation on transcript quantification. BMC Biol 12:42

10. MacKenzie DJ, McLean MA, Mukerji S et al (1997) Improved RNA extraction from woody plants for the detection of viral pathogens by reverse transcription-polymerase chain reaction. Plant Dis 81:222–226

11. Inada N, Wildermuth MC (2005) Novel tissue preparation method and cell-specific marker for laser microdissection of mature leaf. Planta 221:9–16

12. Tang W, Coughlan S, Crane E et al (2006) The application of laser microdissection to in planta gene expression profiling of the maize anthracnose stalk rot fungus *Colletotrichum graminicola*. Mol Plant-Microbe Interact 11:1240–1250

13. Takahashi H, Kamakura H, Sato Y et al (2010) A method for obtaining high quality RNA from paraffin sections of plant tissues by laser microdissection. J Plant Res 123:807–813

Chapter 21

Collection of Phloem Sap in Phytoplasma-Infected Plants

Matthias R. Zimmermann, Torsten Knauer, and Alexandra C. U. Furch

Abstract

Phytoplasmas colonize specifically the phloem sieve elements (SEs) of plants and influence effectively the plant physiology. To study and understand the interaction of phytoplasmas and host plants an access to the cellular, microscale volume of SEs is demanded. Different methods are suitable to collect phloem sap of phytoplasma-infected plants. The two most common methods are the EDTA-facilitated exudation and the stylectomy. For the EDTA-facilitated method, the cut end of a leaf is placed into an EDTA solution. The EDTA prevents and avoids the Ca^{2+} dependent (re-) occlusion of SEs by binding Ca^{2+} ions and the mass flow of SEs is restarted which results in an outflow of the SE content into the EDTA bathing solution. The advantage is on the one hand a simple application and secondly, feasible for all plant species.

The stylectomy method requires piercing-sucking insects like any aphids. During phloem-sap ingestion, the stylet is severed by a microcautery device or a laser from the insect body. Due to the high turgor pressure of the SEs the phloem sap is forced out through the remaining stylet and can be collected with a glass capillary, for example. The stylectomy delivers pure phloem sap, however, the collected volumes are in the range of nano liters and the temporal and staff costs are tremendous. A third method is the spontaneous exudation in phytoplasma-infected apple trees providing only in springtime large volumes of vascular sap after cutting along the bark. For the spontaneous exudation the proportion of phloem sap is unclear. Thus, this third method still needs a closer examination in prospective surveys.

Key words Phloem sap, Stylectomie, EDTA, Exudation, Aphid, Sieve elements, Ca^{2+}

1 Introduction

The phloem as part of the plant vascular system delivers carbohydrates, proteins, amino acids, RNA, and diverse secondary metabolites from source (autotrophic) to sink (heterotrophic) regions of the plant [1] and references therein. All these compounds participate in growth, development, defence, and signaling cascade. It means qualitative and quantitative spatio-temporal analyses of phloem-originated compounds provide significant insights into the physiological status of a plant being also valid for a phytoplasma infection, naturally.

One characteristic of phytoplasmas is the specific colonization of phloem sieve elements (SEs) in diverse plants [2]. Inside SEs,

Rita Musetti and Laura Pagliari (eds.), *Phytoplasmas: Methods and Protocols*, Methods in Molecular Biology, vol. 1875, https://doi.org/10.1007/978-1-4939-8837-2_21, © Springer Science+Business Media, LLC, part of Springer Nature 2019

phytoplasmas proliferate, develop and spread all over the entire plant body. It is a common notice that the inhabitation has an impact (depending on the genetic strand) to the plant development, physiology, and fitness [2]. Phytoplasmas do not have their own genetic equipment for proliferation, but they are able to use the present plant "machineries" and (molecular) resources for a successful development and growth. However, this is only possible if the plant is "convinced" of helping phytoplasmas. It also means that phytoplasmas are able to bypass the plant defence in a specific manner.

All the previous and brief mentioned aspects demonstrate an intensive interaction among the plant especially between the phloem tissue and phytoplasmas. To understand how phytoplasmas manipulate the plant, a specific focus on the interaction inside the phloem cells is mandatory as the molecular exchanges takes place mainly within SEs.

However, there are diverse challenges for the investigation of phloem sap—(1) the phloem is embedded deeply inside the plant tissue covered by diverse cell layers; (2) a second one is the microscale level of the phloem cells; and (3) despite the symplasmatic continuum of the single SEs to a tube, the potential available volume of phloem sap is also limited demanding high sensitive analysis.

So far, different methods have been established to gain the content of SEs. The most prominent methods are the EDTA-facilitated exudation and the stylectomy. These techniques are generally applicable for nearly all plant species. In addition, a brief description for the wound-induced vascular exudation especially observed in phytoplasma-infected apple trees is also shown even if an application is restricted.

2   Materials

2.1   EDTA-Facilitated Exudation

1. Cleaning solution: 5 mM Na_2EDTA (2-Na-ethylenediamine tetraacetic acid), pH 7.0/KOH.
2. Collecting solution: 2.5 mM Na_2EDTA, 1.0 mM MES [2-(N-morpholino)ethanesulphonic acid/KOH (pH 7.0).
3. 15 ml and 2 ml tubes.
4. Razor blades.

2.2   Stylectomy

1. Radio-frequency microcautery [3].
2. Electrolytically sharpened tungsten needle [4].
3. Micromanipulator.
4. Glass microcapillary with a tip diameter of 10–25 μm.
5. 10 ml syringe connected to a silicone tube with a side valve.
6. Binocular.

2.3 The Wound-Induced Vascular Exudation	1. Buffer solution: 1.0 mM MES, pH 7.0/KOH. 2. 2 ml tubes. 3. Scalpel or knife.

3 Methods

3.1 EDTA-Facilitated Exudation

The big advantage of the EDTA-facilitated exudation method is its simple practice and wide application for numerous plant family members (e.g., [5, 6]). Here, the presented protocol has to be seen as a basal suggestion that can be specifically adjusted to the considered plant species, scientific issue, and interesting molecules.

1. Mature leaves are truncated at the base of the petiole with a fresh razor blade (*see* **Note 1**).

2. The cut end of the petiole is recut (about 1–2 cm) under cleaning solution (*see* **Note 2**).

3. The leaves are immediately placed with their cut ends in vials or beakers containing 5 ml of the cleaning solution for approximately 1 h (*see* **Note 3**).

4. The leaves are transferred with their cut ends into 2 ml Eppendorf tubes/cups filled with 1.5 ml collection solution (Fig. 1).

5. The entire preparation is stored in a humid, transparent chamber at 23–27 °C under supplementary lighting or dark conditions (*see* **Note 4**).

6. The preparations are left for several hours, at least (*see* **Note 4**).

7. After exudation the collection solution of each treatment is pooled and sterile filtered through a membrane for example due to the common observed contaminations by the environment and handling.

8. The quite big volumes of the collection solutions need to be concentrated to a final volume of approximately 20 μl. Here, the subsequent proceeding depends on the molecules of interest and one has to choose from the numerous protocols.

Using the example of *Trifolium pratense*, L., the general setup of the EDTA-facilitated exudation is illustrated. (**a**) The cut end of a petiole is placed in the collection solution. Due to the prevented sieve element occlusion the phloem sap runs into the collection solution. (**b**) Several leaves can be pooled in one collection tube and (**c**) diverse sets of leaves can be placed together in the transparent humidity chamber (glass bowl).

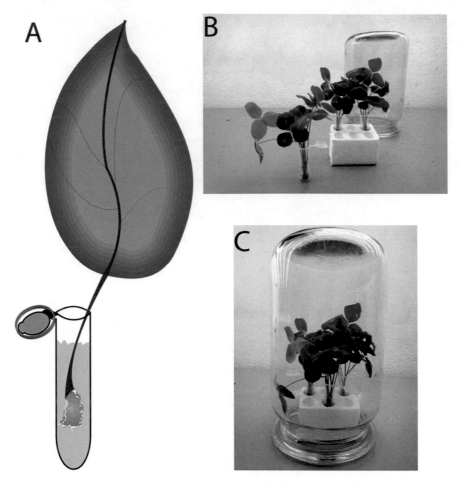

Fig. 1 Setup of EDTA-facilitated exudation

3.2 Stylectomy

The classiness of the stylectomy is the straight access to the phloem and the receipt of pure phloem sap. However, it is a staff and time-consuming method that needs a well-practiced person and the gained volumes are low. It is important to consider these aspects for the design of experiments.

1. Several randomly selected adult and apterous aphids are placed on the plant, preferably on leaves (Fig. 2). The aphids can be freely positioned on the plant or within a clip cage that prevents a potential escape of the aphids. The aphids remain there for 12–24 h for settling and feeding indicated by a non-crabby behavior of the individuals (*see* **Note 5**).

2. The tungsten needle is mounted on a micromanipulator and moved under binocularly observation (Fig. 3a) in the vicinity of the aphid stylet (Fig. 3b, c) (*see* **Note 6**).

3. The tip of the tungsten needle is advanced until it touches the anterior surface of the labium.

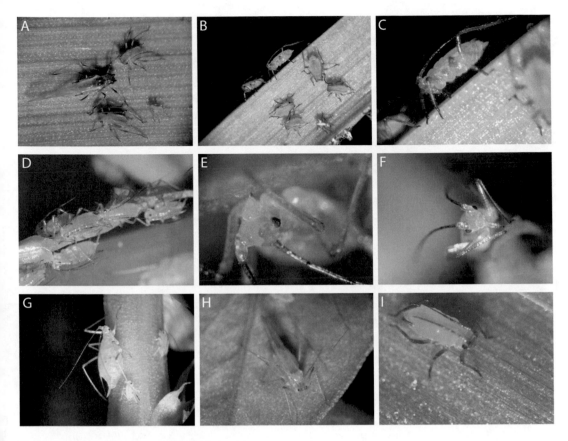

Fig. 2 A small gallery of several aphid species

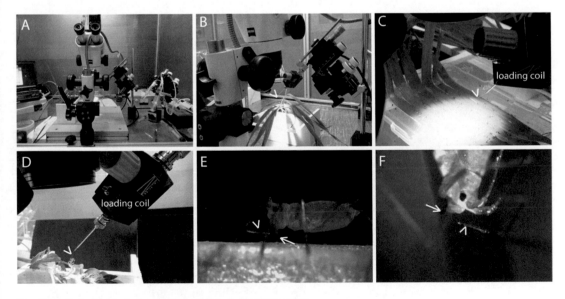

Fig. 3 A photo-illustrated overview of the stylectomy setup

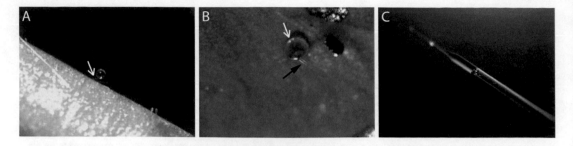

Fig. 4 The collection of the phloem sap droplet. (**a**) A lateral view shows an exuding stylet (white arrow) with a phloem sap droplet on the tip. The phloem sap exudation starts immediately after a successful cut of the stylet. (**b**) A supervision shows the collection of the phloem sap droplet with a glass microcapillary (dark arrow). The tip diameter of the microcapillary is 25 μm. (**c**) The phloem sap is absorbed in the microcapillary due to capillary forces

4. Immediately the pulse of the radio-frequency microcautery is triggered by a foot switch (*see* **Note 7**).

5. In a successful cut, the stylet is severed and starts with exuding (Fig. 4a).

6. The aphid usually run away, or may be brushed off, leaving the cut stylet.

7. The exudates were collected with a borosilicate microcapillary with an inner tip diameter of 25 μm, for example (Fig. 4b, c) (*see* **Note 8**).

8. The phloem sap collection is either executed by suction due to the connection of the microcapillary to a 10 ml syringe via a silicone tube with a side valve or by putting the microcapillary on the stylet.

So far, several specimens of aphids have become quite popular for the application of the stylectomy. (**a–c**) Several views demonstrate apterous aphids of the species *Myzus persicae*, Sulzer 1776 on a *Hordeum vulgare*, L. plant. (**a**) shows also a winged aphid. (**d–g**) Apterous aphids of the species *Acyrtosiphon pisum*, Harris 1776 are shown on a *Vicia faba*, L. plant. (**h**) Winged aphids of the species *Acyrtosiphon pisum* are presented on a *Vicia faba* plant. (**i**) Apterous aphid of the species *Sitobion avenae*, Fabricius locates on a *Triticum aestivum*, L. plant.

The photo illustration of the setup shows an exemplary installation with the monocot plant barley (*Hordeum vulgare*). (**a–d**) The working bench consists of a binocular for the optical surveillance, a movable platform for a suitable arrangement, a micromanipulator for a pinpoint movement, and the microcautery device. (**e** and **f**) The microcauterization of the aphid's stylet (*Sitobium avenae*) is done by means of a hot tungsten needle (arrow head). The stylet is marked with an arrow.

Fig. 5 A visualization of the wound-induced vascular exudation at a tree

3.3 The Wound-Induced Vascular Exudation

The advantage of that method is the possibility of collecting huge volumes of phloem sap in a simple manner. However, only some phytoplasma-infected trees exhibit that phenomenon and thus, make any comparative analysis difficult due to the missing of controls from healthy trees.

1. Choose a phytoplasma-infected tree (*see* **Note 9**).

2. Using a scalpel or knife, a cut is made into the cortex of the stem (Fig. 5).

3. The collection of the exudates is performed using a pipette. 1 ml of the exudate was transferred into a 2 ml tube filled with 1 ml buffer solution (*see* **Note 10**).

    (a) Here, an apple tree infected with "*Candidatus* Phytoplasma mali" shows a wound-induced vascular exudation after an incision of the stem signed with an arrow.

    (b) The great volume of approximately 300 µl per droplet (arrow) is collected by a pipette and transferred into a buffer solution.

4   Notes

1. In common, mature leaves are used as the in vivo translocation direction represents the exudation direction that might increase the yield. It is possible and in parts necessary to pool several leaves in one collection tube. However, the leaves should be of the same age/developmental stage.

2. It is important to prevent an embolism that would inhibit a distribution of EDTA in the plant tissue and disturbs the xylem mass flow. The continuous water supply of the plant tissue is necessary for the reinitialization of the vascular mass flow and thus, for the phloem exudation into the collection solution.

3. The washing step prevents a potential re-embolism, removes material of the damaged cells, and thus, decreases the chance of a contamination with non-phloem material.

4. In the literature, quite wide variations of the parameters are found. The exudation time period reaches up to 12 h with several intervals of 1 h in parts [6, 7]. The quantity of pooled leaves differs. Often an EDTA concentration of 20 mM was used [8–12] and others. Here, an approach with lower EDTA concentrations is proposed [5–7] as EDTA attacks the integrity of membranes [7]. Also different light and dark regimes were still applied—King and Zeevaart [8] and Kovalskaya and colleagues [12] choose darkness, whereas van Bel and Hess [5] and Liu and colleagues [6] used light. Thus, before starting any experiments the conditions should be adjusted to the plant species and scientific issue to achieve optimum results.

5. It is important to choose a suitable aphid/plant combination because some aphid species are restricted to some plant families or species. In addition, the same aphid/host plant combination for rearing should also be used for the experiments to avoid any "adjustments/confusions" of the aphids.

6. A movable flexibility of the single working bench compounds should be considered to place the aphids/plants in an optimum position for a successful cut. Here, a careful handling is necessary.

7. Here, the microcautery device is exemplary presented by a radio-frequency apparatus but also laser-based setups are available [3].

8. The cut stylet can be also covered with silicon oil to prevent evaporation and hardening of the exuding droplet. The phloem sap can be collected after a certain time period. The advantage of the straight collection with a microcapillary is a prevention of (1) contamination by the plant surface and (2) influence of the silicon treatment.

9. Unfortunately, only less reports have been known, so far, dealing with that method [13–15]. However, due to the focus of the current issue on phytoplasmas we want to mention this method.

10. There are any other plant species who exhibit a wound-induced vascular exudation like *Ricinus communis* L. or cucurbits [16]. But, they have not played a role in phytoplasma research until now, and it has to be distinguished the vascular exudation from the latex extrusion [17].

Acknowledgments

This work was supported by the Deutsche Forschungsgemeinschaft (grant FU969/2-1 to ACUF and MRZ) and the Max Planck Society, Germany (TK). We thank Erich Seemüller for providing the pictures about the exudation of apple trees.

References

1. van Bel AJE (2003) The phloem, a miracle of ingenuity. Plant Cell Environ 26:125–149

2. Seemüller E, Garnier M, Schneider B (2002) Mycoplasmas of plants and insects. In: Razin S, Herrmann R (eds) Molecular biology and pathogenicity of mycoplasmas. Springer, Boston, MA, pp 91–115

3. Fisher DB, Frame JM (1984) A guide to the use of the exuding-stylet technique in phloem physiology. Planta 161:385–393

4. Brady J (1965) A simple technique for making very fine, durable dissecting needles by sharpening tungsten wire electrolytically. Bull World Health Org 32:143

5. van Bel AJ, Hess PH (2008) Hexoses as phloem transport sugars: the end of a dogma? J Exp Bot 59:261–272

6. Liu DD, Chao WM, Turgeon R (2012) Transport of sucrose, not hexose, in the phloem. J Exp Bot 63:4315–4320

7. Gaupels F, Knauer T, van Bel AJ (2008) A combinatory approach for analysis of protein sets in barley sieve-tube samples using EDTA-facilitated exudation and aphid stylectomy. J Plant Physiol 165:95–103

8. King RW, Zeevaart JAD (1974) Enhancement of phloem exudation from cut petioles by chelating agents. Plant Physiol 53:96–103

9. Urquhart AA, Joy KW (1981) Use of phloem exudate technique in the study of amino acid transport in pea plants. Plant Physiol 68:750–754

10. Friedman RA, Levin N, Altman A (1986) Presence and identification of polyamines in xylem and phloem exudates of plants. Plant Physiol 82:1154–1157

11. Raps A, Kehr J, Gugerli P et al (2001) Immunological analysis of phloem sap of *Bacillus thuringiensis* corn and of the nontarget herbivore *Rhopalosiphum padi* (Homoptera: Aphididae) for the presence of Cry1Ab. Mol Ecol 10:525–533

12. Kovalskaya N, Owens R, Baker CJ et al (2014) Application of a modified EDTA-mediated exudation technique and guttation fluid analysis for potato spindle tuber viroid RNA detection in tomato plants (*Solanum lycopersicum*). J Virol Meth 198:75–81

13. Kollar A, Seemüller E, Krczal G (1989) Impairment of the sieve tube sealing mechanism of trees infected by mycoplasma-like organisms. J Phytopathol 124:7–12

14. Kollar A, Seemüller E (1990) Chemical composition of phloem exudate of mycoplasma–infected apple trees. J Phytopathol 128:99–111

15. Zimmermann MR, Schneider B, Mithöfer A et al (2015) Implications of *Candidatus* Phytoplasma Mali infection on phloem function of apple trees. Endocytobiosis Cell Res 26:67–75

16. Atkins CA, Smith PM, Rodriguez-Medina C (2011) Macromolecules in phloem exudates—a review. Protoplasma 248:165–172

17. Pickard WF (2008) Laticifers and secretory ducts: two other tube systems in plants. New Phytol 177:877–888

Chapter 22

DAPI and Confocal Laser-Scanning Microscopy for In Vivo Imaging of Phytoplasmas

Rita Musetti and Stefanie Vera Buxa

Abstract

As phytoplasmas are located inside the phloem tissue, always surrounded by numerous layers of other cells, they can result difficult candidates for microscopical investigations. Moreover, the necessity to kill the plant tissues for microscopy observations causes instantaneous and irreversible modifications in the sieve elements, leading to misleading information and erroneous interpretations. Phytoplasmas were here investigated in intact *Vicia faba* host plants using DAPI as fluorescent probe and confocal laser scanning microscopy. The described nondestructive technique may be applied for the imaging of phytoplasmas and of different pathogen-related responses in planta.

Key words Confocal laser scanning microscopy, DAPI, Phytoplasmas, *Vicia faba*

1 Introduction

Fluorescence microscopy operating with the nucleic acid-specific fluorochrome DAPI [4',6-diamidino-2-phenylindole] was extensively used for the detection of phytoplasmas in hand-cut or freezing-microtome sectioned plant materials [1]. A method based on the use of resin-embedded plant tissues, DNA-specific fluorochromes, and fluorescence microscopy was also recently described [2].

All these approaches often evidence technical limitations caused by plant tissue intrinsic properties (phytoplasmas are located inside the phloem tissue, always surrounded by numerous layers of other cells [3]) or by fixation procedures, which reduce the DAPI fluorescent signal-to-noise ratio and cause instantaneous and irreversible reactions of sieve elements to wounding [4]. This may lead to misleading information and erroneous interpretations. For these reasons, the development of protocols for in vivo imaging of fluorescent signal is desirable.

An "in vivo observation" method was developed by Knoblauch and van Bel [5] for the examination of the phloem tissue in

Rita Musetti and Laura Pagliari (eds.), *Phytoplasmas: Methods and Protocols*, Methods in Molecular Biology, vol. 1875, https://doi.org/10.1007/978-1-4939-8837-2_22, © Springer Science+Business Media, LLC, part of Springer Nature 2019

physiological condition, using the confocal laser-scanning microscope (CLSM). This is a nondestructive technique, which allows the observation of different processes in planta [5], including those related to pathogen attack [6].

In this chapter, a CLSM approach is described for imaging phytoplasma DNA-DAPI signal using alive *Vicia faba*. The obtained images could also provide real-time information on structural and biochemical modifications in sieve elements following phytoplasma infection.

2   Materials

1. Plant material: Six-week-old *Vicia faba* plants, healthy and infected with the phytoplasma associated with flavescence dorée (FD) disease [7] (*see* **Note 1**).

2. Apoplasmic buffer solution: 2 mol m^{-3} KCl, 1 mol m^{-3} CaCl$_2$, 1 mol m^{-3} MgCl$_2$, 50 mol m^{-3} mannitol, and 2.5 mol m^{-3} MES/NaOH buffer, pH 5.7 [5].

3. DAPI staining solution: 1 µg/ml in apoplasmatic buffer, pH 5.7 [6].

4. Razor blades.

5. CLSM equipped with a UV laser to detect DAPI fluorescence signal (*see* **Note 2**).

6. To watch the phloem tissue at the observation window without a cover slip, a water immersion objective (i.e., HCX APO L40×0.80 W U-V-l objective, Leica, Heidelberg, Germany) was used in the dipping mode (*see* **Note 3**).

3   Methods

For in vivo observation of sieve tubes, cortical cell layers were removed from the lower side of the main vein of a fully expanded leaf, still attached to an intact plant, to provide a CLSM observation window [5].

1. Mount a leaf attached to an intact *Vicia faba* plant upside down on the stage of a confocal microscope, as described by Knoblauch and van Bel [5] (Fig. 1).

2. Using a razor blade, remove the cortical cell layers down to the phloem from the lower side of the main vein of a full expanded leaf, still attached to the intact plant to create an "observation window" (Fig. 2a).

3. Apply a drop of DAPI solution (1 µg/ml) to the observation window.

Fig. 1 A leaf attached to an intact *Vicia faba* plant is placed downward on the stage of a confocal microscope

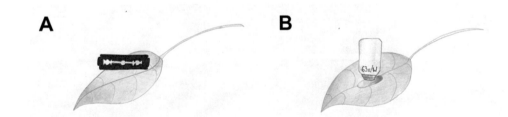

Fig. 2 (a) The cortical cell layers down to the phloem are removed from the lower side of the main vein of a leaf by a razor blade. (**b**) A drop of DAPI solution is applied on the "observation window"

4. Incubate for 15–20 min at room temperature in the dark.

5. Remove DAPI using a pipette and replace it with the apoplasmic buffer.

6. Observe at 63 × under CLSM at 405 nm (Fig. 2b; *see* **Note 2**).

7. To avoid misinterpretations due to autofluorescence, observe in vivo unstained *V. faba* sieve elements at the same excitation wavelength used for DAPI as controls (*see* **Note 4**).

4   Notes

1. *Vicia faba* is a good plant for phloem in vivo observations, sieve elements are well distinguishable thanks to the absence of nuclei and chloroplasts; moreover, the sieve plates are easily discernible. In healthy plants, the presence of the typical Fabaceae forisomes is a further help for sieve element identification.

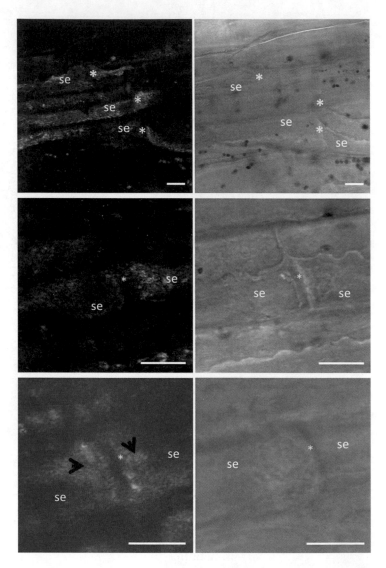

Fig. 3 Confocal laser-scanning microscopy (CLSM) (left half of the panel) and transmission microscopic images (right half) of phloem tissue in intact Flavescence Dorée (FD)-infected *Vicia faba* plants. DAPI fluorescence is massive in infected sieve elements (arrows). Particulate fluorescent structures can be distinguished above all at sieve plates. Bars correspond to 10 μm. (* indicates the sieve plate; se: sieve element)

Fig. 4 Confocal laser-scanning microscopy (CLSM) (left half of the panel) and transmission microscopic images (right half) of phloem tissue in intact healthy *Vicia faba* plants. DAPI signal is absent in the phloem of healthy plants. Bars correspond to 10 μm. (* indicates the sieve plate; f: forisome; se: sieve element)

2. Analysis by CLSM and image processing using DAPI are appropriate tools for the detection and localization of phytoplasmas in the phloem of intact plants. DAPI is excited by 800 nm and the emission is recorded in the spectral window at 405 nm.

3. In the sieve elements of infected plants it is possible to observe fluorescent aggregates, formed by small dot spots (Fig. 3), mainly accumulated at the sieve plate in correspondence of the sieve pores. As DAPI binds to DNA, in the enucleate sieve elements [4], no interference with nuclear staining can occur.

4. The sieve elements in healthy plants test negative to DAPI staining showing any bright dots, as present in the infected ones (Fig. 4).

Acknowledgments

This work was funded with the support of the Deutscher Akademischer Austauschdienst (DAAD) (grant number A/11/04104 for R. Musetti).

The authors are grateful to Elena Zaghini for drawing Fig. 2.

References

1. Andreade N, Arismendi NL (2013) DAPI staining and fluorescence microscopy techniques for phytoplasmas. In: Dickinson M, Hodgett J (eds) Phytoplasma: methods and protocols. Springer Science+Business Media, LLC 2011, New York, pp 115–121

2. Buxa SV, Pagliari L, Musetti R (2016) Epifluorescence microscopy imaging of phytoplasmas in embedded leaf tissues using DAPI and SYTO13 fluorochromes. Microscopie 1:49–56

3. Truernit E (2014) Phloem imaging. J Exp Bot 65:1681–1688

4. van Bel AJE (2003) The phloem, a miracle of ingenuity. Plant Cell Environ 26:125–149

5. Knoblauch M, van Bel AJE (1998) Sieve tubes in action. Plant Cell 10:35–50

6. Musetti R, Buxa SV, De Marco F et al (2013) Phytoplasma-triggered Ca^{2+} influx is involved in sieve-tube blockage. Mol Plant-Microbe Interact 26:379–386

7. Caudwell A, Kuszala C, Larrue J et al (1972) Transmission de la flavescence dorée de la fève à la fève par des cicadelles des genres *Euscelis* et *Euscelidius*. Ann Phytopathol no. hors. série:181–189

Chapter 23

Immunofluorescence Assay to Study Early Events of Vector Salivary Gland Colonization by Phytoplasmas

Luciana Galetto, Marta Vallino, Mahnaz Rashidi, and Cristina Marzachì

Abstract

To visualize phytoplasmas at early stages of vector infection, an immunofluorescence assay was developed. The chapter provides experimental details on dissection of salivary glands, incubation of the dissected organs with phytoplasma suspension, fixation, embedding, sectioning, labeling, and final visualization with confocal microscopy. All the procedure will be described for the leafhopper *Euscelidius variegatus*, natural vector of "*Candidatus* phytoplasma asteris" and laboratory vector of Flavescence dorée phytoplasma.

Key words *Candidatus* phytoplasma asteris, *Euscelidius variegatus*, Leafhopper, Phytoplasma membrane proteins

1 Introduction

Phytoplasmas are plant pathogenic bacteria that cause severe threats to cultivation of different crops worldwide. Phytoplasmas are mainly transmitted by phloem feeding leafhoppers, planthoppers, and psyllids within the Hemiptera [1] in a persistent and propagative manner. Phytoplasma transmission is a complex biological process and interactions between plant, insect, and pathogen [2], as well as insect vector specificity plays pivotal role in spreading of these diseases [3]. Behavioral, environmental, and geographical factors are involved in determining vector specificity of phytoplasma transmission (reviewed in [4]), but interaction at the molecular level between membrane proteins of phytoplasmas and vectors is also involved. Indeed, ingested phytoplasmas must actively multiply and pass from the alimentary canal through the midgut into the haemocoel, before colonizing salivary gland cells (latency period, LP) and being transmitted to a new host plant. During the LP, phytoplasmas closely interact with insect tissues of different organs, and with insect cellular machinery when internalized within insect cells. Therefore, phytoplasmas face diverse challenges to their survival, which they counteract by manipulating host

Rita Musetti and Laura Pagliari (eds.), *Phytoplasmas: Methods and Protocols*, Methods in Molecular Biology, vol. 1875, https://doi.org/10.1007/978-1-4939-8837-2_23, © Springer Science+Business Media, LLC, part of Springer Nature 2019

metabolism [5] and by exploiting the host cellular machinery, as suggested for other vector-borne bacteria [6]. Phytoplasmas lack a cell wall, therefore their plasma membrane is in direct contact with the host. Indeed, phytoplasma membrane proteins with hydrophilic domains exposed on the outer part of the cell are good potential partners for interaction with insect proteins. For some of them, both genome or plasmid encoded, such interaction has been demonstrated [7–11], and shown to be involved in the transmission process [12]. Most of the recent studies on molecular interactions between phytoplasma surface and host cells/organs relay on fluorescence microscopy coupled with the use of specific antibodies. In particular, as these plant pathogens are recalcitrant to cultivation, study of early phases of vector colonization by phytoplasma is not easily achieved. To overcome these technical barriers, coating of latex beads with target phytoplasma membrane proteins has been explored on a vector cell-based assay [7]. On the other hand, phytoplasmas remain viable for short time [13] in fresh suspensions from infected hosts [14], and these can be exploited for the development of organ-based interaction assays. To allow phytoplasma visualization at early stages of vector infection, an immunofluorescence assay was developed, coupling fluorescence microscopy and quantitative PCR [12]. As qPCR protocols for quantification of phytoplasma loads are described elsewhere in this book, the details of this immunofluorescence assay are the focus of this chapter.

The chapter will provide experimental details on dissection of salivary glands, incubation of the dissected organs with phytoplasma suspension, fixation, embedding, sectioning, labeling, and final visualization with confocal microscopy. All the procedure will be described for the leafhopper *Euscelidius variegatus*, natural vector of "*Candidatus* phytoplasma asteris," and laboratory vector of Flavescence dorée phytoplasma [15].

2   Materials

All solutions should be prepared with ultrapure water, stored, and used as detailed in the following sections. All chemicals should be disposed following the appropriate disposal procedures.

2.1 Insects and Phytoplasma Isolate

1. Laboratory healthy population of *Euscelidius variegatus* Kirschbaum, originally collected from Piemonte region (Italy).

2. Phytoplasma-infected insect population, e.g., *E. variegatus* infected with "*Candidatus* Phytoplasma asteris," 16SrI-B, Chrysanthemum yellows isolate (CYp).

3. Plastic and nylon cages for insect rearing.

4. Growth chamber: 24 °C, L:D = 16:8.

5. Mouth aspirator and tubes, CO_2 cylinder and appropriate dispenser, forceps, brush, stereo microscope, wax plate.

2.2 Required Buffers, Reagents, and Labware

1. Ice-cold crushing buffer: 0.3 M glycine, 0.03 M $MgCl_2$, pH 8.0 [13]. Prepare fresh before every use, and sterilize by filtering (0.22 μm).

2. Blocking buffer for organs: 1% Bovine Serum Albumin (BSA) in crushing buffer. Prepare fresh before every use. Use gloves and sterile spatula to weigh the BSA powder. Further sterilization is not required.

3. Fixative solution: 4% paraformaldehyde, 1× phosphate-buffered saline, pH 7.4 (PBS) (*see* **Note 1**), 0.1% Triton X-100. Dissolve paraformaldehyde in PBS in a flask and, under a fume hood, mix on a heated stirrer until the solution (which has a milky appearance) turns colorless. Check the solution regularly to avoid spilling due to over-heating (temperature should be around 65 °C). If the solution appears cloudy, add some drops of 1 N sodium hydroxide: it should then become clear. Let the solution cool down to room temperature and add Triton X-100. Formaldehyde is a hazardous substance: always use fume hood, wear gloves and lab coat, discard the waste in a proper way (*see* **Note 2**).

4. Permeabilization solution: 1× PBS, pH 7.4, 1% Triton X-100.

5. Washing buffer: 1× PBS.

6. Embedding matrix: 8% agarose. Choose a low melting point agarose (e.g., Agarose Type II-A, Sigma). Weigh the agarose, pour into a flask, add deionized water, and melt in a microwave. Check carefully the solution while heating in the microwave to avoid spilling due to boiling (*see* **Note 3**). Keep the embedding matrix warm and melted in a water bath (*see* **Note 4**).

7. Blocking buffer for vibratome sections: 1× PBS, 1% BSA. Prepare fresh before every use.

8. Primary antibody for specific labeling of phytoplasma cells (e.g., A416, polyclonal rabbit antibody raised against the major antigenic membrane protein Amp of CYp; [16]). Make the desired dilution in blocking buffer.

9. Fluorophore-conjugated secondary antibody for specific labeling of primary antibody (any commercial brand, e.g., FITC-conjugated goat anti rabbit). Make the desired dilution in blocking buffer.

10. Nuclear stain, such as 0.1 μg/ml DAPI (4′,6-diamidino-2-phenylindole) in H_2O. DAPI is a mutagen: handle with care and dispose of as hazardous waste.

11. Ice-cold mortar and pestle, syringe filter (0.45 μm), syringe (2 ml), tubes (1.5 and 2.2 ml), plastic mold for embedding, humid chamber, microscope slides and cover glasses, liquid blocker pen, forceps, pipets, antifade mounting medium.

2.3 Further Required Equipment

1. Refrigerated microcentrifuge.
2. Roller drum or equivalent shaker.
3. Vibratome, equipped with blades and glue.
4. Confocal microscope.

3   Methods

3.1 Preparation of Phytoplasma Suspension (See Note 5)

1. On experiment day, prechill two mortars and crushing buffer on ice.
2. Crush 30 *E. variegatus* adults (CYP-infected or healthy insects as negative control) in 900 μl ice-cold crushing buffer, in the cold mortar.
3. Transfer the obtained mixture in an Eppendorf tube using a pipette with a cut tip. The extract should then be clarified by centrifugation at $800 \times g$ for 10 min at 4 °C.
4. Collect the supernatant and filter it through 0.45 μm sterile filter mounted on the appropriate syringe.
5. Store the filtered phytoplasma suspension on ice until needed (*see* **Note 6**).

3.2 Organ Dissection and Incubation with Phytoplasma Suspension

1. To dissect organs from coeval insects, collect newly emerged, healthy *E. variegatus* adults from the rearing colony (*see* **Note 7**) into a glass tube with the mouth aspirator for immediate dissection. For each treatment, at least 10 insects should be dissected (*see* **Note 8**).
2. Before dissection, the leafhoppers should be anesthetized by flushing CO_2 for few seconds. This is achieved by sliding the CO_2 tube into the glass container with the insect and slowly turning the on/off valve to release the CO_2 for 3–5 s. Pay attention not to go full blast, or the insects will go flying around. The anesthesia time should be empirically established according to the CO_2 cylinder and tube diameter.
3. Label two Eppendorf tubes (1.5 ml) as "Treatment" and "Control," place them on ice, and fill them with 200 μl of ice-cold crushing buffer. Place a single anesthetized insect on the wax tray under the stereo microscope, dorsal side down (Fig. 1). Place a drop (20 μl) of crushing buffer close to the insect. With forceps or a needle, detach the head from the thorax by lifting the lower part of gena (Fig. 2a). Insert the

Fig. 1 Ventral view of an anaesthetized *Euscelidius variegatus* adult female ready for dissection

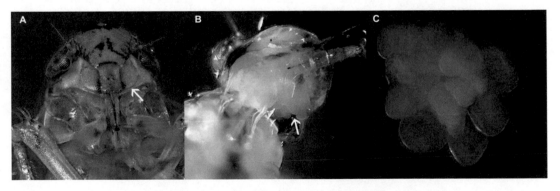

Fig. 2 *Euscelidius variegatus* gland dissection. (**A**) Magnification of insect head; arrow indicates the point for head lifting. (**B**) View of detached head; arrow indicates salivary glands. (**C**) Dissected salivary glands

needle under the globular mass of the glands (Fig. 2b), pull and deposit them in the crushing buffer drop to confirm their integrity (Fig. 2c). Transfer the dissected salivary glands into the appropriate Eppendorf tube, and keep it on ice until all salivary glands of each treatment have been dissected. Proceed with dissection of the successive leafhopper.

4. Slowly spin the tubes containing the dissected salivary glands ($1000 \times g$, $4 \, ^\circ C$), withdraw the supernatant without disturbing the organs (*see* **Note 9**). Add 200 µl ice-cold blocking buffer for organs and incubate at $4 \, ^\circ C$ for 1 h (*see* **Note 5**), with slow shaking (use the lowest speed of a roller drum).

5. Slowly spin the tubes containing the dissected salivary glands ($1000 \times g$, $4 \, ^\circ C$), withdraw the supernatant without disturbing the organs (*see* **Note 9**). Add 200 µl ice-cold phytoplasma suspension or control suspension and incubate at $4 \, ^\circ C$ for 4 h with slow shaking.

6. Perform five successive washes with ice-cold fresh crushing buffer (200 µl each time), to remove unbound phytoplasma.

7. After the last wash, add 200 µl ice-cold fixative solution, and incubate overnight at $4 \, ^\circ C$.

8. Remove fixative solution, add 200 µl of washing buffer and leave 5 min at room temperature. Remove the washing buffer and repeat this washing step twice more.

9. After the last wash, add 200 µl of permeabilization solution and incubate overnight at $4 \, ^\circ C$.

10. Remove permeabilization solution, add 200 µl of washing buffer and leave 5 min at room temperature. Remove the washing buffer and repeat this washing step twice more.

3.3 Vibratome Sectioning

1. Fill a well of an embedding mold with the melted agarose (*see* **Note 10**) and rapidly take out (with a forceps) a salivary gland and insert it in the agarose (*see* **Note 11**). Proceed in this way for all glands. The agarose blocks will solidify in few minutes (*see* **Note 12**).

2. Trim the agarose blocks with a razor blade or a scalpel to remove excess agarose (*see* **Note 13**) and fix them with the glue on specimen plates (one block for each plate) (*see* **Note 14**). Let the glue dry and proceed with the sectioning with a vibrating microtome (follow the instruction of the instrument). In our experience, 100 µm-thick sections work well.

3. As soon as each section is cut and starts floating on the water, collect it with forceps or a small paint brush and transfer it onto a microscope glass (*see* **Note 15**).

4. Observe the sections under a microscope to check the quality: discard damaged or poor-quality sections. When the sections of all samples have been obtained and selected, distribute them to the microscope slides (*see* **Note 16**).

3.4 Labeling and Observation

All the incubations are performed directly on the microscope slide: therefore, to avoid the loss of sections and to minimize the amount of solution to be used, a containment barrier should be created (*see* **Note 17**). Make sure that in each step the solution covers all the sections: usually 100–200 µl should be enough. Always keep the slide in a humid chamber to avoid evaporation of the solutions (*see* **Note 18**).

1. Incubate the sections in blocking buffer for vibratome sections for 30 min at room temperature.

2. Remove the blocking buffer (*see* **Note 19**), add the primary antibody, and incubate at 4 °C overnight (*see* **Note 20**).

3. Remove the primary antibody, add the washing buffer, and let it stand for 5 min at room temperature. Remove the washing buffer and repeat the step twice more.

4. Incubate the sections in blocking buffer for vibratome sections for 30 min at room temperature.

5. Remove the blocking buffer, add the secondary antibody, and incubate at room temperature for 3 h (*see* **Notes 20** and **21**).

6. Remove the secondary antibody, add the washing buffer, and let it stand for 5 min at room temperature. Remove the washing buffer and repeat the step other two times.

7. Remove the last washing buffer, arrange the sections so that they do not overlie and mount in water (add some drops of water and cover with the cover glass). If you decide to perform a nuclear staining, mount directly in DAPI solution (*see* **Note 22**). Instead of using water, a mounting medium with an antifade reagent is recommended.

8. Observe the sections under a confocal microscopy. As a rule of thumb, set laser intensity and detector gain on nontreated samples at first, and then keep the same settings for all the subsequent observations. Take note of the number of observations and positive sections for each treatment (Fig. 3).

4 Notes

1. Prepare a stock solution of 10× PBS and sterilize by autoclave.

2. To avoid being exposed to paraformaldehyde while weighing it, pour the powder in the weigh boat under the fume hood, cover it with another weigh boat, and weigh on the balance in which the tare was previously set. Transfer the weighed paraformaldehyde to the flask inside the fume hood. Alternatively, the paraformaldehyde can be poured under the fume hood directly in a screw cap tube, weighed and dissolved in the

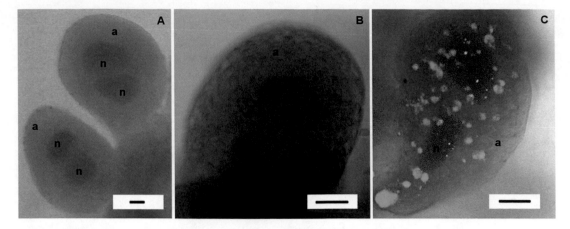

Fig. 3 Confocal microscopy images of sections of *Euscelidius variegatus* salivary glands acini (a). In **A**, antibodies and DAPI stain were omitted. In **B** and **C** glands were labeled with primary and secondary antibodies and stained with DAPI. **B** control sample, in which glands were incubated in suspension from healthy insects; **C** treated sample, in which glands were incubated in suspension from phytoplasma infected insects. Images are the overlay of fluorescence and bright-field acquisition. Nuclei (n) are labeled in red; phytoplasma are labeled in green. Magnification bars correspond to 20 μm

same tube. In this case, the solution can be warmed in a water bath outside the fume hood.

3. This agarose solution is quite viscous, therefore, to avoid spilling, the microwave should be set at the lowest power and stopped as soon as the solution starts boiling; warming can be turned on again after incubation for few seconds at room temperature. The flask should be removed from the microwave from time to time and the solution observed against the light to confirm complete melting of the agarose.

4. Instead of using a water bath, the flask containing the melted agarose may be incubated in a beaker containing preheated water on a heating plate. Do not let the water in the beaker boil. A paper towel between the flask and the beaker bottom may prevent the flask from floating in the hot water.

5. Phytoplasma suspension may be prepared during the 1 h blocking step described in Subheading 3.2, **step 4** paragraph section.

6. Under these conditions, phytoplasmas are viable for few hours if stored at 4 °C.

7. To obtain coeval healthy *E. variegatus* adults, transfer fourth and fifth instar nymphs from the original rearing cage to a new healthy oat in a separate cage 10 days before the experiment. Collect insects from this cage for the experiment.

8. The number of organs for dissection depends on the difficulties that may be encountered during the successive manipulation. Indeed, some organs may be lost during washing, or may break

upon transfer to agarose during embedding. Moreover, sectioning may not be optimal for each embedded organ, and slices may be lost during their transfer to the glass slide. As adequate numbers of observed sections are required to perform appropriate statistical analyses of the results, the number of dissected organs must be sufficient.

9. Use a glass Pasteur pipette for withdrawing the buffer, as glands do not stick to glass. Also, look at the bottom of the tube, to ensure that the tip of the pipette does not touch the organs. Confirm the absence of any salivary gland from the removed buffer by eye inspection. As a strategy, leave some residual buffer at the bottom of the tube.

10. Using a plastic pipet with the tip cut off is an easy way to distribute the embedding matrix into the wells.

11. Dissected organs should be handled carefully. When transferring the gland from the tube to the agarose, pay attention not to squeeze it between the forceps ends. To facilitate picking, transfer the liquid containing the organs from the tube to a small glass plate. If you need to save space and embedding matrix, you can also put more than one organ in each well, given that they are sufficiently far apart to be separated during the block trimming prior to sectioning.

12. After the agarose block solidify, do not let them air-dry; otherwise, it will be difficult to obtain good sections (e.g., sample distortion, sample coming out from the agarose). If you have many samples to prepare, keep the agarose blocks in a humid chamber (see **Note 18**) while preparing new ones. We recommend cutting all the prepared blocks during the same day. Sections, instead, may be stored overnight at 4 °C prior to start labeling. Unused embedding matrix can be stored in a fridge for few weeks, re-melted and used.

13. Trimming the agarose block is also useful to help positioning the specimen according to the blade plane.

14. Use a quick drying glue (glue as Attack Epoxy Resin works well). Use a small drop, not exceeding the basal surface of the agarose block, to avoid waiting many minutes for the glue to dry. After sectioning, the agarose block can be detached from the specimen plate using a wooden or a plastic spatula. Some people use a razor blade, but we do not recommend it, because it can scratch the plate surface.

15. Put some drops of water on the slide: never let the sections dry.

16. Some controls should be envisaged to unmask false positives: (1) a control for autofluorescence of the specimen (primary and secondary antibody omitted); (2) a control for unspecific binding of the secondary antibody (primary antibody

omitted). Controls should be prepared for few slices of each dissected organ. As underlined in **Note 8**, consider having many sections on each slide and, if possible, having more slides for each treatment, to perform appropriate statistical analyses of the results.

17. Containment barrier on the microscope slide can be created drawing an ellipse around the sections with a commercial liquid blocker pen or using a correction fluid pen. An electrical tape can also be used. Within a 25 mm-wide tape stuck to the slide, cut an internal rectangle with a blade and remove it, leaving a tape frame wide enough to contain the sections. Alternatively, stick a piece of tape to a smooth surface, cut an internal rectangle within the tape using a blade, remove the external frame and stick it to the slide.

18. The humid chamber consists of a closed container (glass or plastic) with a moist paper towel at the bottom.

19. The removal of the solutions from the microscope slides can be done with a pipet or filter paper. In the first case, tilt slightly the glass so that the liquid accumulates on one side and then suck with the pipet. In the second case, small pieces of filter paper are placed on the edge of the liquid and allow the liquid to be absorbed. In both cases, sections move and tend to overlie: carefully arrange them with a pipet tip or forceps.

20. Preliminary optimization experiments to decide proper primary and secondary antibody concentrations should be considered.

21. Fluorescent antibodies are light sensitive and fade when exposed to light. Work with a dim light, prepare the dilution in a shaded tube, and place slides in a shaded box (e.g., wrapped in aluminum foil) during incubation times and until observation.

22. If the samples are overstained by DAPI, you can incubate for 5 min in the DAPI solution (prepared in $1\times$ PBS), wash three times, and then mount in water.

References

1. Weintraub PG, Beanland L (2006) Insect vectors of phytoplasmas. Annu Rev Entomol 51:91–111

2. Purcell AH (1982) Insect vector relationships with prokaryotic plant pathogens. Annu Rev Phytopathol 20:397–417

3. Gratz NG (1999) Emerging and resurging vector-borne diseases. Annu Rev Entomol 44:51–75

4. Bosco D, D'Amelio R (2010) Transmission specificity and competition of multiple phytoplasmas in the insect vector. In: Weintraub PG, Jones P (eds) Phytoplasmas: genomes, plant hosts, and vectors. CABI, Wallingford, UK, pp 293–308

5. Toruño TY, Šeruga Musić M, Simi S (2010) Phytoplasma PMU1 exists as linear chromosomal and circular extrachromosomal elements and has enhanced expression in insect vectors

compared with plant hosts. Mol Microbiol 77:1406–1415

6. Dubrana MP, Guegueniat J, Bertin C et al (2017) Proteolytic post-translational processing of adhesins in a pathogenic bacterium. J Mol Biol 429:1889–1902

7. Arricau-Bouvery N, Duret S, Dubrana MP et al (2017) The variable membrane protein VmpA of flavescence dorée phytoplasma interacts with cells of the insect vector. Paper presented at the 3rd international symposium on the Hemipteran-Plant Interactions, University of Madrid, Madrid (S), 4–8 June 2017

8. Galetto L, Bosco D, Balestrini R et al (2011) The major antigenic membrane protein of 'Candidatus phytoplasma asteris' selectively interacts with ATP synthase and actin of leafhopper vectors. PLoS One 6:e22571

9. Ishii Y, Oshima K, Kakizawa S et al (2009) Process of reductive evolution during 10 years in plasmids of a non-insect-transmissible phytoplasma. Gene 446(2):51–57

10. Neriya Y, Sugawara K, Maejima K et al (2011) Cloning, expression analysis, and sequence diversity of genes encoding two different immunodominant membrane proteins in poinsettia branch-inducing phytoplasma (PoiBI). FEMS Microbiol Lett 324:38–47

11. Suzuki S, Oshima K, Kakizawa S et al (2006) Interaction between the membrane protein of a pathogen and insect microfilament complex determines insect-vector specificity. Proc Natl Acad Sci U S A 103:4252–4257

12. Rashidi M, Galetto L, Bosco D et al (2015) Role of the major antigenic membrane protein in phytoplasma transmission by two insect vector species. BMC Microbiol 15:193

13. Bressan A, Clair D, Semetey O et al (2006) Insect injection and artificial feeding bioassays to test the vector specificity of Flavescence dorée phytoplasma. Phytopathology 96:790–796

14. Lefol C, Caudwell A, Lherminier J et al (1993) Attachment of the Flavescence dorée pathogen (MLO) to leafhopper vectors and other insects. Ann Appl Biol 123:611–622

15. Rashidi M, D'Amelio R, Galetto L et al (2014) Interactive transmission of two phytoplasmas by the vector insect. Ann Appl Biol 165:404–413

16. Galetto L, Fletcher J, Bosco D et al (2008) Characterization of putative membrane protein genes of the 'Candidatus Phytoplasma asteris', chrysanthemum yellows isolate. Can J Microbiol 54:341–351

Part IV

Plant-Pathogen Interaction

Chapter 24

Characterization of Phytoplasmal Effector Protein Interaction with Proteinaceous Plant Host Targets Using Bimolecular Fluorescence Complementation (BiFC)

Katrin Janik, Hagen Stellmach, Cecilia Mittelberger, and Bettina Hause

Abstract

Elucidating the molecular mechanisms underlying plant disease development has become an important aspect of phytoplasma research in the last years. Especially unraveling the function of phytoplasma effector proteins has gained interesting insights into phytoplasma-host interaction at the molecular level. Here, we describe how to analyze and visualize the interaction of a phytoplasma effector with its proteinaceous host partner using bimolecular fluorescence complementation (BiFC) in *Nicotiana benthamiana* mesophyll protoplasts. The protocol comprises a description of how to isolate protoplasts from leaves and how to transform these protoplasts with BiFC expression vectors containing the phytoplasma effector and the host interaction partner, respectively. If an interaction occurs, a fluorescent YFP-complex is reconstituted in the protoplast, which can be visualized using fluorescence microscopy.

Key words Bimolecular fluorescence complementation (BiFC), Effector protein, Fluorescence microscopy, Phytoplasma, Protoplast isolation, Protoplast transfection, Protoplast transformation

1 Introduction

Phytoplasma are plant pathogens that reside in the phloem and interact with their host plant at different levels. The phytoplasma-host interactions can be mediated by secreted bacterial proteins, so called effector proteins. These bacterial effector proteins are molecules secreted by plant-associated organisms that alter host-cell structure and function [1]. In the last years, studies unraveling effector gene function have significantly contributed to many important discoveries in plant and biological sciences and have emerged effectors as a central class of molecules in our integrated view of plant-microbe interactions [2]. Characterization of effector protein function and the identification of effector targets will

Katrin Janik and Hagen Stellmach contributed equally to this work.

Rita Musetti and Laura Pagliari (eds.), *Phytoplasmas: Methods and Protocols*, Methods in Molecular Biology, vol. 1875,
https://doi.org/10.1007/978-1-4939-8837-2_24, © Springer Science+Business Media, LLC, part of Springer Nature 2019

greatly enhance our understanding of bacterial and phytoplasma virulence [3, 4]. Plant pathogenic organisms can produce a range of different effector proteins to modulate plant processes and plant defenses for their own benefit.

For most phytoplasma-host interactions information about the exact molecular mechanisms underlying disease development remains scarce. The identification of phytoplasma effectors and the characterization of their interaction with host targets has become, however, an important aspect in phytoplasma research during the last years [4]. Pioneering work has been performed in unraveling the function of 'Candidatus Phytoplasma mali', the causal agent of aster yellows witches' broom (AY-WB) and onion yellows (OY) phytoplasma effectors in the model plants *Arabidopsis thaliana* and *Nicotiana benthamiana* (*N. benthamiana*) [5–10]. Especially, the AY-WB phytoplasma effector protein SAP11 [7] moved to the center of attention, due to the discovery of SAP11-homologues in many other phytoplasma species [11, 12] and its potential role as a key player in phytoplasma disease development [13]. The occurrence of a bacterial effector can be predicted by genome mining and bioinformatic preselection. This in silico identification is then followed by experimental approaches, such as verifying bacterial effector expression in the host during infection, determining its subcellular localization in planta using confocal microscopy, and a subsequent functional characterization [6]. A functional screening often involves the heterologous expression of the candidate effector in a model plant. This allows unraveling whether the effector itself induces symptom development and thus provides the basis to assess its potential role and its importance during infection. To determine the molecular interaction partner, a bacterial protein of interest can be screened against a library of potential proteinaceous interaction partners (e.g., from the natural host species) using a Yeast Two-Hybrid (Y2H) assay [6, 7, 14]. The Y2H principle is based on a positive selection, which enables recombinant yeast expressing the respective "effector-host protein-interaction couple" to grow on selective medium [14]. A detailed protocol for a Y2H screen to identify interaction of a phytoplasma effector protein with its host binding partners has been published [15]. Y2H screen is a convenient technique to screen an effector against hundreds of thousands of potential interaction partners; however, this method is prone to generate false-positive results [16–18]. The physiological relevance of a protein-protein interaction identified in a Y2H must therefore be further analyzed with a technically different method [16]. Since Y2H interactions occur in recombinant yeast cells, it is important to analyze whether the identified interaction of a phytoplasma effector also occurs in planta, i.e., the natural cellular site of the interaction. An in planta interaction can be analyzed by an alternative method, such as bimolecular fluorescence complementation (BiFC). BiFC allows the direct visualization of protein-protein interactions using fluorescence microscopy. This method is based on the principle that two non-fluorescent fragments of a fluorescent protein can coalesce and form a non-covalent fluorescent complex. This fusion is mediated by two proteins that interact with each other. The two interacting proteins are coupled to the two non-fluorescent fragments and the interaction of the proteins and the associated approximation of the fragments leads to the reconstitution of the bimolecular fluorescent complex. BiFC assays are widely used to visualize protein-protein interactions in many different experimental settings and they furthermore allow to track the subcellular locations of the observed interactions [19]. The BiFC method is prone, however, to potential sources of artifacts leading to false positive and false negative results [20]. Therefore, it is important to include appropriate controls to unequivocally demonstrate the specificity of the observed protein-protein interaction [20].

This protocol will provide step-by-step instructions to perform an in planta BiFC analysis of two (potentially) interacting proteins in *N. benthamiana* protoplasts. The interaction partners will be cloned into the two BiFC expression vectors pE-SPYNE and pE-SPYCE, which allow the bicistronic expression of each interaction partner with a non-fluorescent fragment of the yellow fluorescent protein (YFP) [21]. The two plasmids are co-transformed in *N. benthamiana* protoplasts using a modified method based on the protocol from the Sheen lab [22]. The described method is easily feasible and allows analyzing whether a dual protein-protein interaction occurs in an in planta environment and thus provides an important tool for phytoplasma effector studies. With this method we were able to confirm the interaction of the SAP11-like effector protein ATP_00189 from '*Candidatus* Phytoplasma mali' (apple proliferation phytoplasma) with two transcription factors from its natural host *Malus x domestica* [13].

2 Material

All solutions are prepared in double deionized water.

2.1 DNA Preparation and PEG Precipitation

1. pE-SPYCE and pE-SPYNE plasmid constructs with the subcloned effector and target genes (*see* **Note 1**).

2. Midi scale plasmid purification kit to yield about 100 µg plasmid DNA per preparation.

3. PEG-solution 1 (PEG1; for DNA precipitation): 26% (w/v) polyethylene glycol (PEG) 4000, 6 mM $MgCl_2$, 0.6 M Sodium acetate (pH 5.2).

4. Nuclease-free, double deionized water.

2.2 Protoplast Isolation and Transformation

1. Four-week-old *N. benthamiana* plants.
2. Small petri-dishes (5 cm diameter).
3. Razor blades.
4. Enzyme solution: Dilute the stock solutions to a final concentration of 0.4 M mannitol, 20 mM KCl, 20 mM 4-morpholineethanesulfonic acid (MES), 1.5% (w/v) cellulase R-10, 0.4% (w/v) macerozyme R-10, 10 mM $CaCl_2$, 1 mg/mL bovine serum albumin (BSA) (*see* **Note 2**).
5. Desiccator.
6. Vacuum pump.
7. Laboratory shaker.
8. Polyamide-cell strainer (100 μm pores).
9. Centrifuge tubes with round bottom (15 mL) (*see* **Note 3**).
10. Buffer W5: 154 mM NaCl, 125 mM $CaCl_2$, 5 mM KCl, 2 mM MES (pH 5.7) (*see* **Note 4**).
11. MMG: 0.4 M mannitol, 15 mM $MgCl_2$, 4 mM MES (pH 5.7) (*see* **Note 5**).
12. Counting chamber (hemocytometer or any other type with 0.2 mm chamber depth).
13. 2.0 mL microcentrifuge tubes or centrifuge tubes (*see* **Note 6**).
14. PEG-solution 2 (PEG2; for transformation): 0.2 M mannitol, 0.1 M $CaCl_2$, 40% (w/v) PEG 4000 (*see* **Note 7**).
15. Ethanol (70% v/v): pure ethanol (≥96%) diluted to 70% (v/v) in water.
16. Buffer WI: 0.5 M mannitol, 20 mM KCl, 4 mM MES (*see* **Note 8**).
17. Filter paper (grade 6; 3 μm; medium to slow filter paper) (*see* **Note 9**).

2.3 Microscopic Analysis

1. Confocal laser scanning microscope or epifluorescence microscope with appropriate filter sets.
2. Microscopy dishes (glass bottom dishes) or counting chamber depending on type of microscope (*see* **Note 10**).

3 Methods

3.1 DNA Preparation and Precipitation Using PEG

1. Subclone your proteinaceous interaction partners into pE-SPYCE and pE-SPYNE using the Gateway® cloning strategy (*see* **Note 11**).
2. Prepare a midi-scale plasmid preparation of both plasmid constructs (containing your subcloned interaction partners) using a commercially available plasmid preparation kit (*see* **Note 12**).

3. Mix the DNA solution with the same volume of PEG-solution 1 (PEG1) and incubate at room temperature for 20 min.

4. Centrifuge the DNA-PEG1-Mix at 10 °C and $16,000 \times g$ for 20 min and carefully remove the supernatant.

5. Wash the pellet with 1.0 mL Ethanol (70% v/v) and centrifuge at 10 °C and $16,000 \times g$ for another 20 min.

6. Carefully discard the supernatant and dry the pellet in an incubator or thermoblock at 37 °C for some minutes. Dissolve the pellet in 20 μL nuclease-free, double deionized water (*see* **Note 13**).

3.2 Protoplast Isolation

1. Count the leaves from the tip of the shoot and pick the third and the fourth leaf of the *N. benthamiana* plant (*see* Fig. 1) (*see* **Note 14**).

2. Put the leaves on a clean paper sheet. Remove the leaf midrib and cut the leaf lamina into 2 mm broad stripes with a clean razor blade (Fig. 2a). Lay the stripes on top of another and cut them into 2 mm pieces (Fig. 2b).

3. Transfer the leaf pieces into a petri dish containing about 10 mL of the enzyme solution. Use two leaves per 10 mL solution. Vacuum infiltrate the leaf pieces in a desiccator by applying and releasing the vacuum twice. Incubate under vacuum for 30 min (*see* **Note 15**). Leaf pieces should appear darker and slightly transparent after vacuum infiltration (Fig. 3a, b).

Fig. 1 Numbering of leaves of a four-week old *N. benthamiana* plant used for protoplast isolation

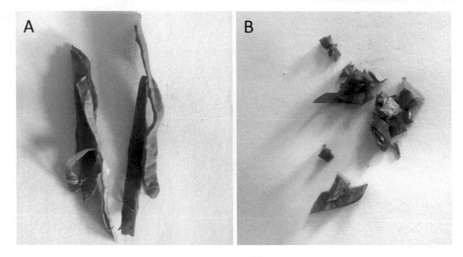

Fig. 2 Leaf sectioning for protoplast isolation from *N. benthamiana* plants. Preparation of 2 mm broad leaf stripes after removal of the midrib (**a**) and cutting the stripes into 2 mm pieces for subsequent infiltration of enzyme solution (**b**)

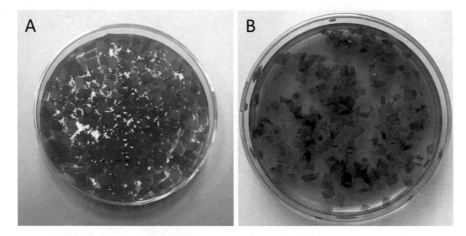

Fig. 3 Appearance of vacuum infiltrated leaf pieces from *N. benthamiana* for protoplast isolation. Shown are the leaf pieces in enzyme solution before (**a**) and after (**b**) vacuum application

4. Release vacuum and incubate the leaf pieces in the dark at room temperature for 4 h (*see* **Note 16**).

5. Carefully shake the petri dishes on a laboratory shaker for 30 min to release the protoplasts (*see* **Note 17**). Set the speed on the shaker in order that the solution will not be spilled.

6. Filtrate 2 × 4.5 mL of the protoplast suspension through a cell strainer into two 15 mL centrifuge tubes (4.5 mL in each tube). Proceed immediately with **step 7**.

7. Centrifuge the two tubes containing the protoplast suspension at 200 × *g* and 4 °C for 1 min and carefully remove the

supernatant. Resuspend the pellet of one tube in 2–3 mL buffer W5 and disperse the protoplasts by carefully swirling the centrifuge tube. Use this suspension to resuspend the pellet in the second centrifuge tube in the same careful manner as described above (*see* **Note 18**).

8. Incubate the protoplast suspension in the dark on ice for 40 min to sediment the protoplasts by gravity.

9. Remove the supernatant, carefully resuspend the pellet in 2–3 mL buffer W5, and repeat **step 8**.

10. Remove the supernatant and carefully resuspend the pellet in 2–3 mL MMG.

11. Count the protoplasts in a counting chamber (*see* **Note 19**).

12. Dilute the protoplast suspension with MMG to a final concentration of 100,000 protoplasts per mL.

3.3 Protoplast Transformation with PEG

1. Prepare microcentrifuge tubes (*see* **Note 20**) with the DNA solution (from Subheading 3.1, **step 6**) use 5–10 μg DNA per 10,000 protoplasts (*see* **Note 21**). That means pipetting of 10–20 μg of each plasmid (pE-SPYCE and pE-SPYNE with the respective inserts) into the microcentrifuge tube and add 200 μL of the protoplast suspension from Subheading 3.2, **step 12**.

2. Add 220 μL (i.e., the 1.1-fold volume of the protoplast suspension) of the PEG-solution 2 (PEG2) and mix by carefully shaking the tube. Incubate the DNA-PEG2-protoplast mix for 5–10 min at room temperature.

3. Add 880 μL buffer W5 (i.e., the 4.4-fold volume of the initial protoplast suspension) and carefully mix the suspension.

4. Centrifuge the mix for 1 min at 4 °C and $200 \times g$. Discard the supernatant (*see* **Note 22**) and carefully resuspend the protoplasts in the same volume as initially used (i.e., 200 μL) buffer WI.

5. Place the tubes horizontally and incubate them overnight in the dark at room temperature.

3.4 Microscopic Analysis

1. After 18 h, carefully resuspend the protoplasts and transfer them to a microscope slide or dish (*see* **Note 10**).

2. Check protoplasts for YFP fluorescence (excitation 514 nm and emission at 516–549 nm) and chlorophyll autofluorescence (excitation 514 nm and emission at 654–735 nm) using either an epifluorescence or a confocal laser scanning microscope (*see* Fig. 4) (*see* **Note 23**).

Fig. 4 Visualization of the interaction of the 'Candidatus Phytoplasma mali' effector ATP_00189 with the Malus x domestica transcription factor MdTCP25 using BiFC in N. benthamiana mesophyll protoplasts. Interaction of ATP_00189 and MdTCP25 leads to the occurrence of YFP fluorescence (**a**). Autofluorescence of chlorophyll in the chloroplasts (**b**). An overlay of both images (**a** and **b**) allows the localization of the YFP signal to the nucleus (indicated by the arrow) of the protoplast (**c**). Pictures were taken using the confocal laser scanning microscope LSM780 (Zeiss GmbH, Jena). Scale bars indicate 10 μm

4   Notes

1. A detailed description of subcloning of the phytoplasma effector and the potential interaction partner is not part of this protocol. For the applied cloning strategy used here *see* Subheading 3.1, **step 1** and **Note 10**.

2. For the enzyme solution mannitol, KCl and MES stock solutions should be prepared [22]: 0.8 M mannitol, 0.1 M KCl, 0.2 M 4-morpholineethanesulfonic acid (MES; pH 5.7). Prepare a fresh working solution as follows: Dilute the above-mentioned mannitol, KCL, and MES stock solutions to the final working dilution in water. Preheat this mix (not containing the enzymes, $CaCl_2$, and BSA) at 55 °C in a water bath for 10 min, then add cellulase R-10 and macerozyme R-10 to the warm solution and leave it at 55 °C for 10 min. Chill the solution on ice and add $CaCl_2$ and BSA. The quality of enzymes is very critical for the protoplast isolation [22]. We obtained the best results with enzymes from Yakult Pharmaceutical (Japan). Filter the solution through filter paper directly into the petri dishes resulting in about 10 mL per 5 cm dish. Alternatively, the solution can be filtered using a 0.45 μm syringe filter unit [22].

3. The use of round bottom centrifuge tubes is highly recommended to avoid the formation of a dense protoplast pellet that is difficult to resuspend.

4. For buffer W5 following stock solutions should be prepared [22]: 5 M NaCl, 1 M $CaCl_2$, 1.1 M KCl, 0.2 M MES (pH 5.7). Dilute the working dilution in water. Buffer W5 at its working concentration can be stored in aliquots at −20 °C or for shorter periods at 4 °C.

5. For buffer MMG following stock solutions should be prepared [22]: 0.8 M mannitol, 0.15 M MgCl$_2$, 0.2 M MES (pH 5.7). Dilute the stock solutions to the final working dilution in water.

6. Whether you need microcentrifuge or centrifuge tubes depends on how much DNA and protoplasts you need to combine for the transformation (*see* Subheading 3.3, **step 1**).

7. Prepare a fresh PEG2 solution. The PEG source for the PEG2 solution is very critical to achieve optimal transformation efficiency. Yoo et al. [22] found the product from Fluka (cat. no. 81240) to work best.

8. For buffer WI mannitol, KCl and MES stock solutions should be prepared [22]: 0.8 M mannitol, 0.1 M KCl, 0.2 M MES (pH 5.7) and diluted to the final working concentrations in water.

9. Alternatively 0.45 µm syringe filter units can be used [22].

10. When using an upright microscope avoid putting a coverslip without some spacers on the protoplasts as this could damage them. Alternatively, use a counting chamber for this task. In case of use of an inverse microscope, the use of a chambered coverslip or a coverslip bottom dish is recommended.

11. It is strongly recommended to subclone the effector and the interactor in both pE-SPYCE and pE-SPYNE BiFC plasmids. The plasmids can then be used in two different transformation combinations, i.e., pE-SPYCE-effector and pE-SPYNE-interactor or pE-SPYNE-effector and pE-SPYCE-interactor. Both vectors are Gateway®-compatible (Invitrogen, Life Technologies, Carlsbad, CA, USA) and the respective cloning strategy must be applied. The pE-SPYCE vector encodes the C-terminal YFP fragment and bicistronically expresses the subcloned insert (effector or interaction partner) together with this YFP fragment. The pE-SPYNE vector expresses the N-terminal fragment of the YFP fluorescent protein together with the respective insert. As an alternative a two-in-one cloning system can be used which allows a ratiometric BiFC (rBiFC) [23].

12. To avoid time constraints, the plasmid preparation should be performed at least the day before the actual protoplast isolation and transformation is planned. Plasmid DNA can be stored at −20 °C until further use. Be sure to have highly pure DNA, because salts and other contaminants from the buffers will strongly affect protoplast survival.

13. Correct drying time depends on how thoroughly the supernatant was removed. Check drying by eye. Do not over-dry the pellet, otherwise resolving the DNA will be cumbersome.

DNA can be also dissolved at 50 °C for 10 min. DNA concentration after dissolving should be 2–5 μg/μL. The DNA solution can be stored at 4 °C. Repeated freeze and thaw cycles should be avoided.

14. Count the leaves as described in Fig. 1. Use only leaves that are well developed and unscathed.

15. During applying the vacuum, bubbles should be visible. Infiltrated leaf pieces develop a darker color, become slightly transparent, and ideally sink to the ground of the petri dish (*see* Fig. 3a, b).

16. Room temperature is considered 22 °C.

17. After 4 h of incubation and careful shaking of the petri dish, the enzyme solution should become greenish and some of the leaf pieces transparent.

18. To facilitate protoplast dispersal a 5 mL serological pipette or a 5000 μL pipette can be used, but protoplasts should be dispensed with extreme caution to avoid their disruption. For all pipetting steps of protoplast-suspension pipette tips should be cut to avoid mechanical stress.

19. Count only undisrupted protoplasts.

20. Use a centrifuge tube instead of a micro-centrifuge tube in case the necessary protoplast-suspension volume exceeds 300 μL due to low protoplast counts. Adapt the subsequent buffer amounts accordingly as mentioned in the protocol.

21. It is recommended to determine the exact ratio between DNA and protoplasts in advance to achieve high transformation efficiency leading to optimal results.

22. Remove the supernatant as completely as possible. A minor protoplast loss is not problematic, but it is important to remove the PEG containing solution, since it harms the protoplasts.

23. An epifluorescence microscope with a proper filter-set for YFP-fluorescence will work for screening protoplasts showing YFP-signal, but better imaging will be obtained using a confocal laser scanning microscope.

References

1. Hogenhout S, van der Hoorn R, Terauchi R et al (2009) Emerging concepts in effector biology of plant-associated organisms. Mol Plant-Microbe Interact 22(2):115–122. https://doi.org/10.1094/MPMI-22-2-0115

2. Win J, Chaparro-Garcia A, Belhay K et al (2012) Effector biology of plant-associated organisms: concepts and perspectives. Cold Spring Harb Symp Quant Biol 77:1–13.

https://doi.org/10.1101/sqb.2012.77.015933

3. Zhou J, Chai J (2007) Plant pathogenic bacterial type III effectors subdue host responses. Curr Opin Microbiol 11:1–7. https://doi.org/10.1016/j.mib.2008.02.004

4. Sugio A, MacLean AM, Kingdom HN et al (2011) Diverse targets of phytoplasma effectors: from plant development to defense

against insects. Annu Rev Phytopathol 49:175–195. https://doi.org/10.1146/annurev-phyto-072910-095323

5. Hoshi A, Oshima K, Kakizawa S et al (2009) A unique virulence factor for proliferation and dwarfism in plants identified from a phytopathogenic bacterium. Proc Natl Acad Sci U S A 106(15):6416–6421. https://doi.org/10.1073/pnas.0813038106

6. Bai X, Correa VR, Toruño TY et al (2009) AY-WB phytoplasma secretes a protein that targets plant cell nuclei. Mol Plant-Microbe Interact 22(1):18–30. https://doi.org/10.1094/MPMI-22-1-0018

7. Sugio A, Kingdom HN, MacLean AM et al (2011) Phytoplasma protein effector SAP11 enhances insect vector reproduction by manipulating plant development and defense hormone biosynthesis. Proc Natl Acad Sci U S A 108(48):E1254–E1263. https://doi.org/10.1073/pnas.1105664108

8. Sugawara K, Honma Y, Komatsu K et al (2013) The alteration of plant morphology by small peptides released from the proteolytic processing of the bacterial peptide TENGU. Plant Physiol 162(4):2005–2014. https://doi.org/10.1104/pp.113.218586

9. MacLean AM, Orlovskis Z, Kowitwanich K et al (2014) Phytoplasma effector SAP54 hijacks plant reproduction by degrading MADS-box proteins and promotes insect colonization in a RAD23-dependent manner. PLoS Biol 12(4):e1001835. https://doi.org/10.1371/journal.pbio.1001835

10. Minato N, Himeno M, Hoshi A et al (2014) The phytoplasmal virulence factor TENGU causes plant sterility by downregulating of the jasmonic acid and auxin pathways. Sci Rep 4 (7399):1–7. https://doi.org/10.1038/srep07399

11. Siewert C, Luge T, Duduk B et al (2014) Analysis of expressed genes of the bacterium 'Candidatus phytoplasma mali' highlights key features of virulence and metabolism. PLoS One 9(4). https://doi.org/10.1371/journal.pone.0094391

12. Sugio A, MacLean AM, Hogenhout SA (2014) The small phytoplasma virulence effector SAP11 contains distinct domains required for nuclear targeting and CIN-TCP binding and destabilization. New Phytol 202(3):838–848. https://doi.org/10.1111/nph.12721

13. Janik K, Mithöfer A, Raffeiner M et al (2017) An effector of apple proliferation phytoplasma targets TCP transcription factors—a generalized virulence strategy of phytoplasma? Mol Plant Pathol 18(3):321–473. https://doi.org/10.1111/mpp.12409

14. Fields S, Song O (1989) A novel genetic system to detect protein-protein interactions. Nature 340(6230):245–246. https://doi.org/10.1038/340245a0

15. Janik K, Schlink K (2017) Unravelling the function of a bacterial effector from a non-cultivable plant pathogen using a Yeast Two-hybrid S screen. J Vis Exp 119. https://doi.org/10.3791/55150

16. Golemis EA, Serebriiskii I, Law SF (1999) The yeast two-hybrid system: criteria for detecting physiologically significant protein-protein interactions. Curr Issues Mol Biol 1 (1–2):31–45

17. Serebriiskii I, Estojak J, Berman M et al (2000) Approaches to detecting false positives in Yeast Two-Hybrid systems. BioTechniques 28 (2):328–336

18. Bruckner A, Polge C, Lentze N et al (2009) Yeast two-hybrid, a powerful tool for systems biology. Int J Mol Sci 10(6):2763–2788. https://doi.org/10.3390/ijms10062763

19. Kerppola T (2006) Design and implementation of bimolecular fluorescence complementation (BiFC) assays for the visualization of protein interactions in living cells. Nat Protoc 1(3):1278–1286. https://doi.org/10.1038/nprot.2006.201

20. Kudla J, Bock R (2016) Lighting the way to protein-protein interactions: recommendations on best practices for bimolecular fluorescence complementation analyses. Plant Cell 28 (5):1002–1008. https://doi.org/10.1105/tpc.16.00043

21. Nagai T, Ibata K, Park E et al (2002) A variant of yellow fluorescent protein with fast and efficient maturation for cell-biological applications. Nat Biotechnol 20(1):87–90. https://doi.org/10.1038/nbt0102-87

22. Yoo S, Cho Y, Sheen J (2007) Arabidopsis mesophyll protoplasts: a versatile cell system for transient gene expression analysis. Nat Protoc 2(7):1565–1572. https://doi.org/10.1038/nprot.2007.199

23. Grefen C, Blatt MR (2012) A 2in1 cloning system enables ratiometric bimolecular fluorescence complementation (rBiFC). Biotech 53 (5):311–314. https://doi.org/10.2144/000113941

Chapter 25

Collection, Identification, and Statistical Analysis of Volatile Organic Compound Patterns Emitted by Phytoplasma Infected Plants

Jürgen Gross, Jannicke Gallinger, and Margit Rid

Abstract

In this chapter, we give an introduction to innovative attempts for the collection, identification, and statistical analysis of volatile organic compound (VOC) patterns emitted by phytoplasma-infected plants compared to healthy plants by the use of state-of-the-art techniques. This encompasses headspace-sampling techniques, gas chromatography coupled with mass spectrometry, and identification of VOC patterns by the "Automated Mass Spectral Deconvolution and Identification System" (AMDIS) followed by appropriate statistical analysis.

Key words Headspace sampling, Thermodesorption, Gas chromatograph/mass spectrometer, AMDIS, Compositional dataset, Principal component analysis, Non-metric multidimensional scaling

1 Introduction

The first two papers showing that phytoplasma vectoring insects like psyllids (Hemiptera: Psylloidea) and planthoppers (Hemiptera: Fulgoroidea) use chemical cues for orientation and host identification were published already 13 years ago [1, 2]. In the subsequent years further studies showed that intraspecific communication in these taxa is mainly mediated by acoustical signals [3, 4], while volatile organic compounds (VOCs) emitted by plants play an important role for interspecific communication [5–8]. These so-called infochemicals (or semiochemicals) can influence target organisms in general (e.g., green leaf volatiles), and may be specific for a particular group of insects (e.g., many kairomones), or even species specific (e.g., pheromones). For instance, the psyllid species *Cacopsylla picta* and *C. melanoneura* are able to distinguish their

Electronic supplementary material: The online version of this chapter (https://doi.org/10.1007/978-1-4939-8837-2_25) contains supplementary material, which is available to authorized users.

Rita Musetti and Laura Pagliari (eds.), *Phytoplasmas: Methods and Protocols*, Methods in Molecular Biology, vol. 1875, https://doi.org/10.1007/978-1-4939-8837-2_25, © Springer Science+Business Media, LLC, part of Springer Nature 2019

specific reproduction and overwintering hosts by means of chemical signals produced by their host plants [2]. Expanded knowledge on the chemically mediated multitrophic interactions of psyllids and planthoppers with their host plants and vectored phytoplasmas may help to design new and innovative methods for their control in integrated production [9–11].

In this chapter, we give an introduction to innovative attempts for the collection, identification, and statistical analysis of VOC patterns emitted by phytoplasma infected plants compared to healthy plants by the use of state-of-the-art techniques.

2   Materials

2.1   Volatile Collection

1. Headspace sampling device, which ensures an exact collection of relative quantities of VOCs in headspace samples. Examples are:

   - Easy-VOC grab sampler for sampling small volumes (50–500 mL; Markes International, Frankfurt, Germany).

   - Single tube pumps.

   - Programmable multi-tube sequential samplers (STS 25, PerkinElmer or MTS 32, Markes International).

   - Micro-chamber/Thermal extractor (Markes International; for fruits, nuts or other cropped small plant parts).

   - Self-constructed quantitative sampling devices: 5-channel stationary system according to [12] or a newly developed portable 6-channel headspace-sampling device with Coriolis mass flow measure and regulation, enabling an exact quantitative measurement of VOCs, which can be operated with a car battery in field experiments (Fig. 1).

Fig. 1 Portable 6-channel headspace sampling device for collecting VOCs on thermal desorption tubes in the field (left photograph). Detailed view of the device (right photograph)

2. Phytoplasma-infected plants and healthy control plants.

3. Bags made of polyethylene terephthalate (PET) (e.g., Top-pits®, Melitta, Minden, Germany).

4. Washing bottles filled with activated charcoal (granulated 4–8 mm) or inline clean air filter cartridges (e.g., Sigma Scientific LLC).

5. Teflon® tubes for connecting of the individual parts.

6. Sample tubes filled with sorbent (e.g., with Tenax® TA60/80; appropriate for many common plant volatiles) with Teflon®--coated brass compression caps (e.g., Swagelok, PerkinElmer).

7. Ethanol 70%.

2.2 Identification

1. AMDIS program (*see* **Note 1**).

2. NIST/EPA/NIH Mass Spectral Library (must be purchased).

3. NIST Mass Spectral Search Program.

3 Methods

3.1 Volatile Collection

Quantitative exact headspace sampling of VOCs emitted by plants can be conducted using the 6-channel headspace-sampling device (Fig. 1) or another mass flow controlled pumping system following this protocol:

1. Carefully wrap parts of or whole target plants in oven plastic bags made of PET, in greenhouse or field (*see* **Note 2**).

2. Purify ambient air by passing through washing bottles filled with activated charcoal or filter cartridges, and stream with defined air flow through the bag until having reached a specified final volume.

3. Collect VOCs from headspace samples on pre-packed sample tubes.

4. Close the sample tubes air-tight after collecting odors by headspace sampling device (*see* **Note 3**).

5. Rinse the washing bottles and all Teflon® tubes with 70% ethanol after each trial and then heat them in an oven at 230 °C for at least 2 h.

3.2 Chemical Analysis

The collected samples on sample tubes should be analyzed using a thermal desorption device (TD) connected to a GC-MS (gas chromatograph—mass spectrometer) instrument for verification of emitted volatiles. For details *see* [13].

3.3 Identification

3.3.1 Identification Criteria

The volatile compounds could be identified by comparing the characteristic ion fragmentation pattern (mass spectrum) with data from mass spectrum libraries (e.g., NIST Mass Spectral Library, National Institute of Standards and Technology) [14]. To verify the identification, it is recommended to additionally compare with either the retention times (RT) of standard compounds obtained by the equal analytical system and protocol or the retention indices (RI) with RI from literature obtained by similar analytical systems. The identification of components only via mass spectra or only via RI may lead to false identifications. Mass spectra may be very similar and RI values may be overlapping. The peak areas of single compounds can be integrated and relative proportions can be calculated [13]. Coupling of mass spectra with retention indices using free AMDIS software (Automated Mass Spectral Deconvolution and Identification System) from the National Institute of Standards and Technology (NIST), U.S. Department of Commerce, is very helpful for the identification and quantification of VOC patterns obtained by a GC-MS. This saves time and labor by automatic subtraction of noise (by deconvolution) and calculation of net match factors based on mass spectra and RI. Proper chromatograms from clean samples without contaminations and noise (as low as possible) enable a fast and less complicated analysis procedure.

3.3.2 Workflow with AMDIS

1. *Set your alkane run into the program*: Open the chromatogram of the alkane standard mixture (*File → Open...*) (*see* **Note 4**).

2. Open the "Analyze GC/MS Data" window (*Analyze → Analyze GC/MS Data*). In this window set the desired type of analysis, check and select the correct calibration files and libraries.

3. Select *RI Calibration/Performance* as type of analysis. The *Calib/Stds. Lib, Intern. Std. Lib* and *RI Calib. Data* buttons invoke the *Analysis Settings* window, where library and calibration files can be selected (*Select New...*).

4. Set the *Calibration/Standards Library*: select the ONSITE. CSL from the AMDIS folder as calibration library. This calibration standard library file contains information about alkanes from C_5 to C_{30}.

5. Set the *RI Calibration Data*: Open the *Select New...* dialog in the *Analysis Settings* for RI Calibration Data. Give an appropriate location and name for the .CAL file, which will be created.

6. *Run* the analysis.

7. Check the performance: (*see* **Note 5**) Ensure that every expected alkane is detected and contains the right RI value. The targets can also be checked by inspecting the .CAL file, which can be found in the selected folder (*see* **step 5**).

The .CAL file can be opened in a text editor (for example Notepad from Windows). It contains the RT, RI, the net match value and the signal-to-noise ratio of every alkane detected in the analysis as plain text. If necessary, correct the . CAL file manually in a text editor.

8. *Create your target library:* By creating your target library you define all components AMDIS will cast for in every sample in the subsequent analysis.

9. Open the chromatogram of the respective sample.

10. Change the type of analysis to *Use Retention Index Data* in the "Analyze GC/MS Data" window. This is an important step after setting the alkane run; otherwise, AMDIS will keep running *RI Calibration/Performances* and overwrites the created calibration file.

11. Selecting *Use the Retention Index Data* will enable choosing a target library and a calibration data file.

12. Select the correct .CAL file and *Run* the analysis.

13. Choose peaks that you would like to include in your analysis (*see* **Note 6**).

14. If one peak is selected, click *Analyze → Go to NIST MS Program,* which will search through NIST MS Library and included user libraries (*see* **Note 7**).

15. Create your target library file: *Library → Build One Library → Files → Create New Library.* Give an appropriate file location and name and save your target library as *Simple MS File* (.MSP). The .MSP file format is required to import mass spectra from NIST MS Search Program.

16. Insert mass spectra into your target library: Import the selected peak as reference in your target library: *Library → Build One Library → Add:* This enables you to include the so-called "known unknown" components in your analysis, by saving the mass spectra (together with RT and RI values) of components in your target library, even without a reliable identification. To import a mass spectrum from the NIST MS or user libraries right click on the component in the NIST MS Search Program choose *Export Selected* and select a target library, in which the component should be included. Ensure that the mass spectrum will be appended to the .MSP file; otherwise, the created target library will be overwritten.

17. Check the RI for included components from NIST MS Library and correct or add RI values. The information about the components can be edited in the library window. The NIST chemistry WebBook provides information on RI values for different column types and temperature programs (http://webbook. nist.gov/chemistry/).

18. It is recommended to go randomly through some of your samples to select all components for your target library.

19. *Set Identification and Deconvolution Settings for subsequent analysis*: Configuration of the identification and deconvolution parameters is always a trade-off between reducing the number of false positives to a minimum and the proper identification and detection of the components of interest (especially for small peaks). Therefore settings should be adjusted precisely prior to analysis (*see* **Note 8**). For the final analysis it is important to always use the same alkane run (.CAL), target library (.MSP), and the same settings (.INI), for every sample.

20. Set proper *Analyze Settings → Identification* (*see* **Note 9**).

21. *Set proper Analyze Settings → Deconvolution* (*see* **Note 10**).

22. *Export your data (File → Generate Report)*: This generates a report from the open chromatogram (after the run). It is possible to append reports of further runs to the same report file. The file is saved as .TXT and can be opened with Microsoft Excel (*see* **Note 11**).

23. For most calculations extract the following columns from the report: FileName, Name, and Area. AMDIS will only report information about targets, which are detected in the chromatogram. Therefore it is necessary to add a row for each target, which was not detected. Every sample must contain a row for each target. If components were not present in one of the samples, add a row to your data table with an area value of 0. To transpose the data and for adding zero values either use the pivot-table function in Microsoft excel or the provided *R*-script in the online supporting material.

3.4 Statistical Analysis

Different methods can be used to visualize and calculate relationships between compositions of VOCs of infected and uninfected plants. Multivariate methods try to project the differences (resp. similarities) unbiased as possible in a multidimensional space. For every dataset figure out, which analysis is valid and addresses your research question. Here, a workflow with some possible discriminatory methods is presented briefly. Further methods, explanations, and R-scripts for multivariate statistical analyses can be found in [15, 16].

1. Generate compositional dataset (*see* **Note 12**).

2. Transform your data (*see* **Note 13**).

3. Choose a proper ordination method (*see* **Notes 14** and **15**).

4. Create a biplot, for visualization of whole odor bouquets (*see* **Note 16**).

5. Verify (dis)similarity by statistical tests (*see* **Note 17**).

6. Identify single (or groups of) components responsible for separation (*see* **Note 18**).

4   Notes

1. Download the "*Full Program Installation*" item from http://chemdata.nist.gov/dokuwiki/doku.php?id=chemdata:downloads:start#amdis. Unzip the folder, run the .EXE file, and follow installation wizard setup.

2. Strictly avoid any injuries of leaves, fruits, or shoots to prevent an emission of green leaf volatiles (GLV) due to damage of the plants.

3. Samples should be stored for a maximum of 1 week before getting thermodesorbed and analyzed by a thermodesorption device coupled to a gas chromatograph connected to a mass spectrometer (TD-GC-MS).

4. Ideally, the alkane run and the chemical analysis of the components are done on the same GC-MS system (device, separation column, temperature program, etc.).

5. AMDIS labels all detected peaks with an arrow and identified targets with a "T" on the top of the chromatogram. Furthermore, an "Information List" is provided in an extra window.

6. Right click into the chromatogram area gives the opportunity to show the selected component as a black line in the chromatogram (*Show Component on Chromatogram*).

7. Alternative: select the component in the "Information List," right click will also give the opportunity to *NIST Library* → *Go to NIST MS Program*.

8. Settings can be saved as .INI file. Nevertheless, we recommend recording your analysis settings in your documentation. AMDIS uses the default ONSITE.INI automatically for analysis. Other .INI files cannot be loaded manually. Therefore, restoring configurations from saved .INI files is cumbersome. Additionally, the ONSITE.INI will be overwritten after changes were made. Saved .INI files are needed to generate analysis in batch mode with your settings.

9. A high *Minimum match factor* will reduce the number of false positive identified components. For "clean" samples it is best practice to set the match factor about 80. This value defines the required percentage of similarity between the spectra of the sample with the spectra saved in the target library (including additional information as RI). Components, which are less similar, will not be identified as targets. However, it is not always possible to prevent contaminations and samples are not as clean as desired. Lowering the minimum match factor is a possibility to nevertheless identify components, but it may increase the number of false identifications. In this case we recommend a manual check of identification quality. If the RI

data is used as type of analysis you can specify the *RI window*. The RI window can be scaled linearly, resulting in increasing size with increasing RI value. The *net match* value for components with RI values outside of this window is reduced. The applied penalty value, by which the net match is reduced, depends on the selected penalty level and the distance of the RI value from the library RI value for this component. For more details please consult the AMDIS manual.

10. In the *Deconv.* tab it is possible to set the expected width of components (*Component width*), omit single m/z values from your analysis, and further specify the parameters used by the algorithm for deconvolution. You can select how AMDIS should handle interfering Ions (*Adjacent peak subtraction*), how peaks should be separated (*Resolution*), how peak width and noise are considered (*Sensitivity*), and how important an equal peak shape (*Shape requirements*) is. For more detailed information about the deconvolution algorithm and settings the AMDIS manual is recommended.

11. It is recommended to compare the reports manually with chromatograms in AMDIS to ensure that the targets are identified correctly. Especially, when desired mass spectra are very similar, the automatic identification can cause false targeting.

12. Although it is possible that phytoplasmas promote their host plants to produce new compounds, it is more likely that the quantities of some compounds change due to phytoplasma infection [12]. Furthermore, phytoplasmas rely on the spread by insect vectors and in some systems the odor of infected plants attracts their vectors [5, 6]. In many systems the respective ratios of the compounds of an odor bouquet act as biological active signal for insects [17]. Thus, the influence of phytoplasma infection on the composition of the entire odor bouquet is important. Therefore, the generation of a compositional dataset is vital, especially when an internal standard cannot be applied.

13. Calculating relative proportions of compounds causes a constant-sum constraint and prevents values of individual variables to vary independently. An appropriate method to un-constrain data is the centered log-ratio transformation [18]. Furthermore, because PCA uses the Euclidian distance between samples for ordination, care should be taken if the data contain many zero values. To prevent that the absence of compounds is interpreted as similarity, data can be transformed by the method of Chord or Hellinger [19].

14. A classical multivariate method is the principal component analysis (PCA). The PCA provides a simple way of low-dimensional projection of complex compositional data. A

similarity matrix is generated by calculating the Pearson correlation coefficients between the components. Therefore, the PCA is a parametric method and the data should be normalized prior to analysis.

15. Another opportunity is the non-metric multidimensional scaling (NMDS), which is a more flexible procedure, because different ordination methods can be applied. NMDS is a robust nonparametric method, in which ranks of differences between samples are projected to a low-dimensional space. For chemical compositions it is common to use the Bray-Curtis-Dissimilarity matrix [20].

16. For interpretation of PCA commonly the two principal components (PCs), which explain most of the variance, are visualized in a biplot. The NMDS plots visualize the dissimilarity (Bray-Curtis) and elucidate patterns in compositional data sets by finding grouping or gradients of the data. As an example compare with [21]. The stress-value describes the goodness of fit of the projection of a NMDS plot. A low stress-value (<0.3) provides a good representation of the difference between the distances in the reduced dimension compared to the complete multidimensional space.

17. For example, permutational multivariate analysis of variance (PERMANOVA) is a common nonparametric test for discrimination of groups. PERMANOVA can still be calculated if there are less replicates than variables [22], because the test statistic is estimated for a desired number of sample reorderings (permutations). As the data must not fulfill further assumptions to distribution or dependency, the variability within groups can influence the analysis. Therefore, a permutational analysis of multivariate dispersion (PERMDISP) [23] should be applied to elucidate in case statistical significant PERMANOVA results rely on differences in group dispersion and not only on separation by location (attributes) of groups. As an example compare with [24]. PERMANOVA together with PERMDISP help to interpret ordinations, such as PCA and NMDS.

18. The calculated loadings of a PCA represent the correlation of variables (volatile compounds) with the associated PC. Therefore, the amount of a compound with a high loading will increase into the direction of the PC and compounds, which are important for the separation of groups, can be identified. Comparable to the loadings of a PCA, compounds, which influence the grouping in the NMDS, can be identified by fitting vectors of variables. For the identification of the gradients' direction these vectors can be projected as arrows onto the ordination. Additionally, the distance between the origin and the endpoints of arrows indicates the strength of the gradient.

Acknowledgment

The authors are grateful to Eva Gross (Schriesheim, Germany) for language editing.

References

1. Soroker V, Talebaev S, Harari AR, Wesley SD (2004) The role of chemical cues in host and mate location in the pear psylla *Cacopsylla bidens* (Homoptera: Psyllidae). J Insect Behav 17(5):613–626. https://doi.org/10.1023/B:JOIR.0000042544.35561.1c

2. Gross J, Mekonen N (2005) Plant odours influence the host finding behaviour of apple psyllids (*Cacopsylla picta*; *C. melanoneura*). IOBC WPRS Bull 28(7):351–355

3. Eben A, Mühlethaler R, Gross J, Hoch H (2015) First evidence of acoustic communication in the pear psyllid *Cacopsylla pyri* L. (Hemiptera: Psyllidae). J Pest Sci 88 (1):87–95. https://doi.org/10.1007/s10340-014-0588-0

4. Lubanga U, Guédot C, Percy D, Steinbauer M (2014) Semiochemical and vibrational cues and signals mediating mate finding and courtship in Psylloidea (Hemiptera): a synthesis. Insects 5 (3):577

5. Mayer CJ, Vilcinskas A, Gross J (2008a) Pathogen-induced release of plant allomone manipulates vector insect behavior. J Chem Ecol 34(12):1518–1522. https://doi.org/10.1007/s10886-008-9564-6

6. Mayer CJ, Vilcinskas A, Gross J (2008b) Phytopathogen lures its insect vector by altering host plant odor. J Chem Ecol 34 (8):1045–1049. https://doi.org/10.1007/s10886-008-9516-1

7. Mayer CJ, Jarausch B, Jarausch W et al (2009) *Cacopsylla melanoneura* has no relevance as vector of apple proliferation in Germany. Phytopathology 99(6):729–738. https://doi.org/10.1094/Phyto-99-6-0729

8. Mayer CJ, Vilcinskas A, Gross J (2011) Chemically mediated multitrophic interactions in a plant–insect vector-phytoplasma system compared with a partially nonvector species. Agr Forest Entomol 13(1):25–35. https://doi.org/10.1111/j.1461-9563.2010.00495.x

9. Gross J (2016) Chemical communication between phytopathogens, their host plants and vector insects and eavesdropping by natural enemies. Front Ecol Evol 4:104. https://doi.org/10.3389/fevo.2016.00104

10. Gross J, Gündermann G (2016) Principles of IPM in cultivated crops and implementation of innovative strategies for sustainable plant protection. In: Horowitz A, Ishaaya I (eds) Advances in insect control and resistance management. Springer, Cham, pp 9–26. https://doi.org/10.1007/978-3-319-31800-4_2

11. Eben A, Gross J (2013) Innovative control of psyllid vectors of European fruit tree phytoplasmas. Phytopathogenic Mollicutes 3 (1):37–39. https://doi.org/10.5958/j.2249-4677.3.1.008

12. Rid M, Mesca C, Ayasse M, Gross J (2016) Apple proliferation phytoplasma influences the pattern of plant volatiles emitted depending on pathogen virulence. Front Ecol Evol 3:152. https://doi.org/10.3389/fevo.2015.00152

13. Weintraub P, Gross J (2013) Capturing insect vectors of phytoplasmas. In: Dickinson M, Hodgetts J (eds) Phytoplasma, Methods in molecular biology (methods and protocols), vol 938. Humana Press, Totwa, NJ, pp 61–72. https://doi.org/10.1007/978-1-62703-089-2_6

14. Stein SE (2010) NIST chemistry WebBook, NIST Standard Reference Database http://webbook.nist.gov/chemistry

15. Brückner A, Heethoff M (2017) A chemo-ecologists' practical guide to compositional data analysis. Chemoecology 27(1):33–46. https://doi.org/10.1007/s00049-016-0227-8

16. Paliy O, Shankar V (2016) Application of multivariate statistical techniques in microbial ecology. Mol Ecol 25(5):1032–1057. https://doi.org/10.1111/mec.13536

17. Bruce TJ, Pickett JA (2011) Perception of plant volatile blends by herbivorous insects—finding the right mix. Phytochemistry 72 (13):1605–1611. https://doi.org/10.1016/j.phytochem.2011.04.011

18. Kucera M, Malmgren BA (1998) Logratio transformation of compositional data: a resolution of the constant sum constraint. Mar Micropaleontol 34(1):117–120. https://doi.org/10.1016/S0377-8398(97)00047-9

19. Legendre P, Gallagher ED (2001) Ecologically meaningful transformations for ordination of species data. Oecologia 129(2):271–280. https://doi.org/10.1007/s004420100716

20. Bray JR, Curtis JT (1957) An ordination of the upland forest communities of southern wisconsin. Ecol Monogr 27(4):325–349. https://doi.org/10.2307/1942268

21. Gallinger J, Gross J (2018) Unraveling the host plant alternation of *Cacopsylla pruni* – adults but not nymphs can survive on conifers due to phloem/xylem composition. Front Plant Sci 9:484. https://doi.org/10.3389/fpls.2018.00484

22. Anderson MJ (2001) A new method for non-parametric multivariate analysis of variance. Austral Ecol 26:32–46. https://doi.org/10.1111/j.1442-9993.2001.01070.pp.x

23. Anderson MJ (2006) Distance-based tests for homogeneity of multivariate dispersions. Biometrics 62(1):245–253. https://doi.org/10.1111/j.1541-0420.2005.00440.x

24. Rid M, Markheiser A, Hoffmann C, Gross J. (2018). Waxy bloom on grape berry surface is one important factor for oviposition of European grapevine moths. J Pest Sci 91 (4):1225–1239. https://doi.org/10.1007/s10340-018-0988-7

Quantification of Phytohormones by HPLC-MS/MS Including Phytoplasma-Infected Plants

Marilia Almeida-Trapp and Axel Mithöfer

Abstract

There is strong evidence that phytohormones such as abscisic acid, auxin, salicylic acid, and jasmonates might play a role in defense of the host plants during phytoplasma infections. However, these compounds are usually present at low concentration in complex matrixes, requiring a sensitive and selective method to analyze and quantify them. Here, we present a HPLC-MS/MS method to quantify phytohormones in different infected and noninfected plant tissues.

Key words Phytohormones, Mass spectrometry, Quantification, HPLC-MS, Jasmonates, Salicylic acid, Abscisic acid, Auxin, OPDA, Jasmonic acid

1 Introduction

Phytohormones represent a diverse group of low molecular weight compounds that are involved in different cellular processes in plants. Among many other functions they can regulate developmental processes (e.g., plant growth, tissue differentiation) but can also be involved in plant communication, or act as signal molecules in order to mediate and coordinate responses to biotic and abiotic stresses [1, 2].

Although the general changes in plant metabolism during phytoplasma infection are still not very well understood, changes in the level of phytohormones have already been reported during phytoplasma infection for some plants. For example, higher levels of abscisic acid (ABA), auxin (IAA), and salicylic acid (SA) were observed in phytoplasma-infected plants [3–5]. In more detail, level of jasmonates seems to decrease during the symptomatic phase of the infection [4, 5] and increases again in the recovering plants [5]. Recently, Paolacci and coworkers have shown that during development of symptoms in "bois noir" disease after infection in grapevine (*Vitis vinifera*) the initiation of SA-mediated signaling takes place while at the same time a decrease of JA-mediated

Rita Musetti and Laura Pagliari (eds.), *Phytoplasmas: Methods and Protocols*, Methods in Molecular Biology, vol. 1875, https://doi.org/10.1007/978-1-4939-8837-2_26, © Springer Science+Business Media, LLC, part of Springer Nature 2019

defense is detected [6]. However, JA-mediated pathways counter-act salicylate-triggered responses while the plants recover [6]. These examples show that very likely it is important to understand the role of phytohormones in order to understand the plants' responses to phytoplasma infections.

One of the main issues in measuring phytohormones present in plant tissues is that they are usually present at low concentration (nM to μM) in very complex matrixes. Therefore, selective and sensitive methods for a precise and reliable quantification of the phytohormones are necessary. Here, we present a straightforward protocol for analysis of six stress-related signaling compounds from four phytohormone classes (indole-3-acetic acid, abscisic acid, salicylic acid, and the jasmonates: 12-oxophytodienoic acid, jasmonic acid, and jasmonic acid isoleucine conjugate) in plant tissues using a HPLC-MS/MS method [7]. Both high sensitivity and specificity are conferred to this method due to the Single Reaction Monitoring experiments (SRM), which can specifically select precursor and fragment ions related to each phytohormone [8]. After optimization and adjustment to the particular plant and tissue, this method can be employed in phytoplasma-infected as well as in non-infected samples.

2   Materials

- Equipment: tube rotator for 1.5 mL tubes, cooled centrifuge, high-performance liquid chromatograph (HPLC), mass spectrometer (able to perform MS/MS experiments).

- Solvents and additives: acetonitrile (HPLC), methanol (MeOH, HPLC grade), di-deionized water, formic acid (HPLC grade).

- High purity standards: indole-3-acetic acid (IAA), (+)-cis, trans-abscisic acid (ABA), salicylic acid (SA), 12-oxophytodienoic acid (OPDA), jasmonic acid (JA), jasmonic acid isoleucine conjugate (JA-Ile). Deuterated standards: $[^2H_5]$ indole-3-acetic acid (d5-IAA), $[^2H_4]$ salicylic acid (d4-SA), $[^2H_6]$ (+)-cis, trans-abscisic acid (d6-ABA), $[^2H_6]$ jasmonic acid (d6-JA).

2.1 Preparing Original Standard and Internal Standard Stock Solutions

2.1.1 Original Standards Stock Solutions

The original standard stock solution will be used to prepare the working solutions (*see* Subheading 2.2).

1. Weight 10 mg of each standard (ABA, IAA, JA, JA-Ile, OPDA, and SA).

2. Dissolve each of these compounds in 10 mL of MeOH (do it in a 15 mL falcon tube).

3. Make 1.0 mL (1 mg) aliquots in 1.5 mL tubes.

4. Dry all the tubes in a speedvac and keep the tubes at −80 °C.

5. To prepare the stock solution (1.0 mg/mL) add 1.0 mL of acetonitrile into one tube containing each original standard and vortex it to ensure that the compound was solubilized. Keep the solutions at −80 °C.

2.1.2 Internal Standards Stock Solutions

The internal standard stock solutions will be used to prepare the internal standards working solutions (*see* Subheading 2.2). The internal standards (d6-ABA, d5-IAA, d6-JA, and d4-SA) are sold in small amounts or in solutions. Avoid weighting them. Dissolve and/or dilute each internal standard in acetonitrile to a final concentration of 100 µg/mL. Keep them at −80 °C.

2.2 Preparing Original Standard and Internal Standard Working Solutions

2.2.1 Original Standard Working Solution

The *original standard working solution (OSWS)* will be used to prepare the calibration curves (*see* Subheading 2.5), and must be kept at −20 °C. In order to prepare each of them, mix (in 1.5 mL tubes) the volumes of original standard stock solutions (Subheading 2.1.1) and MeOH:water (7:3) described in Table 1.

2.2.2 Internal Standard Working Solution

The *internal standard working solutions (ISWS)* will be used to prepare extraction solutions (*see* Subheading 2.4). They also must be kept in −20 °C. For d-ABA and d-JA, take 100 µL of stock solution (Subheading 2.1.2) and add 800 µL of MeOH:Water (7:3). For d-IAA and d-SA, take 200 µL of stock solutions (Subheading 2.1.2) and add 900 µL of MeOH:Water (7:3).

2.3 Preparing Solution for Optimizations

- *Ionization and fragmentation*: Use original standard working solution (Subheading 2.2.1) and internal standard working solution (Subheading 2.2.2) to prepare a solution containing 1 µg/mL of each standard (*see* **Note 1**).

- *Limit of detection and quantification*: Prepare the testing solutions adding 1 µL of each Original Standard Working Solution (Subheading 2.2.1) to 994 µL of MeOH:Water (7:3). Make serial dilutions (from TP01 to TP5) to generate the Test Points described in Table 2.

2.4 Preparing Extraction Solution

The extraction solution contains all the internal standards (d6-ABA, d5-IAA, d6-JA, and d4-SA), and will be used to prepare the calibration curve in matrix (Subheading 3.6.2) as well as to extract the samples (Subheading 3.5).

The extraction solution might be freshly prepared.

For each sample 1 mL of extraction solution is needed, which is prepared adding 1 µL of each Internal Standard Working Solution (Subheading 2.2, ISWS) to 996 µL of MeOH:Water (7:3).

Table 1
Preparation of standard working solutions

	ABA	IAA	JA	JA-Ile	OPDA	SA
Volume of stock solution (μL)	80	10	40	2	40	200
Volume MeOH:Water (7:3) (μL)	920	960	960	998	960	800
Final concentration (ng/mL)	80	10	40	2	40	200

Table 2
Test points used to determine limits of detection and quantification of each phytohormone

	Concentration in ng/mL					
	ABA	IAA	JA	JA-Ile	OPDA	SA
Test Point 01 (TP01)	80	10	40	2	4	200
Test Point 02 (TP02)	20	2.5	10	0.5	1	50
Test Point 03 (TP03)	4	0.5	2	0.1	0.2	10
Test Point 04 (TP04)	2	0.25	0.5	0.05	0.1	5
Test Point 05 (TP05)	1	0.125	0.25	0.025	0.05	2.5

2.5 Preparing the Points of Calibration Curve

2.5.1 Calibration Curve in Solvent

1. Label ten 1.5 mL tubes from C10-C1.
2. In C10 add 360 μL of MeOH:Water (7:3), 4 μL of each Original Standard Working Solution (Subheading 2.2.1), and 4 μL of each Internal Standard Working Solution (Subheading 2.2.2).
3. In a 1.5 mL tube add 960 μL of MeOH:Water (7:3) and 10 μL of each Internal Standard Working Solution (Subheading 2.2.2).
4. Add the volumes of internal standard solution prepared in three into each tube (C9-C1), as described in Table 3.
5. Make the serial dilution transferring the volumes of previous solution (C10-C2) into the label tubes as described in Table 3 (i.e., add 210 μL of C10 in to C9, 160 μL of C9 into C8, and so on).

2.5.2 Calibration Curve in Matrix

To prepare the spiking solutions, which will be used to spike the plant samples used for the calibration curve in matrix, follow the same steps described in Subheading 2.5.1 with small changes:

- In **item 2** add and 4 μL of each Original Standard Working Solution (Subheading 2.2.1) in 376 μL of MeOH:Water, and NO internal standard (*see* **Note 2**).
- Skip **item 3**.

Table 3
Serial dilution scheme to prepare the calibration curve

	C10	C9	C8	C7	C6	C5	C4	C3	C2	C1
Vol of internal standard Sol (µL)		70	80	125	125	125	100	120	75	60
Vol previous sol (µL)		210	160	125	125	125	100	80	75	40

- In **item 4**, use MeOH:Water 7:3 instead of Internal Standard Solution (*see* **Note 2**).

3 Methods

The implementation of any quantification method involves the following steps, which will be described in detail in a sequence shown in Scheme 1.

3.1 Optimization of Compounds' Ionization and Fragmentation

The first step of the method implementation comprises the optimization of ionization and fragmentation for each phytohormone, which can be done according to the following steps.

1. Infuse a solution containing 100 ng/mL of each original and internal standards into the ESI-MS (at a flow rate of 2–10 µL/min) in combination with 100–200 µL/min of MeOH:Water (7:3) with 0.05% of formic acid (*see* **Note 3**).

2. Optimize for each compound the source parameters such as source voltage, capillary voltage, source lens (or similar ones depending on the type of spectrometer being used).

3. Optimize source parameters such as temperature and gas flow based of the flow rate used during the measurements (*see* step 4 below).

4. Optimize fragmentation parameters (i.e., isolation window, collision energy, Activation Q, or similar parameters depending on the mass spectrometer) for each compound in other to reach a high intensity for the transitions described in Table 4 (*see* **Note 4**).

3.2 Chromatographic Separation

1. Determine the retention time of each original and internal standard, analyzing 10 µL of the same solution used in 3.1 by HPLC-MS (in full scan mode). Use a Luna Phenyl-Hexyl column (150 × 4.6 mm, 5 µm; Phenomenex, Aschaffenburg, Germany), formic acid (0.05%, v/v), and MeOH with 0.05% (v/v) of formic acid as mobile phases A and B, respectively (*see* **Note 5**).

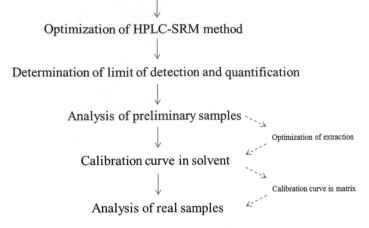

Optimization of ionization and fragmentation parameters
for each phytohormone

Optimization of HPLC-SRM method

Determination of limit of detection and quantification

Analysis of preliminary samples

Optimization of extraction

Calibration curve in solvent

Calibration curve in matrix

Analysis of real samples

Scheme 1 Pipeline for implementing a quantification method based in HPLC-MS/MS-based measurements

Table 4
SRM transitions for phytohormone quantifications

Compound	*m/z* Precursor Ion	Fragments
ABA	263.0	153.0
d6-ABA	269.0	159.0
IAA	176.0	130.0
d5-IAA	181.0	135.0
JA	209.0	59.0
d6-JA	215.0	59.0
JA-Ile	322.0	130.0
OPDA	291.0	165.0
SA	137.0	93.0
d4-SA	141.0	97.0

2. The elution profile is: 0–10 min, 42–55% B in A; 10–13 min, 55–100% B; 13–15 min 100% B; 15–15.1 min 100–42% B in A; and 15.1–20 min 42% B in A. The mobile phase flow rate is 1.1 mL/min.

3. Using a splitter reduces the flow from the HPLC to send only 150–250 μL to the mass spectrometer (*see* **Note 6**).

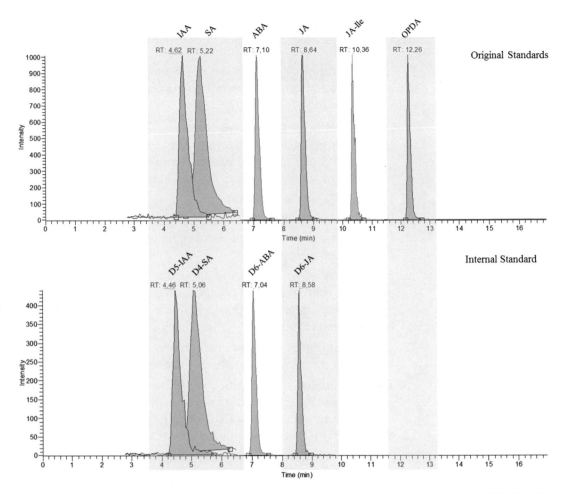

Fig. 1 HPLC-SRM chromatogram from a sample containing phytohormones original standards (ABA, IAA, JA, JA-Ile, OPDA) and internal standards (D5-IAA, D6-ABA, D6-JA, and D4-SA)

4. Once you have determined the retention time of each phytohormone (original and internal standard), set up the MS events using Selected Reaction Monitoring (SRM) transitions with the optimized parameters (Subheading 3.1) and retention times.

5. Run a sample containing 100 ng/mL of each original and internal standard phytohormone using this method. The chromatograms will look similar to those shown in Fig. 1.

3.3 Determining Limit of Detection and Quantification in Solvent

Before measuring plant samples, it is necessary to determine the limits of detection and quantification that can be reached in the setup that is being used.

Analyze the solutions described in Subheading 2.3 (Limit of Detection and Quantification), and determine for each original standard the concentration that provides a signal/noise ratio of

3 (limit of detection), and the one, which provides a signal/noise ratio of 10 (limit of quantification) (*see* **Note 7**).

3.4 Extracting the Samples

The phytohormone content varies a lot upon different types of stress (i.e., wounding). It is important to ensure that the sample collection is performed carefully and fast.

1. Prepare the material for sample collection depending on the amount of plant tissue that will be harvested. It can be, for instance, 1.5 mL tubes, 15 mL falcon tubes, or small envelops made out of aluminum paper or weighing paper.

2. Prepare a Dewar with liquid nitrogen where you can store the tubes/envelops containing the samples.

3. Right before the sample collection, precool the tubes or envelops in liquid nitrogen.

4. Harvest the plant tissue and immediately place it into the precooled tubes/envelops. After collection, the plant material must be kept frozen in liquid nitrogen, or stored in -80 °C until the extraction.

5. Grind the samples using a tissue lyzer or mortar and pestle. In both cases it is very important that the samples do not thaw.

6. Using precooled 1.5 mL tubes, weigh around 100 mg of plant material, and record the exact amount of each sample. Be sure once again that it does not thaw during the weighting process. Keep the samples in liquid nitrogen until the extraction.

7. Add 1.0 mL of extraction solution (Subheading 2.4) into each tube containing the samples.

8. Shake the samples for 30 min in a tube rotator.

9. Centrifuge the samples for 5 min at $15,000 \times g$.

10. Transfer the supernatant into a new 1.5 tube.

11. Dried the samples in a speedvac.

12. Once the samples are dried, resuspend them in 100 μL of methanol.

13. Vortex the samples vigorously.

14. Centrifuge them for 5 min at $15,000 \times g$.

15. Transfer them into vials containing small inserts. In this step, small pieces of washed cotton can be used to avoid transferring some solid material.

3.5 Analyzing Preliminary Samples

Before extracting and analyzing the real samples, few preliminary samples must be analyzed in order to confirm that the method is sensitive enough to detect and quantify the phytohormones levels present in the tissues, which will be analyzed (*see* **Note 8**). Perform the extraction of some samples according to the procedure described in Subheading 3.4 and analyze them using the optimized HPLC-SRM method.

If the sensitivity is sufficient for quantification of desired compounds, the next steps will be to prepare and analyze a calibration curve followed by the samples (Subheadings 3.6 and 3.7). However, if the signals of some phytohormones cannot be seen in the blank sample, an extraction optimization can be carried out in order to improve the extraction as described in Subheading 3.8.

3.6 Calibration Curve

3.6.1 Calibration Curve in Solvent

1. In order to draw the calibration curve in solvent, the samples described in Subheading 2.5 (C1-C10) might be analyzed from the lowest to the highest concentration using the optimized HPLC-SRM method. The final concentration for each point of the calibration curve is described in Table 5, for each phytohormone.

2. Once all the points were analyzed, optimize the integration parameters such as peak detection algorithm, number of smoothing points, area/noise factor, S/N threshold.

3. Integrate the peaks corresponding to each concentration of each phytohormone using these parameters and draw the calibration curve using the values described in Table 5.

4. Optimize parameters as regression type (i.e., linear, quadratic, etc.), and the weight of the curve (i.e., equal, $1/\times$, $1/\times2$, etc.) (*see* **Note 9**).

3.6.2 Calibration Curve in Matrix

Since the chemical and physical properties of plant tissues have an impact on the extraction of phytohormones, a more precise quantification can be done when the calibration curve is prepared in matrix (plant tissue where the phytohormones will be quantified). It can be done according to the following steps (*see* **Note 10**).

1. Harvest around 2.0 g of plant material (control plants) in a precooled container. Keep it in liquid nitrogen.

2. Homogenize and grind this sample in a mortar or in a tissue lyzer. Ensure that it does not thaw.

3. Aliquot the grinded and homogenized plant sample in 12 tubes containing 100 mg each. Keep them in liquid nitrogen.

4. Add 100 μL of each solution (C1-C10, described in Subheading 2.5.2) into one tube containing the plant material (prepared in **step 3**). Keep the samples on ice.

5. Add 1.0 mL of extraction solution containing the internal standards (Subheading 2.3).

6. Proceed with the extraction as described in Subheading 3.4 (**steps 8–13**).

7. Analyze and process the data as described in Subheading 3.6.1.

8. Compare the calibration curves in matrix (Subheading 3.6.2) and in solvent (Subheading 3.6.1) to evaluate how much the

Table 5
Phytohormone concentrations in the calibration curve

	Concentration in ng/mL or ng/g FW									
	C10	C9	C8	C7	C6	C5	C4	C3	C2	C1
ABA	800	600	400	200	100	50	25	10	5	2
IAA	100	75	50	25	12.5	6.25	3.12	1.25	0.62	0.25
SA	2000	1500	1000	500	250	125	62.5	25	12.5	5
JA	400	300	200	100	50	25	12.5	5	2.5	1
JA-Ile	20	15	10	5	2.5	1.25	0.625	0.25	0.125	0.05
OPDA	400	300	200	100	50	25	12.5	5	2.5	1

plant tissue influences the quantification of each phytohormone (*see* **Note 10**).

3.7 Analyzing the Samples

Once all the optimization was done and the method is sensitive enough to quantify the phytohormones in the intend tissues, the samples can be extracted and analyzed. It should be done in batches, which includes the samples and the calibration curve (either in matrix or solvent).

3.8 Optimizing the Extraction of Phytohormones

The optimization of phytohormones extraction can be performed according to the following steps (*see* **Note 11**):

1. Harvest certain amount of the plant material in a precooled container. The amount of plant depends on how many parameters will be optimized. Keep it in liquid nitrogen.

2. If you want to test dry material, dry the sample in a freeze dryer (or go directly to Subheading 3).

3. Homogenize and grind the plant material in a mortar or in a tissue lyzer. Ensure that it does not thaw.

4. Aliquot the grinded and homogenized plant material in 1.5 mL containing 100 mg each. Prepare 5 replicates for each condition, which will be evaluated. If working with fresh tissue keep the samples always in liquid nitrogen.

5. Add 1.0 mL of different extraction solutions (the ones that will be evaluated) and proceed with the extraction as described in Subheading 3.4 (**steps 8–13**).

6. Analyze the samples using the HPLC-SRM method and compare the average area and standard deviation to determine which extraction procedure works the best for the plant tissue.

4   Notes

1. When working with very sensitive mass spectrometers it is better to infuse solutions with lower concentrations (i.e., 100 ng/mL of each phytohormone).

2. When the calibration curve is prepared in matrix, the internal standard will be added through the extraction solution, in the same way than during sample extraction.

3. During the optimization of ionization parameters, it is important to use the conditions (i.e., solvent composition and flow rate) as closed as possible to those which will be used during the HPLC-MS/MS analysis. Therefore, it is highly recommended that the optimization of ionization parameters is carried out combining the infusion of phytohormones and HPLC flow.

4. The parameters related to fragmentation vary according to the analyzer, which is being used. For ion trap, it is important to set up the Activation Q value properly. This parameter influences the mass range of the fragments that can be detected. The default value is usually 0.250. However, in the case of JA, it is not possible to detect the fragment ion with m/z 59 (which is the most specific fragment of this compound) with this Activation Q value. In this particular case the value must be set to 0.210.

5. If the column described in Subheading 3.2 is not available, other chromatographic columns (with different size or similar stationary phase) can also be used. However, a more extensive optimization will be needed.

6. When using most of electrospray sources, high flow can decrease the sensitivity. Therefore, to reach high sensitivity, it is highly recommended to split the HPLC flow before sending it to the mass spectrometer. So, the 1.1 mL flow (coming from the HPLC) might be reduced to 150–250 μL before going into the MS.

7. Depending on the sensitivity of the mass spectrometer, it might be necessary to analyze solutions with lower concentrations than those presented in Table 2, or even concentrations that are in between the test points.

8. The amount of phytohormones can vary greatly depending on the plant which is being analyzed, as well as on the growth stage, tissue, culture conditions, and so on. Therefore, analyze few preliminary samples before starting with the extraction of real samples.

9. In HPLC-MS/MS-based methods, the type of regression and the weight applied to the points of calibration curves are very important. In Fig. 2 graphs for residual errors from jasmonic

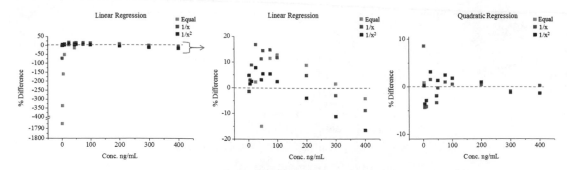

Fig. 2 Residual errors from jasmonic acid (JA) calibration curve using different regressions and weight

acid calibration curve with different regression and different weights are presented. Although all processing (linear or quadratic regression, with different weights) lead to a high r^2 (between 0.9928 for linear regression with equal weight and 0.9998 for quadratic regression with equal weight), the residues generated from these different processing methods are very diverse, especially for the low concentration points. In this case, the quadratic regression is the one with lower residual errors, which are distributed randomly throughout the x-axis. These data show the importance of choosing a good fit model to the calibration curve, depending on which type of mass spectrometer is being used, as well as on the range of the calibration curve and the target compound.

10. The chemical composition and the physical properties are very diverse depending not only on the plant that is being analyzed (Fig. 3) but also on the type of tissue (leaf, steam, roots, etc.). Therefore, it is essential to evaluate the effect of the matrix on the extraction and ionization of the phytohormones, especially when working with absolute quantification. Moreover, if the matrix effect is not so relevant, as occurs for quantification of JA in *Arabidopsis* tissues, the calibration curve can be done in solvent. However, calibration curves in matrix are preferred for plant tissues or compounds where the effect of the matrix is prominent.

11. Due to the great variability in the physical properties of the plant tissues and its chemical composition, quite often it is necessary to optimize the extraction of phytohormones according to the plant material, which is being analyzed. In order to optimize the extraction, parameters such as nature of the plant material (dry or fresh tissue), composition of extraction solution, extraction-accelerating processes (i.e., sonication, heating) can be modified. Figure 4 shows how extraction solution and the nature of the plant material influence the extraction of JA in different plant tissues.

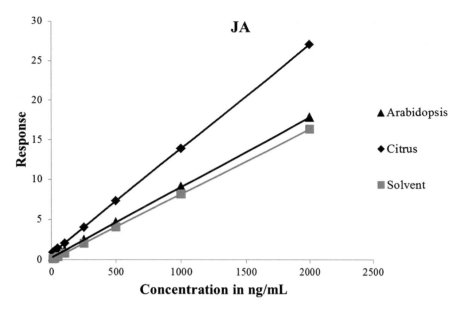

Fig. 3 Matrix effect of different plants in the quantification of jasmonic acid (JA)

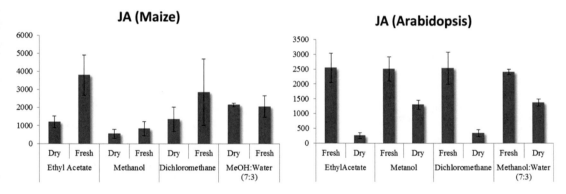

Fig. 4 Extraction of jasmonic acid (JA) in Maize and Arabidopsis using different extraction conditions

References

1. Shan X, Yan J, Xie D (2012) Comparison of phytohormone signaling mechanisms. Curr Opin Plant Biol 15:84–91. https://doi.org/10.1016/j.pbi.2011.09.006

2. Chaiwanon J, Wang W, Zhu J-Y et al (2016) Information integration and communication in plant growth regulation. Cell 164:1257–1268. https://doi.org/10.1016/j.cell.2016.01.044

3. Gai Y-P, Han X-J, Li Y-Q et al (2014) Metabolomic analysis reveals the potential metabolites and pathogenesis involved in mulberry yellow dwarf disease. Plant Cell Environ 37:1474–1490. https://doi.org/10.1111/pce.12255

4. Patui S, Bertolini A, Clincon L et al (2013) Involvement of plasma membrane peroxidases and oxylipin pathway in the recovery from phytoplasma disease in apple (*Malus domestica*). Physiol Plant 148:200–213. https://doi.org/10.1111/j.1399-3054.2012.01708.x

5. Janik K, Mithöfer A, Raffeiner M et al (2017) An effector of apple proliferation phytoplasma targets TCP transcription factors—a generalized virulence strategy of phytoplasma? Mol Plant Pathol 18:435–442. https://doi.org/10.1111/mpp.12409

6. Paolacci AR, Catarcione G, Ederli L et al (2017) Jasmonate-mediated defence responses, unlike

salicylate-mediated responses, are involved in the recovery of grapevine from bois noir disease. BMC Plant Biol 17:118. https://doi.org/10.1186/s12870-017-1069-4

7. Almeida Trapp M, De Souza GD, Rodrigues-Filho E et al (2014) Validated method for phytohormone quantification in plants. Front Plant Sci 5:417. https://doi.org/10.3389/fpls.2014.00417

8. Hird SJ, Lau BP-Y, Schuhmacher R et al (2014) Liquid chromatography-mass spectrometry for the determination of chemical contaminants in food. Trends Anal Chem 59:59–72. https://doi.org/10.1016/j.trac.2014.04.005

INDEX

Rita Musetti and Laura Pagliari (eds.), *Phytoplasmas: Methods and Protocols*, Methods in Molecular Biology, vol. 1875,
https://doi.org/10.1007/978-1-4939-8837-2, © Springer Science+Business Media, LLC, part of Springer Nature 2019

Printed in the United States
By Bookmasters